VORTEX DOMINATED FLOWS

A Volume Celebrating

Lu Ting's 80th Birthday

VORTEX DOMINATED FLOWS

A Volume Celebrating Lu Ting's 80th Birthday

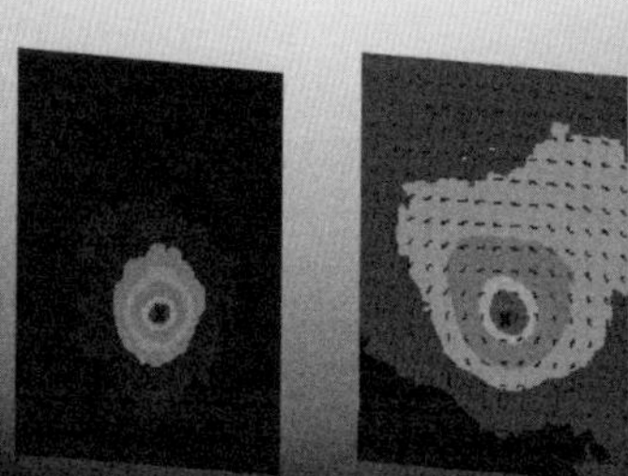

edited by

Denis Blackmore
New Jersey Institute of Technology, USA

Egon Krause
RWTH Aachen, Germany

Chee Tung
Ames Research Center, USA

World Scientific

NEW JERSEY • LONDON • SINGAPORE • BEIJING • SHANGHAI • HONG KONG • TAIPEI • CHENNAI

Published by

World Scientific Publishing Co. Pte. Ltd.

5 Toh Tuck Link, Singapore 596224

USA office: 27 Warren Street, Suite 401-402, Hackensack, NJ 07601

UK office: 57 Shelton Street, Covent Garden, London WC2H 9HE

Library of Congress Cataloging-in-Publication Data

Vortex dominated flows : a volume celebrating Lu Ting's 80th birthday / edited by Denis Blackmore, Egon Krause & Chee Tung.

p. cm.

Includes bibliographical references

ISBN 981-256-320-2 (alk. paper)

1. Viscous flow--Mathematics. 2. Vortex-motion--Mathematics. I. Ting, L. (Lu), 1925– II. Blackmore, Denis L. III. Krause, E. (egon), 1933– IV. Tung, C. (Chee), 1936–

TA357.5.V56V67 2005

532'.0595--dc22

2005041848

British Library Cataloguing-in-Publication Data

A catalogue record for this book is available from the British Library.

Printed in Singapore by World Scientific Printers (S) Pte Ltd

To Lu Ting on his eightieth birthday: In your four score years you have enriched mathematics, engineering and science, and the lives of all who know you. Long may you continue to do so.

Contents

Preface

This is a volume of papers dedicated to Lu Ting on the occasion of his eightieth birthday. The theme is vortex dominated flows and related fields - areas in which Lu Ting's work has made an indelible mark. A Foreword by Joseph Keller precedes the papers, each of which comprises a chapter. Authors of the papers are Shizuko Adachi, Syed Assad, Denis Blackmore, Jyoti Champanerkar, Guido Dietz, Max Gunzburger, Katsuya Ishii, Nicholas Kevlahan, Michael Klaas, Egon Krause, Chjan C. Lim, Joon W. Lim, Goucan Ling, Paul Newton, Tomonori Nihei, Janet Peterson, Wolfgang Schröder, J. S. Shang, Houman Shokraneh, Oliver Thomer, Chee Tung and Claus Weiland - all of whom are admirers of Lu and his work, and some of whom are close friends, scientific collaborators, students, or students of students. The work in this volume represents investigations on the leading-edge of research on vortical flows and their applications, and the editors have made every effort to insure that the contributions meet the high standards characterizing the scientific papers of Lu Ting. Not only do the papers constitute a fitting tribute to Lu Ting on his eightieth birthday, they comprise a collection of papers that should prove useful to graduate students and more advanced researchers interested in vortex dominated flows and related fields.

Professor Ting was born in 1925 in Kiangsi in China. He obtained the degree of bachelor of science in mechanical engineering from Shanghai Jiao-Tong University at the age of 21. Immediately thereafter, in 1946 he was awarded a scholarship by the Chinese Government for studies in the USA, where he earned the degree of master of science in mechanical engineering at Harvard University in 1949. Only two years later, in 1951, he was awarded the degree of doctor of philosophy at New York University under the supervision of H. F. Ludloff - a student of L. Prandtl. After working in industry for four years he joined the Aerodynamics Laboratory of the Polytechnic Institute of Brooklyn in Freeport, Long Island, as a Research Associate in 1955. In 1957 Dr. Ting was appointed Professor at the Polytechnic Institute of Brooklyn. He developed a fruitful collaboration with Professor Antonio Ferri, the founder of the Aerodynamics Laboratory. During the time between 1955 and 1977 they coauthored a large number of publications on gasdynamics, supersonic and hypersonic flows, and on sonic-boom problems. When Professor Ferri left Polytechnic Institute of Brooklyn in 1964 to become the Head of the Department of Aeronautics and Astronautics of New York University, Professor Ting followed him. For four years he taught and did research at the former Daniel Guggenheim School of Aeronautics. He then was appointed Professor in the Mathematics Department of New York University and joined the Courant Institute of Mathematical Sciences, where he remained until his retirement in 2002. He is now Professor Emeritus at the Courant Institute, where he continues to pursue his research.

Professor Ting has been engaged in theoretical research since 1951, initially in mechanics, later on in space mechanics, and then primarily in fluid mechanics and applied mathematics. Most recently, he has been focusing on vortex dominated flows. During his time in industry in the early fifties, he worked on problems of asymmetric deformation of thin shells, at that time one of the most important problems in elasto-mechanics. Ten years later, at the beginning of the Space Age, he studied problems in space mechanics. He constructed solutions for determining reentry trajectories and analyzed problems of optimal orbital transfer. His many investigations in these fields as well as in fluid mechanics, gasdynamics, wave propagation, multi-phase flows, acoustics, applied mathematics, matched asymptotic expansions, and numerical methods are amply demonstrated in his impressive list of publications.

Professor Ting continued collaborating with other scientists during his time at the Courant Institute, as for example with Professor Joseph B. Keller, who later moved on to Stanford University. Together they investigated general problems of perturbation theory, and also analyzed problems in fluid mechanics and acoustics using the method of matched asymptotic expansions. One of their more recent publications is on weak shock diffraction and singular rays, which summarizes their joint work over a period of more than ten years, while there latest joint effort is a paper on geometry involving extremal planar sets. Professor Ting presented some of the major results at the GAMM Annual Meeting in Regensburg, Germany, in 1997. Another example of his cooperation with other scientists is the joint work with Professor Michael J. Miksis of Northwestern University. Since 1983, when Professor Miksis was a Postdoctoral Fellow at the Courant Institute, they have been working together on problems of multi-phase flows and wave propagation.

Professor Ting was also actively engaged in collaborative scientific research in Germany. For more than two decades he participated in scientific investigations of the Aerodynamisches Institut. Almost every year he contributed to the research program of the institute with advice, discussions and lectures. During his time in Aachen, he initiated cooperation with Professor Dr. Rupert Klein, who obtained the degree of Dr.-Ing. in mechanical engineering at the RWTH Aachen in 1987, and who is now Professor at the Freie Universität in Berlin, and a recent recipient of the Leibniz prize. Together they wrote several papers on vortex dynamics, which have received international recognition. They published the book "Viscous Vortical Flow" in the Lecture Notes of Physics of the Springer Verlag, No. 374. This publication resulted from a series of ten lectures Professor Ting presented in Aachen on the occasion of the 75th anniversary of the Aerodynamisches Institut in 1988. They are now working on a second edition of this book with Professor Omar Knio. Professor Ting and Professor Klein have continued studying vortical flows, and their research team expanded to include Professor Omar Knio of the John Hopkins University, a recent recipient of the Bessel prize known for his research on the vortex-element method, and later Professor Denis Blackmore of the New Jersey Institute of Technology, known for his work on dynamical systems, joined the team - which they affectionately refer to as the "Gang of Four".

For almost twenty years, Professor Ting was a consultant at the Institute for Computer Applications to Science and Engineering (ICASE) at NASA Langley Research Center, Hampton, Virginia, USA, where he worked with Dr. Lucio Maestrolla on problems of aero-acoustics and of interactions of aero-acoustics with structural dynamics.

In 1980 –1981 Professor Ting spent his sabbatical year in China as a guest scientist, funded by the National Research Council. He presented two series of lectures on perturbation theory and its applications at the Institute of Mechanics of the Chinese Academy of Sciences in Beijing. The lecture notes were published in two volumes. Three of the four members of the Institute of Mechanics who helped prepare the notes, later achieved the rank of professor at the Institute. During his time in China, Professor Ting supervised four master's students at his Alma Mater, the Shanghai Jiao-Tong University. Two of them later obtained Ph.D. degrees in the

United States.

In October 1996, the Alexander von Humboldt Foundation conferred its Research Award on Professor Dr. Lu Ting of the Courant Institute of Mathematical Sciences of New York University, honoring him with this prestigious award for his contributions in theoretical fluid mechanics, for which he has received worldwide recognition. He penetrated the most difficult problems and pioneered solutions for many of them. His extraordinary, wide-ranging work is documented in more than 135 publications.

Most of Professor Ting´s students are university professors or have research related positions: Professor Max Gunzburger is now the Francis Eppes Distinguished Professor of Mathematics at Florida State University; Dr. C. H. Liu was at NASA Langley Research Center in Hampton, Virginia, Dr. Arje Nachmann is a program director at the Air Force Office of Scientific Research: Dr. Chee Tung is at the Aeroflight Dynamics Directorate of the US Army Aviation and Troop Command, NASA Ames Research Center, Moffet Field, California; Dr. Michael Siclari is at Grumman Aircraft Research Center, Farmingdale N. Y.; Dr. Sun Shen is at Westinghouse Research Center, Orlando, Florida; and Professor Egon Krause, whose doctoral thesis was supervised by Professor Ferri and Professor Ting, went on to become the Director of the Aerodynamisches Institut at the RWTH Aachen.

Any meeting with Professor Ting at international conferences is always an extraordinary scientific experience. For example the 8th International Conference on Numerical Methods in Fluid Mechanics in Aachen in 1982, the AIAA Conference on Computational Fluid Dynamics in 1983, the 4th International Conference on Applied Numerical Modeling in Tainan in 1984, the 4th International Conference on Nonlinear Mechanics, Shanghai in 2002, and the always well attended Mini-Symposia on Vortical Flows, organized by Professor Ting and his students at the International Congresses on Industrial and Applied Mathematics in Paris 1987, in Washington 1991, and in Hamburg 1995.

Professor Ting's life was strongly influenced by his father, Kao-Ying Ting, who was a professor of Chinese literature and devoted his life to the study of classical literature. He revised many classical works and edited them in modern print. Originally the books were written without punctuation and printed with wooden blocks only in limited number. Professor Kao-Ying Ting corrected errors in the prints, divided the text into chapters and included punctuation marks in the writings. The new books he edited were easy to read, printed in large numbers, and made available to the general public. Lu Ting is definitely his father's son when it comes to his abiding interests in literature and history.

Professor Ting also had a special relationship with his uncle, Professor Yu-Tai Yao, who probably influenced him in his choice of career even more than his father. In 1913, the young Yu-Tai Yao was awarded a scholarship by the Chinese government to study physics in the USA. He earned a B. S. degree in physics at the University of Chicago, and then a Ph.D. degree with Professor K. T. Compton at Princeton University. After returning to China, Dr. Yao established the Physics Department at Nankai University, and together with three other physicists who had returned from abroad almost at the same time, he established the Chinese Institute of Physics, in which generations of young scientists were educated. Professors of physics at Beijing University belong to the fifth and sixth generation of these scholars. In 1929 Professor Yao was awarded a scholarship for further studies in Germany by the Chinese Foundation for Education and Culture (The China Institute). He was as a guest scientist at Leipzig University and later at the Institute for Astrophysics in Potsdam. In 1932 he returned to China. A year later he was appointed Head of the Physics Department of Beijing University; a position he held until the early 1950's. He died tragically in 1968 during the cultural revolution.

At Nankai University, two of Professor Yao's students, T. Y. Wu and S. S. Chern, quickly advanced in their careers and later became his best friends. T. Y. Wu is today a world renowned theoretical physicist, who only a few years ago stepped down as President of Academia Sinica in Taiwan. Professor Wu obtained his doctor's degree at the University of Michigan. Later he became a Professor and one of the leading scientists at Beijing University. He educated the second generation of Chinese physicists: Professor S. S. Chern, who died recently after an extraordinary career at the University of California at Berkeley and the Nankai Institute, was one of the greatest differential geometers of the twentieth century. Two of the third generation of Professor Yao's students, Professor Chen-Ning Yang of SUNY Stony Brook and Professor Tsung-Dao Lee of Columbia University, were awarded Nobel prizes. Professor Yang was Professor Wu's teaching assistant, and Professor Lee was educated by Professor Wu as an undergraduate.

Clearly, mathematics and science are integral parts of Lu Ting's heritage - a tradition he has carried on in a very distinguished career spanning over half a century. Professor Ting can still be found almost every day in his office at the Courant Institute, deeply involved in writing a scientific paper or working on the latest edition of his book on viscous vortical flows. His research interests have even broadened in recent years to include areas of geometry and dynamical systems. Time has not diminished his quest for knowledge, his creative brilliance, nor his love for solving challenging problems.

List of Publications

Ting, L., Ludloff, H. F.: Aerodynamics of Blasts, J. Aeronautical Science, Vol. 19, pp. 317-328, 1952.

Ting, L.: The Shock Strength in a Two-Dimensional Nonsteady Flow, J. Aeronautical Science, Vol. 19, pp. 351-352, 1952.

Ting, L.: Diffraction and Reflection of Weak Shocks by Structures, J. Math. and Phys., Vol. XXXII, pp. 102-116, 1953.

Yuan, S. W., Ting, L.: On Radial Deflections of a Cylinder Subjected to Equal and Opposite Concentrated Radial Loads, J. Applied Mechanics, Vol. 24, pp. 278-282, 1957.

Ting, L.: Diffraction of Disturbances Around a Convex Right Corner with Application in Acoustics and Wing-Body Interference, J. Aeronautical Science, Vol. 24, pp. 821-830, 1957.

Ferri, A., Clarke, J. H., Ting, L.: Favorable Interference in Lifting Systems in Supersonic Flow, J. Aeronautical Science, Vol. 24, pp. 791-804, 1957.

Ting, L., Yuan, S. W.: On Radial Deflection of a Cylinder of Finite-Length with Various End-Conditions, J. Aeronautical Science, Vol. 25, pp. 230-234, 1958.

Ting, L.: Some Integrated Properties of Solutions of the Wave Equation with Nonplanar Boundaries, Quart. Appl., Math, Vol. XVI, pp. 373-384, 1959.

Ting, L.: Pressure Distribution on Dihedral Wings at Supersonic Speed, J. Fluid Mechanics, Vol. 6, pp. 302-312, 1959.

Ting, L.: On the Mixing of Two Parallel Streams, J. Math. and Phys., Vol. 38, pp. 153-165, 1959.

Ting, L.: On the Integral of Moment of Pressure in Supersonic Flow, J. Aeronautical Science, Vol. 26, pp. 758-759, 1959.

Ting, L.: On the Diffraction of an Arbitrary Pulse by a Wedge or a Cone, Quart. Appl. Math., Vol. XVIII, pp. 89-92, 1960.

Ting, L.: Boundary Layer over a Flat Plate in the Presence of Shear Flow, Phys. Fluids, Vol. 3, pp. 78-81, 1960.

Ting, L., Wang, K.: An Approximate Analytic Solution of Re-Entry Trajectory with Aerodynamic Forces, ARS J., Vol. 30, pp. 565-566, 1960.

Ting, L., Libby, P. A.: Remarks on the Eddy Viscosity in Compressible Mixing Flow, J. Aero. Sci., Vol. 27, p. 797, 1960.

Ting, L.: Optimum Orbital Transfer by Impulses, ARS J., Vol. 30, pp. 1013-1018, 1960.

Ting, L.: Optimum Orbital Transfer by Several Impulses, Astronautica Acta, Vol. VI, pp. 256-265, 1960.

Kobayakawa, S., Ting, L.: Generalized Integral Relationship in Flow Fields with Vortex Sheets, J. Aeronautical Science, Vol. 27, pp. 872-873, 1960.

Wang, K., Ting, L.: Aerodynamic Heating of Re-Entry Vehicle, ARS J., Vol. 30, pp. 1180-1181, 1960.

Ting, L.: Approximation Technique for Variational Problems, Aero. Eng., Vol. 20, pp. 32-33, 89-90, 1961.

Moskowitz, S. E., Ting, L.: An Approach to Problems of Optimum Rocket Trajectories, ARS J., Vol. 31, pp. 551-553, 1961.

Chow, R. R., Ting, L.: Higher Order Theory of Curved Shock, J. Aero. Sci., Vol. 28, pp. 428-430, 1961.

Ferri, A., Zakkay, V., Ting, L.: Blunt Body Heat Transfer at Hypersonic Speed and Low Reynolds Number, J. Aero. Sci., Vol. 28, pp. 962-972, 1961.

Wang, K., Ting, L.: Approximate Solutions for Re-Entry Trajectories with Aerodynamic Forces, Astronautica Acta, Vol. VIII, pp. 28-41, 1962.

Ferri, A., Ting, L.: Practical Aspects of Re-Entry Problems, Astronautica Acta, Vol. VIII, pp. 63-81, 1962.

Ferri, A., Zakkay, V., Ting, L.: On Blunt Body Heat Transfer at Hypersonic Speed and Low Reynolds Numbers, J. Aero. Sci., Vol. 29, pp. 882-883, 1962.

Ting, L., Pierucci, M.: Optimum Launching of a Satellite by Two Impulses of Unequal Specific Impulse, Astronautica Acta, Vol. IX, pp. 174-183, 1963.

Ting, L., Brofman, S.: On Take-Off From Circular Orbit by Small Thrust, Zeitschrift für Angewandte Mathematik and Mechanik, Vol. 44, pp. 417-428, 1964.

Ting, L.: Effect of Shock Induced Vorticity on the Compressible Boundary Layer along a Flat Plate, AIAA J., Vol. 2, pp. 490-493, 1964.

Ting, L., Ruger, C.: Oblique Injection of a Jet into a Stream, AIAA J., Vol. 3, pp. 534- 535, 1965.

Ting, L., Tung, C.: On the Motion and Decay of a Vortex in a Non-Uniform Stream, Physics of Fluids, Vol. 8, pp. 1039-1051, 1965.

Ting, L.: On the Initial Conditions for Boundary Layer Equations, J. Math. Phys., Vol. XLIV, pp. 353-367, 1965.

Ting, L.: Pressure Distribution on a Surface with Large Normal Injection, AIAA J., Vol. 4, pp. 1573-1579, 1965.

Keller, J. B., Ting, L.: Periodic Vibrations of Systems Governed by Nonlinear Partial Differential Equation, Communications in Pure and Applied Math. Vol. XIX, pp. 371-420, 1966.

Tung, C., Ting, L.: Motion and Decay of a Vortex Ring, Physics of Fluids, Vol. 10, pp. 901-910, 1967.

Ting, L., Ruger, C.: A Note on the Mixing of Two Parallel Subsonic Streams, J. Math. Phys. Vol. XLVI, pp. 343-348, 1967.

Ting, L., Chen, S.: Perturbation Solutions and Asymptotic Solutions in Boundary Layer Theory, J. Engineering Math. Vol. I, pp. 327-340, 1967.

Ting, L.: Asymptotic Solutions of Wakes and Boundary Layers, J. Engineering Math., Vol. 2, pp. 23-38, 1968.

Ting, L., Pan, Y. S.: Incidence of N-Waves on Structures, Second Conference on Sonic Boom Research, NASA SP-180, May 1968.

Ismail, A., Ting, L.: Large Rate of Normal Injection to Surface in High Speed Flow, Astronautica Acta, Vol. 14, pp. 653-663, 1969.

Ting, L., Gunzburger, M.: Diffraction of Shock Wave by Moving Thin Wing, J. Fluid Mechanics, Vol. 42, pp. 585-608, 1970.

Ting, L., Ismail, A.: Large Rates of Normal Injection near Trailing Edge, Quarterly of Applied Math., Vol. 28, pp. 181-190, 1970.

Kleinstein, G., Ting, L.: Optimum Solution for Heat Conduction Problems, Zeitschrift für Angewandte Mathematik und Mechanik, Vol. 51, pp. 1-16, 1971.

Ting, L.: On the Motion and Decay of Vortices, Journal of Aircraft, Wake Turbulence and Detections, edited by J.R. Olsen, A. Goldburg and M. Rogers, Plenum Press, New York, pp. 11-40, 1971.

Ting, L., Kung, F.: Diffraction of a Pulse by a Three-Dimensional Corner, Journal of Eng. Math., Vol. 6, pp. 225-241, 1972.

Ting, L., Liu, C. H., Kleinstein, G.: Interference of Wing and Multi-Propellers, AIAA Journ., Vol. 10, pp. 906-914, 1972.

Ferri, A., Ting, L., Werner, J.: Supersonic Jet Engine Noise Analysis,

Proc. of Interagency Symposium on University Research in Transportation Noise, US Department of Transportation, Vol. 1, pp. 39-47, March 1973.

Ferri, A., Siclari, M., Ting, L.: Sonic Boom Analysis for High Altitude Flight at High Mach Number, Progress in Astronautics and Aeronautics, Vol. 38, (editor, Nagamtsu, E.), American Institute of Aeronautics and Astronautics, New York, pp. 301-320, 1975.

Ting, L., Keller, J. B.: Planning of a Flat Plate at High Froude Number, Phys. of Fluids, Vol. 17, pp. 1080-1086, 1974.

Ting, L.: Transmission of Singularities through a Shock Wave and the Sound Generation, Phys. of Fluids, Vol. 17, pp. 1518-1526, 1974.

Buckmaster, J., Nachman, A., Ting, L.: The Buckling and Stretching of a Viscida, J. Fluid Mech., Vol. 69, pp. 1-20, 1975.

Ting, L.: Spherical Means of Solutions of a Partial Differential Equation in a Conical Region, SIAM Journal of Applied Math, Vol. 29, pp. 597-623, 1975.

Ting, L.: Interaction of Two Vortices in a Fluid with Lower Stagnation Pressure, Quarterly of Applied Math., Vol. 34, pp. 353-363, 1977.

Ting, L.: The Passage of a Vortex Sheet through a Shock Wave, Modern Developments in Shock Tube Research, edited by G. Kaminoto, Kyoto University, Kyoto, Japan, pp. 378-385, July 1975.

Gunzburger, M., Liu, C. H., Maestrello, L., Ting, L.: Mathematical Formulation for the Propagation of Sound through a Turbulent Jet, J. Eng. Math., Vol. 10, pp. 219-230, 1976.

Ting, L., Keller, J. B.: Optimum Shape of a Planning Surface at High Froude Number, Journal of Ship Research, Vol. 21, pp. 40-43, 1977.

Lo, K. C. R., Ting, L.: Studies of Merging of Vortices, Phys. of Fluids, Vol. 19, pp. 912- 913, 1976.

Ferri, A., Ting, L., Lo, R.: Non-linear Sonic Boom Propagation Including the Asymmetric Effects, AIAA Journ., Vol. 15, pp. 653-658, 1977.

Ting, L.: Non-linear Periodic Waves, NASA CP-2001 Advances in Engineering Sciences, Vol. III, pp. 837-850, November 1-3, 1976.

Ting, L., Yu, H. W.: Interaction of Vortices Submerged in a Quiescent Fluid of Lower Stagnation Pressure, presented at XIV International Congress of Theoretical and Applied Mechanics, Delft, Netherlands, August 30 - September 3, 1976.

Ting, L., Keller, J. B.: Radiation from the Open End of a Cylindrical or Conical Pipe, J. of Acoustic Society of America, Vol. 61, pp.

1438-1444, 1977.

Ting, L.: Periodic Solutions of Non-linear Wave Equation in n-Dimensional Space, SIAM Journal on Applied Math., Vol. 34, pp. 504-514, 1978.

Callegari, A., Ting, L.: Motion of a Curved Vortex Filament with Decaying Vortical Core and Axial Velocity, SIAM Journal on Appl. Math., Vol. 35, pp. 148-175, 1978.

Ting, L.: Sound Propagation through a Subsonic Jet due to a Source near the Duct Exit, Journal of Sound and Vibration, Vol. 57, pp. 523-534, 1978.

Ting, L.: Radiation of Acoustic Source near the Trailing Edge of a Wing in Forward Motion, AIAA Journ., Vol. 18, pp. 262-268, 1980.

Ting, L.: Sound Propagation through Parallel Jets Exiting from Ducts, J. Acoustical Society of America, Vol. 67, pp. 782-791, 1980.

Ting, L.: Studies on the Motion and Decay of Vortex Filaments, Lecture Notes in Physics, Vol. 148, Advances in Fluid Dynamics, Springer-Verlag, pp. 67-105, 1981.

Stickler, D., Ahluwalia, D. S., Ting, L.: Application of Ludwig's Uniform Progressing Wave Ansatz to a Smooth Caustic, J. Acoust. Soc. Am., Vol. 69, pp. 1673-1681, 1981.

Ting, L.: On the Applications of Integral Invariants and Decay Laws of Vorticity Distributions, J. Fluid Mechanics, Vol. 127, 497-506, 1983.

Liu, C. H., Ting, L.: Numerical Solutions of Viscous Flow in Unbounded Fluid, Lecture Notes in Physics Vol. 170, E. Krause (ed), Springer-Verlag, pp. 357-363, 1982.

Ting, L., Ahluwalia, D. S., Miksis, M. J.: Solutions of a Class of Mixed Free Boundary Problems, SIAM J. of Appl. Math., Vol. 43, pp. 759-775, 1983.

Miksis, M. J., Ting, L.: Wave Propagation through Slender Curved Tube, J. of Acoustical Society of America, Vol. 74, pp. 631-639, 1983.

Ting, L., Ling, G. C.: Studies of the Motion and Core Structure of a Geostrophic Vortex, Proc. 2nd Asian Congress or Fluid Mechanics, Science Press, Beijing China, pp. 900-905, October 1983.

Miksis, M. J., Ting, L.: Surface Waves Induced by an Impinging Jet, Physics of Fluids, Vol. 26, pp. 2378-2384, 1983.

Maestrello, L., Ting, L.: Analysis of Active Control by Surface Heating, AIAA Journ., Vol. 23, pp. 1038-1045, 1985.

Miksis, M. J., Ting, L.: Nonlinear Radial Oscillations of a Gas Bubble

Including Thermal Effects, J. Acoust. Soc. Am., Vol. 76, pp. 897-905, 1984.

Caflisch, R. E., Miksis, M. J., Papanicolaou, G. C., Ting, L.: Effective Equations for Wave Propagation in Bubbly Liquids, J. Fluid Mechanics, Vol. 153, pp. 959-973, 1985.

Liu, C. H., Tavantzis, J., Ting, L.: Numerical Studies of Motion of Vortex Filaments- Implementing the Asymptotic Analysis, AIAA Journ., Vol. 24, pp. 1290-1297, 1986.

Miksis, M. J., Ting, L.: Wave Propagation in a Bubbly Liquid with Finite Amplitude Asymmetric Oscillations, Phys. Fluids, Vol. 29, pp. 603-618, 1986.

Ting, L.: Applied Mathematics and Modeling of Wave Propagation in Bubbly Media, Proc. 4th Intl. Conf. Appl. Numerical Modeling (keynote address), National Cheng Kung University, Tainan, Taiwan, pp. 1-18, 1984.

Liu, C. H., Ting, L.: Motion and Interaction of Decaying Trailing Vortices in Spanwise Shear Wind, NASA Tech. Paper No. 2599, 1986.

Krause, E., Liu, C. H., Ting, L.: Vortex Dominated Flow with Diffusive Core Structures, AIAA Paper No. 851556, Invited paper, 18th AIAA Fluid Dynamics Conference, Cincinnati OH, Sept. 1985.

Ting, L.: Theoretical and Numerical Studies of Vortex Interaction and Merging, Invited paper, Intl. Symp. Comp. Fluid Dynamics, Tokyo, September 1985, Numerical Methods in Fluid Dynamics I, Editor, K. Oshima, Japan Soc. Compu. Fiuid Dyn., pp. 218-229, 1986.

Caflisch, R. E., Miksis, M. J., Papanicolaou, G. C., Ting, L.: Wave Propagation in Bubbly Liquids at Finite Volume Fraction, J. Fluid Mech., Vol. 160, pp. 1-14, November 1985.

Blackmore, D., Ting, L.: Surface Integral of Its Mean Curvature Vector, SIAM Review, Vol. 27, pp. 569-572, December 1985.

Ng, K. C., Ting, L.: Wave Propagation through a Thin Bubbly Layer, J. Amer. Acoust. Soc., Vol. 79, pp. 924-926, April 1986.

Caflisch, R. E., Miksis, M. J., Papanicolaou, G. C., Ting, L.: Wave Propagation in a Bubbly Liquid, Amer. Math. Soc. Lectures in Appl. Math., Vol. 23, pp. 327-343, 1986.

Liu, C. H., Ting, L., Weston, R. P.: Boundary Conditions for N-S Solutions of Merging of Vortex Filaments, Intl. Symp. Comp.Fluid Dynamics, Tokyo, September 1985, Numerical Methods in Fluid Dynamics II, Editor, K. Oshima, Japan Soc. Compu. Fluid Dynamics, pp. 253-264, 1986.

Weston, R. P., Ting, L., Liu, C. H.: Numerical Studies of the Merging of Vortex Filaments, AIAA Paper No. 86-0558, Presented at the AIAA 24th Aerospace Sciences Meeting, Reno NV, January 1986.

Miksis, M. J., Ting, L.: A Numerical Study of Thermal Effects on Nonlinear Bubble Oscillations, J. Acoust. Soc. Am., Vol. 81, pp. 1331-1340, May 1987.

Liu, G. C., Ting, L.: Numerical Studies of the Merging of Vortices with Decaying Cores, AIAA Paper No. 86-1072, presented at the AIAA/ASME 4th Joint Fluid Mechanics, Plasma Dynamics and Lasers Conference, Atlanta GA, May 1986.

Ting, L., Liu, G. C.: Merging of Vortices with Decaying Cores and Numerical Solutions of Navier-Stokes Equations, Lecture Notes in Physics 246, Springer Verlag, pp. 612-616, 1986.

Caflisch, R. E., Miksis, M. J., Papanicolaou, G. C., Ting, L.: Waves in Bubbly Liquids, IMA Vols. in Math. and its Application Vol. I, Homogenization and Effective Moduli of Materials and Media, Springer Verlag, pp. 175-181, 1986.

Ting, L., Miksis, M. J.: Exact Boundary Conditions for Scattering Problems, J. Acoust. Soc. Amer., Vol. 80, pp. 1825-1827, Dec. 1986.

Liu, C. H., Ting, L.: Interaction of Decaying Trailing Vortices in Spanwise Shear Flow, Computers and Fluids, Vol. 15, pp. 77-92, Jan. 1987.

Miksis, M. J., Ting, L.: Viscous Effects on Wave Propagation in a Bubbly Liquid, Phys. Fluids, Vol. 30, pp. 1683-1689, June 1987.

Tavantzis, J., Ting, L.: The Dynamics of Three Vortices - Revisited, Physics of Fluids, Vol. 31, pp. 1392-1409, 1988.

Ling, G. C., Ting, L.: Two-time Scales Inner Solutions and Motion of a Geostrophic Vortex, Sci. Sinica, Ser. A 31, pp. 806-817, 1988.

Miksis, M. J., Ting, L.: Wave Propagation in a Multiphase Media with Viscous and Thermal Effects, Amer. Nuclear Soc. Proc., National Heat Transfer Conference, pp. 145-154, 1987.

Miksis, M. J., Ting, L.: A Numerical Method for Long Time Solution of Integro- Differential Systems in Multiphase Flow, Computers and Fluids, Vol. 16, pp. 327-340, 1988.

Miksis, M. J., Ting, L.: Surface Waves Produced by the Scattering of an Incident Wave from an Interface, Proc., 12th World Congress on Scientific Computation, IMACS, Paris, July 18-22, Vol. II, pp. 298-301, 1988.

Cheng, L., Ting, L.: Void Distribution in a Shallow Bubbling Pool, Proceedings of the 26th National Heat Transfer Conference, pp. 101-109, August 1989.

Miksis, M. J., Ting, L.: Scattering of an Incident Wave from an Interface Separating Two Fluids, Wave Motion, Vol.11, pp. 545-557, Nov. 1989.

Miksis, M. J., Ting, L.: Effects of Bubbly Layers on Wave Propagation, J. Acoust. Soc. Amer., Vol.86, pp. 2349-2358, Dec. 1989.

Ting, L., Miksis, M. J.: On Vortical Flow and Sound Generation, SIAM J. Appl. Math., Vol. 50, pp. 521-536, April 1990.

Klein, R., Ting, L.: Far Field Velocity Potential Induced by an Exponentially Decaying Vorticity Distribution, ZAMP, Vol.41, pp. 395-418, May 1990.

Ting, L., Keller, J. B.: Slender Jets and Thin Sheets with Surface Tension, SIAM J. Appl. Math., Vol. 50, pp. 1533-1546, Dec. 1990.

Miksis, M. J., Ting, L.: Effective Equations for Multiphase Flows - Waves in Bubbly Liquid, Advances in Applied Mechanics Vol. 28, Eds: J. W. Hutchinson and T. Y. Wu, Academic Press, pp. 141-260, 1991.

Ting, L., Klein, R.: Viscous Vortical Flows, Lecture Notes in Physics 374, 222 pages, ISBN 0-387-53713-9, Springer-Verlag, 1991.

Miksis, M. J., Ting, L.: Wave propagation in a bubbly liquid at low volume fraction, Chem. Engrg. Commu., Vol. 118, pp. 59-73, 1992.

Klein, R., Ting, L.: Vortex filament with axial core structure variation, Appl. Math. Letters, Vol. 5, pp. 99-103, 1992.

Keller, J. B., Ting, L.: Singularities in Semi-linear Waves, Commu. Pure Appl. Math., Vol. XLVI, pp. 341-352, 1993.

Ting, L., Bauer, F.: Viscous Vortices in Two- and Three-dimensional Space, Computers & Fluids, Vol. 22, pp. 565-588, 1993.

Frendi, A., Maestrello, L., Ting, L.: An Efficient Model for Coupling Structure Vibrations with Acoustic Radiation, ICASE report 93-18, April 1993, J. Vibration and Sound, Vol. 182, pp. 741-757, 1995.

Ting, L.: On Surface Conditions for Structural Acoustic Interaction in Moving Media, presented in Workshop on Perturbation Methods in Physical Mathematics, Rensselaer Polytechnic Institute, June 23-26, 1993, SIAM J. Appl. Math., Vol. 55, pp. 369-389, 1995.

Miksis, M. J., Ting, L.: Panel Oscillations and Acoustic Waves, Appl. Math. Letters, Vol. 8, No. 1, pp. 37-42, Jan. 1995.

Keller, J. B., King, A., Ting, L.: Blob Formation, Phys. Fluids, Vol. 7, pp. 226-228, 1995.

Klein, R., Ting, L.: Theoretical and Experimental Studies of Slender Vortex Filaments, Appl. Math. Letters, Vol. 8, No. 2, pp. 45-50, March 1995.

Bauer, F., Maestrello, L., Ting, L.: Acoustic Wave Propagation in Unsteady Moving Media, ICASE Rep. No. 95-39, J. Acoust. Soc. Amer., Vol. 99, pp. 1291-1305, March 1996.

Miksis, M. J., Ting, L.: Structural Acoustic Interactions and On Surface Conditions, to appear in IMA Volumes in Mathematics and its Applications, Vol. 86, Computational Wave Propagation, Editors, B. Engquist and G. A. Kriegsmann, pp.165-178, Springer-Verlag, 1997.

Knio, O., Klein, R., Ting, L.: Vortical Flow outside a Sphere and Sound Generation, SIAM J. Appl. Math., Vol. 57, No. 4, August 1997.

Knio, O., Klein, R., Ting, L.: Interaction of a slender vortex filament with a rigid sphere: dynamics and far-field Noise, presented in the mini-symposium on Aero- and hydroacoustics', 19th International Congress of Theoretical and Applied Mechanics, Kyoto, Japan, August 25-31, 1996.

Klein, R., Knio, O., Ting, L.: Representation of Core Dynamics in Slender Vortex Filament Simulations, Phys. Fluids, 8, pp. 2415-2425, 1996.

Ting, L., Keller, J. B.: Weak Shock Diffraction and Singular Rays, to be presented as an invited talk in the Annual Scientific Conference, GAMM 97 at the University Regensburg, March 24-27, 1997.

Ting, L., Klein, R., Knio, O.: Asymptotic Theory of Slender Vortex Filaments - Old and New, invited lecture, IUTAM-Symposium on Dynamics of Slender Vortices, RWTH Aachen, August 31 - September 3, 1997, in Proceedings of the Symposium, Kluwer Academic Publ., pp. 3-20, 1998.

Knio, O., Ting, L., Klein, R.: Interaction of a slender vortex with a rigid sphere: Dynamics and far-field sound. Journal of the acoustical Society of America, 1998 V103 pp. 83-98.

Ting, L., Klein, R., Knio, O.: Asymptotic theory of slender vortex filaments - old and new, Fluid Mechanics and its Applications, Vol. 44 pp. 3-29, 1998.

Knio, O., Ting, L.: Noise emission due to slender vortex solid body interactions - Ring/sphere interaction at high Reynolds number, Fluid Mechanics and its Applications Vol. 44 P369-378, 1998.

Ting, L.: On the dynamics of vortex filaments and their core structure, Journal of Shanghai University, Vol. 3, No. 2 p87-94, 1999.

Blackmore, D., Ting, L.: Higher order conditions for weak shocks: modified Prandtl relation, PAMM, Vol. 1, pp. 397-398, 2000 (online).

Blackmore, D., Ting, L.: Interaction of shocks and weak vorticity fields, Proceedings of Fifth World Congress on Computational Mechanics, Vienna, Austria, July 2002 (online).

Ting, L., Blackmore, D.: Higher order shock conditions for curved shocks in unsteady flows, Proceedings of Fourth International Conference on Nonlinear Mechanics, Shanghai, China, 2002, pp. 160-164.

Blackmore, D., Ting, L.: Vorticity jumps across shock surfaces, Proceedings of Second MIT Conference on Computational Fluid and Solid Mechanics, Vol. 1, K. J. Bathe, ed., Elsevier, Amsterdam, 2003, pp. 847-849.

Ting, L., Keller, J. B.: Extremal convex planar sets, Discrete and Computational Geometry, Vol. 33, pp. 369-393, 2005.

Blackmore, D., Knio, O., Ting, L.: Perturbations of three point vortex problems, Proceedings of Third MIT Conference on Computational Fluid and Solid Mechanics. (to appear in 2005)

Acknowledgment

The editors would like to thank Yuriy Mileyko for his assistance in preparing the LaTeX files for the manuscript.

The Editors

Denis Blackmore
Egon Krause
Chee Tung

Foreword by Joeseph B. Keller

I am very pleased to say a few words about my good friend and scientific collaborator Lu Ting in this impressive volume of papers on vortex dominated flows celebrating his eightieth birthday. His scientific contributions speak for themselves, but I would like to offer some insights based on my own experiences with Lu, as well as my congratulations to Lu for reaching this milestone.

Lu Ting and I have worked together for over 40 years. Our first joint work was entitled "Periodic vibrations of systems governed by nonlinear partial differential equations." In it, we use perturbation analysis and bifurcation theory to determine the frequencies and modes of vibration of various physical systems. The novelty was the application to partial differential equations of methods, which, previously, had been used almost exclusively on ordinary differential equations.

Most of our subsequent work concerned problems in fluid dynamics, and employed singular perturbation theory and matched asymptotic expansions. Lu is an expert in both fluid dynamics and the use of matched asymptotic expansions, so it was natural that we worked on these topics. Whenever I encountered problems in these areas, I consulted him. His physical insight into fluid flows enabled us to find the appropriate mathematical simplifications to solve many flow problems.

Joe Keller
Stanford University

Chapter 1

Circular Discrepancy and a Monte Carlo Algorithm for Generating a Low Circular Discrepancy Sequence

Syed M. Assad, Physics Department, National University of Singapore
and
Chjan C. Lim, Mathematical Sciences, Rensselaer Polytechnic Institute
Troy, NY 12180, USA

Dedicated to Professor Lu Ting on his 80th Birthday

Abstract

We present a numerical method for generating a low discrepancy point sequence inside the circle. The method is based on physically minimizing the energy of a vortex gas system and naturally incorporates the random fluctuations in a simulated system at positive temperatures.The quality of this distribution of points is compared against two sequences obtained from the Hammersley sequence and the Sobol sequence, using the half-plane notion of discrepancy also known as circular discrepancy.

1.1 Introduction

This paper deals with discrepancy of N points in the disc and a new method for generating low discrepancy sequences. Numerous algorithm [Niederreiter (1991)] have been developed for generating low discrepancy points on the cube $I^s = [0,1]^s$. On the cube, low discrepancy points are of importance in performing quasi Monte-Carlo methods. For example to numerically evaluate the proper integral of a function of bounded variation $f(\vec{x})$, we can estimate this integral of f by sampling f at N points. Then Koksma's inequality bounds the integration error by $V(f)D_N^*$, where $V(f)$ reflects the regularity of f and D_N^* describes the regularity (discrepancy) of the points [Morokoff and Caflisch (1995); Niederreiter (1978)].

Consider the simple problem of approximating the area of a disc segment (an intersection of a disc with a half plane) by choosing N random points in the disc, and counting the number of points that falls in this disc segment. The area of the segment can then be approximated by x/N, where x is the number of points falling in the disc segment. As N gets larger, we expect that the approximation would get better. It is well known that the error of this random or Monte Carlo method is $O(N^{-1/2})$.

Quasi Monte Carlo methods on the other hand use deterministic sequences of points to perform the numerical integration of sufficiently smooth functions. In principle the integration error of these methods is $O(N^{-1}(\log N)^d)$ where d is the dimension of the domain of integration. There is evidence that the quasi Monte Carlo methods outperform the traditional Monte Carlo methods when the effective dimension of the integrands are not too high. See [Morokoff and Caflisch (1995); Owen (1991)] for a definition of the effective dimension of a large class of integrands.

Hybrid Monte Carlo quasi Monte Carlo methods where the integration mesh from a quasi MC method has been perturbed by random fluctuations, are known to have better asymptotic error of $O(N^{-3/2}(\log N)^d)$ [Owen (1991)]. Our method is based on the equilibrium statistical mechanics of point vortex logarithmic interactions and naturally incorporates the random fluctuations of a simulated system at positive temperatures.

This paper is organized as follows. In Section 1.2, we shall define the circular discrepancy of N points in a disc. This is defined as the maximum possible error in approximating the area of an arbitrary disc segment given N fixed points. This shall be used to quantify the quality of a sequence of points in the disc.

We show in Section 1.3 that given an arbitrary sequence of N points, its circular discrepancy can be measured in $O(N^3)$ time. Section 1.4 presents a method for generating N well distributed points in the disc that achieves a low circular discrepancy. This method is based on Monte-Carlo simulations of a vortex gas system.

In Section 1.5 we introduce the square discrepancy. Square discrepancy measures the 'uniformity' of a sequence of points on the square. We shall use two quasi-random sequences with low square discrepancy, namely the Hammersley sequence and the Sobol sequence to provide numerical comparisons with the vortex gas sequence.

Finally Sections 1.6 and 1.7 presents the numerical results.

1.2 Circular discrepancy

Definition: *Let $\mathcal{P}$ be an arbitrary distribution of N points in the circular disc D of unit area. For each disc segment A (i.e. an intersection of D with a half-plane), let $Z[\mathcal{P}; A]$ denote the number of elements of $\mathcal{P}$ which lies in A. We define the circular discrepancy of $\mathcal{P}$ with respect to A as*

$$\Delta[\mathcal{P}; A] = \left| \frac{Z[\mathcal{P}; A]}{N} - \mu(A) \right|$$

where $\mu(A)$ denotes the usual area of A. The circular discrepancy of $\mathcal{P}$ is then defined as

$$\Delta[\mathcal{P}] = \sup_{A} \Delta[\mathcal{P}; A]$$

where the supremum is taken over all disc segments A of D. Finally the best discrepancy (for N points) is defined by

$$\Delta[N] = \inf_{\mathcal{P}} \Delta[\mathcal{P}]$$

$\Delta[N]$ is bounded as follows [Beck (1987)]:

$$c_1 N^{\frac{-3}{4}} (\log N)^{-\frac{7}{2}} < \Delta[N] < c_2 N^{\frac{-3}{4}} (\log N)^{\frac{1}{2}},$$

where c_1 and c_2 are constants.

1.3 Measuring $\Delta[\mathcal{P}]$

In this section, we shall show that given a distribution of N points $\mathcal{P}$, its circular discrepancy $\Delta[\mathcal{P}]$ can be measured in finite time.

We can replace the infinite set of half-plane A in the supremum with a finite subset without affecting the value of $\Delta[\mathcal{P}]$. We first identify a finite set of half-planes that would have maximum local discrepancy as follows: If let's say the edge of the half-plane A does not contain a point $p \in \mathcal{P}$, then unless this edge is a diameter, it can be translated slightly to make $\mu(A)$ bigger. This edge can also be translated in the opposite direction to make $\mu(A)$ smaller. As long as either of these two translations is small enough so that it does not cross any points in $\mathcal{P}$, the counting function $Z[\mathcal{P}; A]$ does not change. So one of the two translation results in an increase in circular-discrepancy $\Delta[\mathcal{P}; A]$.

If the edge of the half-plane A is a diameter of D, then we can rotate this edge about the center of D without affecting the value of $\mu(A)$. If the rotation is small such that it does not cross any points in $\mathcal{P}$, then the counting function $Z[\mathcal{P}; A]$ does not change. Hence all this family of rotated half-planes would have the same circular-discrepancy $\Delta[\mathcal{P}; A]$ with a particular half-plane that has a point $p \in \mathcal{P}$ on its edge.

Hence the half-plane we are looking for in maximizing the circular-discrepancy $\Delta[\mathcal{P}; A]$ has a point of p on its boundary. Thus:

Theorem 1.1 *Let $\mathcal{P}$ be a set of N points in the circular disc D of unit area. The disc segment A that achieves the maximum discrepancy with respect to $\mathcal{P}$ is one of the following types:*

(i) A contains one point $p \in \mathcal{P}$ on its boundary,
(ii) A contains two or more points of $\mathcal{P}$ on its boundary.

Theorem 1.2 *For type (i), the disc segment A that contains one point $p_i \in \mathcal{P}$ locally extremize $\mu(h)$ when the boundary of A is perpendicular to the line that contains p_i and the center of the disk. See figure 1.1(a).*

Proof: Let p be a point in $\mathcal{P}$. Without loss of generality, we can set up our coordinate axis such that p lies on the positive y-axis, with coordinates $(p_x, p_y) = (0, y)$. Consider the disc segment $A_p(\phi)$, whose boundary $l(\phi)$ contains p and makes the angle ϕ with the positive y-axis. The disc segment contains all point directly above p when $0 < \phi < \pi$, and it contains all points below p when $\pi < \phi < 2\pi$.

When $0 < \phi < \pi$, the area of $A_p(\phi)$ is:

$$\mu(A_p(\phi)) = R^2 \cos^{-1}\left(\frac{d_1}{R}\right) - d_1\sqrt{R^2 - {d_1}^2},$$

where R is the radius of D and $d_1 = y \sin\phi$ is the closest distance between $l(\phi)$ and the center of D.

When $\pi < \phi < 2\pi$, the area of $A_p(\phi)$ is:

$$\begin{aligned}\mu(A_p(\phi)) &= \pi R^2 - \mu(A_p(\phi - \pi)) \\ &= \pi R^2 - \left(R^2 \cos^{-1}\left(\frac{d_2}{R}\right) - d_2\sqrt{R^2 - {d_2}^2}\right),\end{aligned}$$

where $d_2 = y\sin(\phi - \pi) = -y\sin\phi$ is the closest distance between $l(\phi)$ and the center of D.

We want to find the value of ϕ^* which would locally extremizes $\mu(A_p(\phi))$. Scaling the unit disc D to the disc of unit radius would not change the value of ϕ^*. Hence setting $R = 1$, we get

$$\mu(A_p(\phi)) = \begin{cases} \cos^{-1}(y\sin\phi) - y\sin\phi\sqrt{1 - y^2\sin^2\phi} & \text{, if } 0 \le \phi < \pi \\ \pi - \cos^{-1}(-y\sin\phi) - y\sin\phi\sqrt{1 - y^2\sin^2\phi} & \text{, if } \pi \le \phi < 2\pi \end{cases}$$

$\mu(A_p(\phi))$ as parameterized by ϕ is continuous and smooth since it represents the area of the disc segment $A_p(\phi)$. To find the local extremas of $\mu(A_p(\phi))$, we differentiate $\mu(A_p(\phi))$ with respect to ϕ.

$$\begin{aligned}\frac{d\mu(A_p(\phi))}{d\phi} &= -\frac{y\cos\phi}{\sqrt{1 - y^2(\sin\phi)^2}} + \frac{y^3\cos\phi(\sin\phi)^2}{\sqrt{1 - y^2(\sin\phi)^2}} - y\cos\phi\sqrt{1 - y^2(\sin\phi)^2} \\ &= -y\cos\phi\sqrt{4 - 2y^2 + 2y^2\cos(2\phi)}\end{aligned}$$

Setting $d\mu(A_p(\phi))/d\phi$ to zero, we get two real solutions, $\phi = \pi/2$ and $\phi = 3\pi/2$. •

Having identified the disc segments A that achieves maximum local discrepancy for type (i), now we can proceed to calculate this discrepancy, $\Delta[\mathcal{P}; A]$.

Remark 1.1 *Let $p_i = (x_i, y_i)$ be an arbitrary point in $\mathcal{P}$. If p_i lies on the center of the disc, then the disc segment of type (i) is not uniquely defined. However, all this family of disc segment have the same area, and one could rotate the disc segment such that they contain two points of $\mathcal{P}$. Then these disc segments could be dealt with as type (ii).*

If p_i is not exactly on the center, then the disc segment A of type (i) that achieves maximum local discrepancy would have its boundary on the line $l(x_i, y_i)$, where l passes through p_i with slope $-x_i/y_i$. Two disc segments share this boundary. Let A_1 be the disc segment with smaller area. We count $Z[\mathcal{P}; A_1]$ as follows: if $y_i \geq 0$, then $Z[\mathcal{P}; A_1] = \sum_{j \neq i}^{N} g(p_j)$, where

$$g(p_j) = \begin{cases} 1 \text{ , if } p_j \text{ lies above } l(x_i, y_i) \\ 0 \text{ , otherwise.} \end{cases}$$

If $y_i < 0$, then $Z[\mathcal{P}; A_1] = \sum_{j \neq i}^{N} (1 - g(p_j))$.

The possible candidates for the discrepancy would then be:

$$\left\{ \left| A_1 - \frac{Z}{N} \right|, \left| A_1 - \frac{Z+1}{N} \right|, \left| (1 - A_1) - \frac{N-Z}{N} \right|, \left| (1 - A_1) - \frac{N-Z-1}{N} \right| \right\}$$

The second candidate arises if we include p_i inside A_1. The third and fourth candidates are due to the other (larger) disc segment, with area $1 - A_1$, that shares the boundary $l(x_i, y_i)$ with A_1.

Simplifying the third and fourth candidates shows that they are exactly the same as the first and second candidates respectively. Hence we can simplify the discrepancy $\Delta[\mathcal{P}; A]$ to

$$\max \left\{ \left| A_1 - \frac{Z}{N} \right|, \left| A_1 - \frac{Z+1}{N} \right|, \left| (1 - A_1) - \frac{N-Z}{N} \right|, \left| (1 - A_1) - \frac{N-Z-1}{N} \right| \right\}$$
$$= \max \left\{ \left| A_1 - \frac{Z}{N} \right|, \left| A_1 - \frac{Z+1}{N} \right| \right\}$$
$$= \begin{cases} A_1 - \frac{Z}{N} & \text{, if } Z < A_1 N - \frac{1}{2} \\ \frac{Z+1}{N} - A_1 & \text{, otherwise.} \end{cases}$$

As for type (ii), there are $2({}^N C_2)$ ways of choosing A such that it contains two points of P on its boundary. And for each A, we have to check $N - 2$ points to see if they lie inside A. Hence the half plane discrepancy of a set of N points in the unit disk can be computed by brute force in $O(N^3)$ time. The details for calculating $\Delta[\mathcal{P}, A]$ for type (ii) disc segments are very similar as that for type (i) and shall not be repeated here.

Point sets with three or more collinear points can be ignored by the following argument since we are really interested in low circular discrepancy sequences: The discrepancies of point sets with three or more collinear points can always be reduced by moving one of these points slightly so that it is now no longer collinear with the rest. By the same token, disc segments with three or more points on its boundary are very rare (and can be handled

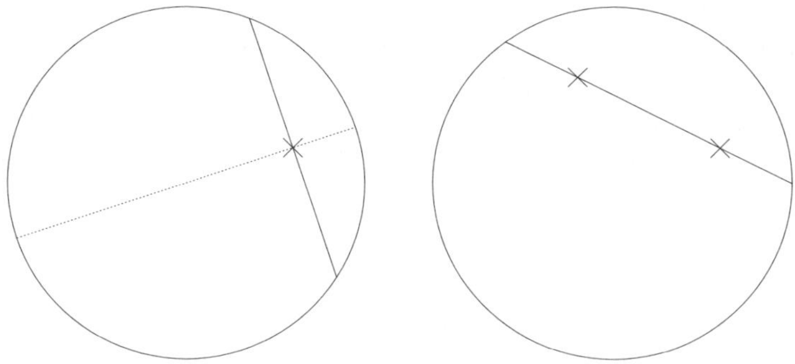

(a) (b)

Fig. 1.1 Half planes that would extremize the discrepancy. (a) The dotted line shows the line passing through the center of the circle and through a point p. The half-plane with its edge through p and perpendicular to the dotted line would extremizes the local discrepancy. (b) The half-plane with its edge passing through two points would also extremizes discrepancy locally.

separately) when computing the discrepancy of well-known sequences such as the Halton and Sobol sequences.

1.4 The vortex points

In this section, we describe a method for generating sequences with a finite number of points in D using Monte-Carlo simulations of a physical vortex gas system.

Given a finite set of N points, P_N lying on some surface, consider the functional $E(P_N) = \sum_{ij} f(r_{ij})$ where the sum is taken over all pairs of points. r_{ij} is the separation between the i-th and j-th points. If the surface is a closed n-sphere (like points on the circumference of a circle or points on surface of the globe) and if f is monotonously decreasing on that sphere, then by symmetry arguments, the set P_N that would minimizes E is one in which all the points are most spread apart. Hence we can use any arbitrary monotone function f to obtain a 'uniformly distributed' set of points on the closed spherical surfaces by numerically minimizing E on that surface.

To find such 'uniformly distributed' points inside the unit disc, we shall use the functional $E(P_N) = \sum_{ij} f(r_{ij}) + \beta \sum_i (r_i{}^2)$, where the first sum is over all pairs of points. The second sum is over all individual points and r_i is the distance of the i-th point from the origin. The second sum in E

ensures that the set P_N that minimizes E would have to be bounded in size. Also since f depends only on the separation of two points, then by symmetry arguments, the minimizing set must be radially symmetric. So the minimizing set would be bounded inside a disc of radius $R(f, \beta, N)$. The set can then be rescaled to the unit disc by dividing the coordinates of each point by R.

However unlike the previous closed n-sphere case, an arbitrary monotone f would not cause the minimizing set to be 'uniformly distributed'. The distribution of the minimizing set from such a function would depend on r. On the two dimensional plane when f is the log function, the minimizing set was shown in [Lim and Assad (2005)] to be 'uniformly distributed' inside a radius that has a simple dependence on β and N. Outside this radius, the distribution drops to zero.

This log interaction arises in the physical interaction of N point vortices of equal circulation and signature [Newton (2000)]. Their interaction is governed by the logarithmic Hamiltonian:

$$H_N(\vec{s}) = -\sum_{j<k}^{N} \ln |z_j - z_k|, \tag{1.1}$$

where $z_j \in \mathcal{C}$ is the position of the j-th vortex in the complex plane. $\vec{s}$ represents the state of the entire vortex system.

Rotational invariance of the Hamiltonian implies that the moment of vorticity:

$$\Gamma(\vec{s}) = \sum_{j=1}^{N} |z_j|^2$$

is a first integral of motion.

Hence the functional $E(P_N) = -\sum_{ij} \log(r_{ij}) + \beta \sum_i ({r_i}^2)$ corresponds to the augmented energy of an N identical point vortex gas system.

The equilibrium distribution of point vortex gas on the plane in thermal contact with an energy reservoir is described by the partition function:

$$Z(\beta, \mu) = \int_{R^2} dz_1 \ldots \int_{R^2} dz_N \exp(-\beta H) \exp(-\mu \Gamma),$$

where β is the inverse temperature, and μ is the Lagrange multiplier associated with the moment of vorticity of the system.

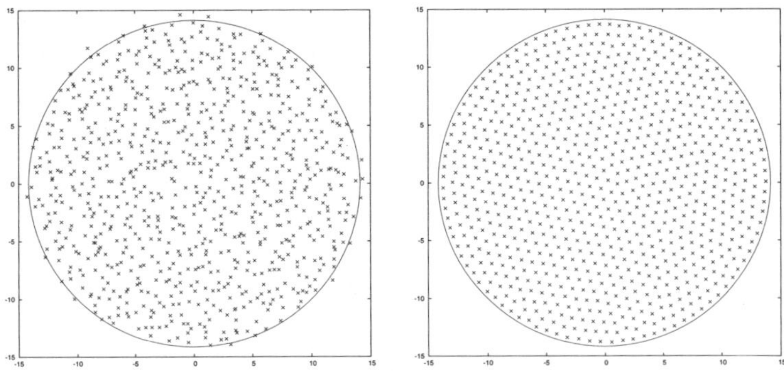

(a) (b)

Fig. 1.2 (a) Irregular vortex distribution of 400 vortices at $\beta = 1, \mu = 2$. Most of the points lies in a circle of radius 14.14. (b) Regular vortex distribution of 400 vortices at $\beta = 1000, \mu = 2000$. All the points are well spaced from each other and they all fall inside a circle of radius 14.14.

The exact partition function remains unsolved. However, to obtain the equilibrium vortex gas distribution, a Monte-Carlo simulation of the system is carried out [Assad and Lim (2005)]. When the Lagrange multipliers β and μ are large enough, the vortex gas distribution becomes crystalline [Assad and Lim (2005)]. They are regularly spaced apart in a virtual bounding circle of radius $\sqrt{\beta N/(2\mu)}$. We shall refer to this distribution as the regular vortex gas distribution.

Figure 1.2(a) shows a particular vortex distribution obtained by Monte Carlo simulations with $\beta = 1$ and $\mu = 2$. Figure 1.2(b) shows a vortex distribution at a higher value of $\beta = 1000$ and $\mu = 2000$. We are interested to find the circular discrepancy, $\Delta[\mathcal{P}]$ of these point vortices. For this, the vortices will be scaled from the circle $D_{R=\sqrt{\beta N/(2\mu)}}$ onto D by a uniform scaling about the origin of the system.

1.5 Other sequences

A lot of studies have been done to obtain low square discrepancy point sets on the unit square, $I_2 = [0,1) \times [0,1)$. While it is possible to define the square discrepancy using half planes (as in the circular discrepancy), we shall not do that here. The square discrepancy of point sets on I_2 is more commonly defined using subintervals whose sides are parallel to the edge of

I_2. Many sequences have been developed so as to achieve a low discrepancy with respect to this definition, so we shall use this definition in this paper.

Definition: *Let $\mathcal{P}$ be an arbitrary distribution of N points in I_2. For each subinterval $J = (x_1, x_2) \times (y_1, y_2) \cap I_2$, let $Z[\mathcal{P}; J]$ denote the number of elements of $\mathcal{P}$ which lies in J. We define the square discrepancy of $\mathcal{P}$ with respect to J as*

$$\Delta_s[\mathcal{P}; J] = \left| \frac{Z[\mathcal{P}; J]}{N} - \mu(J) \right|$$

where $\mu(J)$ denotes the usual area of J. The square discrepancy of $\mathcal{P}$ is then defined as

$$\Delta_s[\mathcal{P}] = \sup_J \Delta[\mathcal{P}; J]$$

where the supremum is taken over all subinterval J.

As for the circular discrepancy, we can identify a finite set of subintervals J that would have maximum local square discrepancy. Thus the square discrepancy of a sequence can also be measured in finite time.

The Hammersley sequence and the Sobol sequence are quasi-random sequences with low square discrepancy. They are generated using algorithms from number theory, rather than randomly. We shall map them onto D and then measure the corresponding circular discrepancy. To obtain a sequence in D from a sequence in I_2, we first need a suitable mapping between these two areas.

1.5.1 *Mapping the square to a disc*

We shall use two different ways of obtaining a sequence in D from one on I_2.

Method 1: Let p be a point in the unit square with Cartesian coordinates (x, y). We map it to the point p^* on the disc of unit area D, where p^* has the polar coordinates $(\sqrt{(x/\pi)}, 2\pi y)$.

Remark: *Under this mapping, an area element $dx\,dy$ in the unit square maps to an area of the same measure on D.*

Proof: On D the area element is $rdr\,d\theta$. Using $dr = dx/(2\sqrt{\pi x})$ and $d\theta = 2\pi dy$, we get the area element $rdr\,d\theta = dx\,dy$. •

A second method of obtaining points in D from a distribution is as follows. We shall pick only those points that lies in the circle inscribed in

I_2. Points outside this circle will be discarded. We make this precise with the following definition.

Method 2: Let $\mathcal{P}$ be a distribution of N points in the unit square. Then $\mathcal{P}^* = \{(x, y) \in \mathcal{P} : (x - 1/2)^2 + (y - 1/2)^2 \leq 1/4\}$ would be a distribution of $N^* \leq N$ points in the disc of radius 1/2. The distribution $\{(2(x - 1/2)/\sqrt{\pi}, 2(y - 1/2))/\sqrt{\pi} : (x, y) \in \mathcal{P}^*\}$ then consists of N^* points in D.

We shall see in section 1.6, that $\Delta_s[P_1] < \Delta_s[P_2]$ does not imply $\Delta[P_1] < \Delta[P_2]$.

Figures 1.4, 1.5 and 1.6 shows three transformations from the I_2 to D using method 1. Figures 1.7, 1.8 1.9 shows three mappings from D to the I_2 also using method 1. In figure 1.9(a), we obtained a distribution of 400 points in D by method 2 on 1024 points of the Hammersley sequence on I_2.

1.5.2 *Hammersley sequence*

The Hammersley point set for a two-dimensional point x_n is defined as [Niederreiter (1991)]:

$$x_n = \left(\frac{n}{N}, \phi_2(n)\right) \text{ for } n = 0, 1, \ldots N - 1,$$

where $\phi_2(n)$ is the base two radical inverse function. The base two radical inverse function for an integer n is given by:

$$\begin{aligned} &\text{If } n = a_0 2^0 + a_1 2^1 + \ldots + a_m 2^m, \text{ where } a_i = 0 \text{ or } 1, \\ &\text{then } \phi_2(n) = a_0 2^{-1} + a_1 2^{-2} + \ldots a_m r^{-(m+1)} \end{aligned}$$

1.5.3 *Antonov-Saleev variant of the Sobol sequence*

The Sobol sequence generates numbers between zero and one directly as binary fractions of length w bits from a set of special binary fractions, v_i, $i = 1, 2, \ldots, w$, called *direction numbers*. We shall not go into the generation of v_i except to say that it is obtained recursively from the coefficients of an arbitrary primitive polynomial in the field Z_2 [Bratley and Fox (1988); Sobol (1976)].

With the direction numbers, the Antonov-Saleev variant of the Sobol sequence for a two-dimensional point $x_n = (x_n^1, x_n^2)$ for $n = 1, 2, \ldots, m$ is obtained by [Antonov (1979)]:

$$x_n^i = g_1^n v_1^i \oplus g_2^n v_2^i \oplus \ldots \text{ for } i = 1, 2 \tag{1.2}$$

where $G(n) = \ldots g_3^n g_2^n g_1^n$ is the Gray code binary representation of n. Here $\oplus$ denotes the bitwise XOR operator.

1.5.3.1 *Example*

Suppose in the first dimension we have $v_1^1 = 0.1, v_2^1 = 0.11$ and $v_3^1 = 0.111$. The Gray code binary representation $G(n)$, for n up to six is:

$$\begin{aligned}
G(1) &= 1 \oplus 0 = 1 \\
G(2) &= 10 \oplus 01 = 11 \\
G(3) &= 11 \oplus 01 = 10 \\
G(4) &= 100 \oplus 010 = 110 \\
G(5) &= 101 \oplus 010 = 111 \\
G(6) &= 110 \oplus 011 = 101
\end{aligned}$$

To compute x_6^1, we first note from $G(6)$ that $g_1^6 = 1, g_2^6 = 0$ and $g_3^6 = 1$. Then from equation (1.2) we have,

$$\begin{aligned}
x_6^1 &= v_1^1 \oplus v_3^1 \\
&= 0.100 \oplus 0.111 \\
&= 0.011
\end{aligned}$$

In practice, we would calculate x_n^1 more efficiently using an equivalent recursive algorithm [Bratley and Fox (1988)]:

$$x_{n+1}^1 = x_n^1 \oplus v_c^1 \tag{1.3}$$

where c is the position of the right-most zero bit in the binary representation of n (adding a leading zero to n if necessary), and $x_0^1 = 0$. With this algorithm we find the first six non-zero terms of x_i^1:

$$
\begin{aligned}
x_0^1 &= 0 \ , \ c = 1; \\
x_{01}^1 &= x_0^1 \oplus v_1^1 = 0.0 \oplus 0.1 = 0.1 \ , \ c = 2; \\
x_{10}^1 &= x_{01}^1 \oplus v_2^1 = 0.10 \oplus 0.11 = 0.01 \ , \ c = 1; \\
x_{011}^1 &= x_{10}^1 \oplus v_1^1 = 0.01 \oplus 0.10 = 0.11 \ , \ c = 3; \\
x_{100}^1 &= x_{011}^1 \oplus v_3^1 = 0.110 \oplus 0.111 = 0.001 \ , \ c = 1; \\
x_{101}^1 &= x_{100}^1 \oplus v_1^1 = 0.001 \oplus 0.100 = 0.101 \ , \ c = 2; \\
x_{110}^1 &= x_{101}^1 \oplus v_2^1 = 0.101 \oplus 0.110 = 0.011 \ , \ c = 1.
\end{aligned}
$$

The values for the second dimension x_i^2 are computed in a similar manner by using a different set of direction numbers v_i^2. The implementation of the Antonov-Saleev variant of the Sobol sequence used in this paper was obtained from [Press *et al.* (1992)].

1.6 Numerical results 1

Table 1.1 shows the discrepancy values for the sequences in figures 1.4 to 1.9. All these sequences has $N = 400$. An explanation on how each sequence are generated are given as captions following each figure.

Table 1.1 Discrepancies of the 400 points in figures 1.4 to 1.9.

	Square Discrepancy	Circular Discrepancy
Fig 1.4: Uniform	0.0975	0.05
Fig 1.5: Hammersley 1	0.0136035	0.0169734
Fig 1.6: Sobol	0.0182048	0.021981
Fig 1.7: Vortex regular ($\beta = 1000, \mu = 2000$)	0.126578	0.0176981
Fig 1.8: Vortex irregular ($\beta = 1, \mu = 2$)	0.0467964	0.0248931
Fig 1.9: Hammersley 2	0.051529	0.0242217

Clearly, a sequence with low square discrepancy need not have a low circular discrepancy and vice versa. In particular, the 'regular vortex' distribution has the highest square discrepancy, $\Delta_s[\mathcal{P}]$, however its circular discrepancy $\Delta[\mathcal{P}]$ is second lowest after the 'Hammersley 1' distribution.

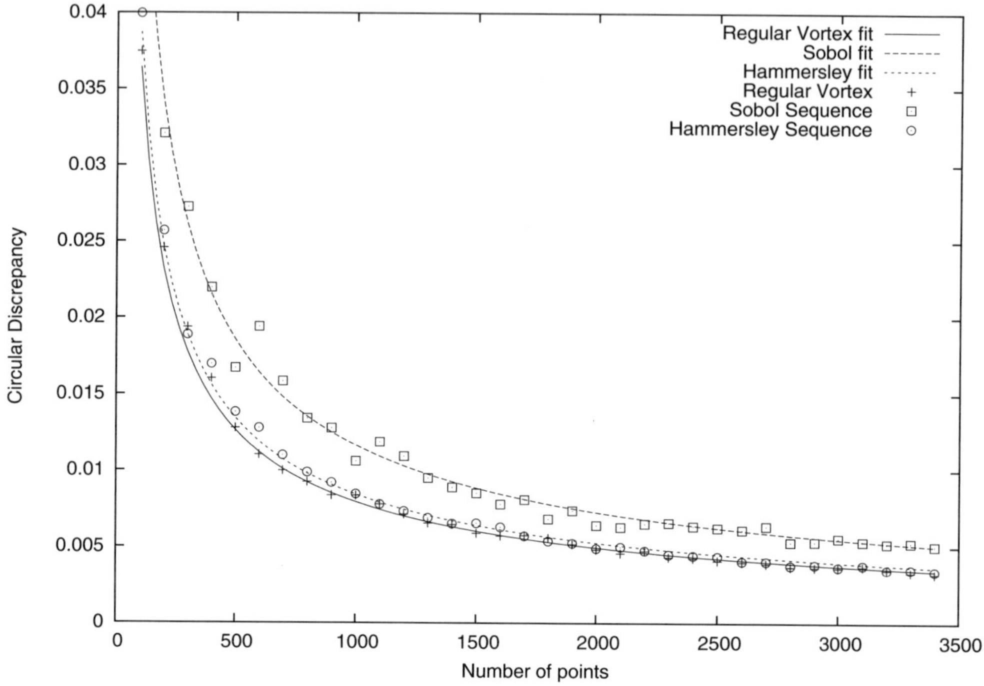

Fig. 1.3 Graph of circular discrepancy with the number of points. The data points greater than $N = 500$ are fitted to the function $f(x) = cN^{-3/4}(logN)^{1/2}$.

1.7 Numerical results 2

Here, we look at the behavior of circular discrepancy with N.

Figure 1.3 shows the circular discrepancy for the regular vortex sequence, the Sobol sequence and the Hammersley sequence for N up to 3400. The data points for the 'regular vortex' are obtained using $\beta = 2000$ and $\mu = 1000$. The distribution for the 'Hammersley sequence' and 'Sobol sequence' are obtained by applying the transformation (method 1) in section 1.5.1 to the respective sequence on the square.

The general trend is that the discrepancy decreases with more points. However this is not always the case.

For almost all N, the 'regular vortex' distribution has the lowest circular discrepancy among the three sequences, with the Hammersley sequence coming closest to it.

The data points greater than $N = 500$ are fitted to the function $f(x) =$

$cN^{-3/4}(logN)^{1/2}$ where c is a constant, using the least square fit. The constants c for each sequence are given in table 1.2.

Table 1.2 The constant for the least square fit to $f(x)$.

Sequence	c
Vortex regular ($\beta = 2000, \mu = 1000$)	0.536733
Sobol	0.787307
Hammersley	0.576499

1.8 Conclusion

The equilibrium vortex gas distribution at high β and μ can be used to provide a convenient low circular discrepancy point set. These point sets would be useful if one wishes to sample a fixed number of data on a circular domain.

These methods can be generalized to higher dimensions, $d > 2$, by modifying the Hamiltonian in equation 1.1 to:

$$H_N(\vec{s}) = \sum_{j<k}^{N} \frac{1}{|z_j - z_k|^{4-d}}.$$

The expression for the moment of vorticity and the partition function remains unchanged. At high β and μ, one would obtain a regular distribution of N points in the hyper-sphere of radius $\sqrt[d]{(d-2)\beta N/(2\mu)}$.

Acknowledgement

The work of Professor Chjan Lim is supported in part by ARO grant W911NF-05-1-0001 and DOE grant DE-FG02-04ER25616.

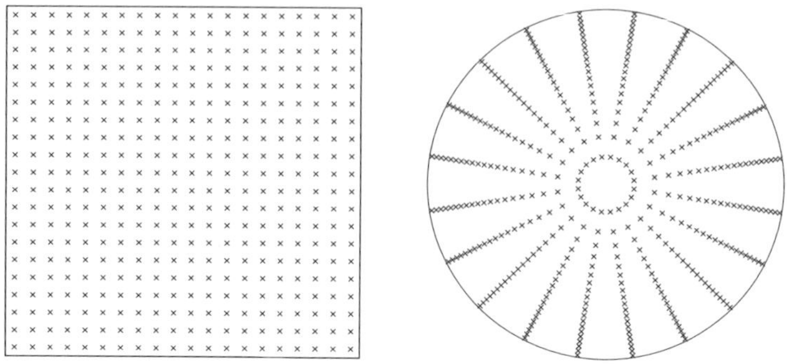

(a) (b)

Fig. 1.4 (a) Uniform distribution of 400 points in I^2. (b) Using the transformation in method 1 of section 1.5.1, the distribution becomes rays on a disc.

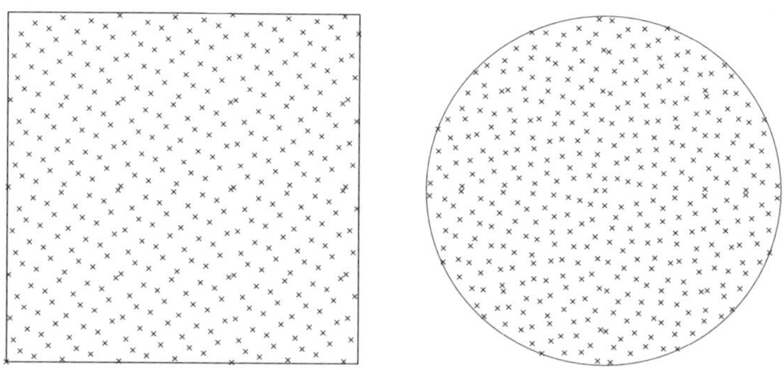

(a) (b)

Fig. 1.5 (a) 400 points from the Hammersley sequence. (b) Transformation of the Hammersley sequence onto the disc using method 1 of section 1.5.1.

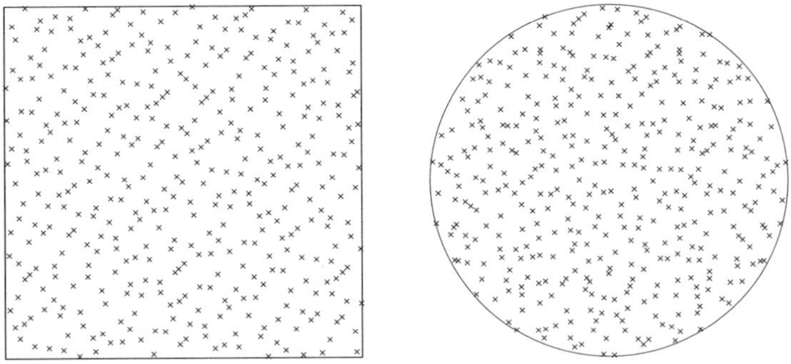

(a) (b)

Fig. 1.6 (a) 400 points from the Antonov-Saleev variant of the Sobol sequence. (b) Transformation of the Antonov-Saleev variant of the Sobol sequence onto the disc using method 1 of section 1.5.1.

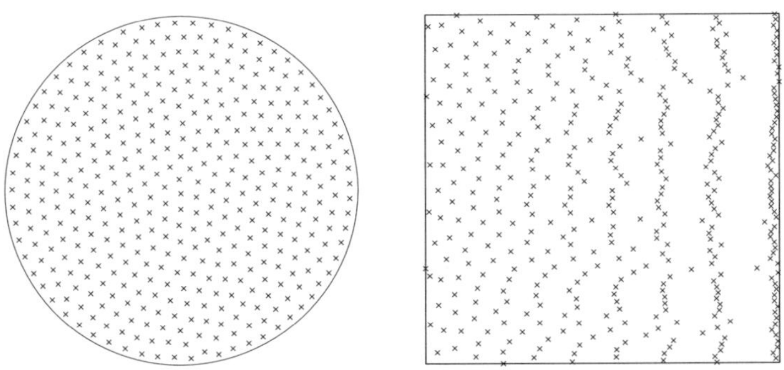

(a) (b)

Fig. 1.7 (a) A regular vortex distribution of 400 points from a particle Monte Carlo simulations at $\beta = 1000$ and $\mu = 2000$. The boundary of the sequence is set to be $\sqrt{\beta N/(2\mu)}$. (b) The regular vortex distribution is mapped onto I_2 using method 1 of section 1.5.1.

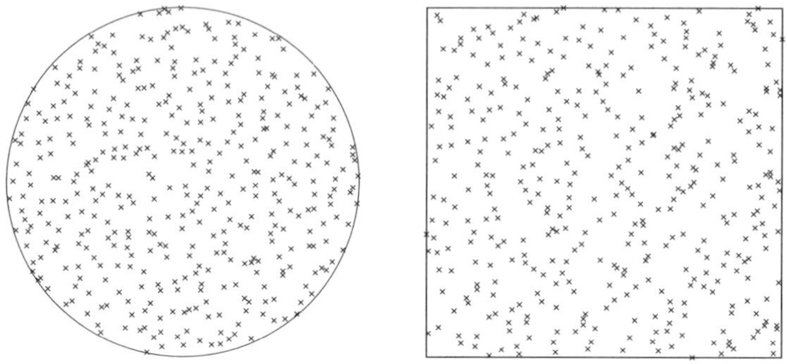

(a) (b)

Fig. 1.8 (a) Irregular vortex distribution of 400 points obtained by taking the closest 400 points from the center of a particle Monte Carlo simulation with 800 particles ran at $\beta = 1$ and $\mu = 2$. The boundary is chosen to be half the sum of the distance of the 400th particle and 401st particle. (b) The irregular vortex distribution is mapped onto I_2 using method 1 of section 1.5.1.

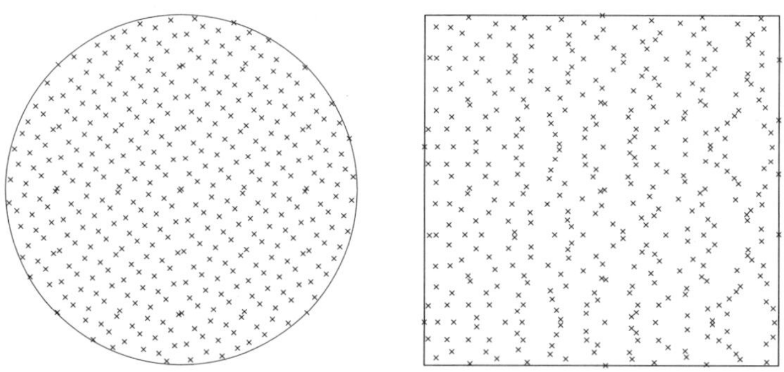

(a) (b)

Fig. 1.9 (a) The first 400 points closest to the center obtained from the Hammersley sequence of 1024 points. The boundary of the sequence is chosen at halfway between the 400th and 401st point. (b) The Hammersley sequence is mapped onto I_2 using method 1 of section 1.5.1.

Bibliography

I. A. Antonov and V. M. Saleev, *An economic method of computing LP-sequences*, USSR Computational Mathematics and Mathematical Physics **19** (1979), 252–256.

S. M. Assad and C. C. Lim, *Statistical equilibrium of the Coulomb / vortex gas on the unbounded 2-dimensional plane*, Discrete and Continuous Dynamical Systems-Series B (2005), vol. 5(1), 1–14.

J. Beck and W. W. L. Chen, *Irregularities of distribution*, Cambridge University Press, 1987.

P. Bratley and B. L. Fox, *Algorithm 659 implementing Sobol's quasirandom sequence generator*, ACM Transactions on Mathematical Software **14** (1988), no. 1, 88–100.

C. C. Lim and S. M. Assad, *A variational principle for the statistical equilibrium of single-species vortex gas and cylindrical electron plasmas*, Submitted to Regular and Chaotic Dynamics, 2005.

W. J. Morokoff and R. E. Caflisch, *Quasi-Monte Carlo Integration*, Journal of Computational Physics **122** (1995), no. 2, 218–230.

P. K. Newton, *The N-vortex problem: Analytical techniques*, Springer-Verlag New York, 2000.

H. Niederreiter, *Quasi-Monte Carlo methods and pseudo-random numbers*, Bulletin of the Mathematical Society **84** (1978), no. 6, 957–1041.

H. Niederreiter, *Construction of low-discrepancy point sets and sequences*, Sets, Graphs and Numbers (Budapest), vol. 60, 1991, 529–559.

A. B. Owen, *Monte Carlo extension of quasi-Monte Carlo*, Proceedings of the 1998 Winter Simulation Conference, 1991, 571–577.

W. H. Press, S. A. Teukolsky, W. T. Vetterling, and B .P. Flannery, *Numerical recipes in C: The art of scientific computing*, Cambridge University Press, 1992.

I. M. Sobol, *On the distribution of points in a cube and the approximate evaluation of integrals*, USSR Computational Mathematics and Mathematical Physics **16** (1976), 236–242.

Chapter 2

Periodic and Quasiperiodic Motion of Point Vortices

Denis Blackmore
Department of Mathematical Sciences and Center for Applied Mathematics and Statistics
New Jersey Institue of Technology, Newark, New Jersey 07102-1982, USA

Jyoti Champanerkar
Department of Mathematics
William Paterson University, Wayne, New Jersey 007470, USA

Dedicated to Lu Ting on his eightieth birthday

Abstract

A number of generalizations and recent extensions of the Poincaré-Birkhoff fixed point theorem are presented, along with applications to some problems in vortex dynamics. The problem of four or more point vortices in a plane is analyzed in considerable detail for the case where all the vortex strengths have the same sign. This is accomplished using a combination of KAM theory and a recent version of the Poincaré-Birkhoff fixed point theorem. For example, it is proved that if the diameter of the initial system of vortices is sufficiently small, there exists an ample set of starting configurations that produce dynamics exhibiting quasiperiodic flows on invariant KAM tori, and periodic orbits of arbitrarily large period near these tori. It is also shown that analogous dynamical behavior occurs for configurations of any finite number of point vortices in a half-plane.

2.1 Introduction

In this paper, our chief aims are essentially twofold: First, we shall describe some recent and new results concerning the existence of periodic orbits for both continuous and discrete Hamiltonian dynamical systems. Then we shall apply some of these results, together with Kolmogorov-Arnold-Moser (KAM) theory, to show that certain configurations of any finite number of point vortices moving in an *ideal* (= incompressible, inviscid) fluid in a plane or in a half-plane produce motions that are periodic or quasiperiodic and confined to invariant tori. In the course of our exposition, we shall also provide a brief introduction to Hamiltonian dynamics and describe some other vortex motion related applications in the realm of fluid dynamics.

Hamiltonian dynamics and the closely linked fields of symplectic geometry and symplectic topology have been the focus of an exceedingly active and steadily growing research concentration over the last eighty years or so. There are several reasons for the intensity of interest in Hamiltonian dynamics: The theory involves a rich blend of algebra, algebraic topology, differential equations, differential geometry, differential topology and variational methods variational that has a mathematical and esthetic elegance that makes it compelling in its own right; recent advances, especially in infinite-dimensional Hamiltonian systems with soliton and soliton-like solutions, have kept this field at the leading edge of current pure and applied research; and it has a vast and still expanding array of important applications in such areas as mechanics, optics, fluid dynamics, and quantum field theory. Two of the landmark results in Hamiltonian dynamics are the Poincaré-Birkhoff fixed point theorem and the KAM theorem (see, e.g. [2]-[4], [6]-[8], [12], [19]-[24], [27], [38], [39] and [42]). Many generalizations and extensions of these two fundamental theorems have been proved (even going so far as extending them to infinite-dimensional systems) since they were introduced, and many significant applications of them to physically based problems of current interest have been found and studied. We shall make a further contribution to this process of theoretical generalization and application of this extraordinary pair of theorems in this paper.

Applications of the theory of differential equations to vortex dominated ideal fluid flows trace back to the pioneering work of Helmholtz [26] (cf. [32] and [47]). The equations governing the flow of an ideal fluid have an intrinsic (stream function) structure that leads naturally to a variety of Hamiltonian dynamical systems, both of the finite-dimensional and infinite-dimensional types (see, e.g. [1], [4], [9], [10], [12]-[17], [28]-[30], [33]-[36],

[40], [41], [48]-[50], [52] and [53]). For example, the motions of point vortices on a surface and coaxial vortex rings in space are governed by finite-dimensional Hamiltonian systems (see e.g. [12], [15], [17], [34], [35] and [49]), while the dynamics of vortex filaments moving in an ideal fluid in space can be modeled by an infinite-dimensional Hamiltonian system (see [9] and [18]). The applications part of this paper is focused on obtaining analogs for four or more point vortices in the plane and half-plane of recent results for three or more coaxial vortex rings ([10], [12]) concerning the persistence of invariant tori and the existence of periodic solutions.

The differential equations of motion of point vortices in an ideal fluid in the plane were first formulated in detail by Gröbli [25]. He also identified some constants of motion, and indicated that the behavior of three or fewer vortices should be regular (integrable). Much later, Synge [45] was able, using trilinear coordinates, to give an almost complete characterization of the dynamics of three vortices moving in a plane. Forty years thereafter, Tavantzis & Ting [46] introduced a new integral in trilinear coordinates, which effectively combined two of the integrals used by Synge, and used it to tie up most of the loose ends in Synge's work. They also were able to give precise characterizations of the stability of certain types of initial configurations of the three point vortices. Moreover, although they did not study the vortex system from a Hamiltonian perspective, they essentially proved - in combination with the results of Gröbli and Synge - the complete integrability of the motion of three vortices in a plane, and identified three independent integrals in involution (see also Aref [1]). Ting and his collaborators have continued their investigation of point vortex and vortex ring dynamics in a recent series of papers, which include [14], [48], and [49].

It was through the work of Kirchhoff [29], Lin [36], Novikov [41], and Aref [1] that the Hamiltonian formulation of point vortex dynamics came to the fore, and this approach has produced many outstanding advances. Among these are the complete integrability of the motion of three vortices on a sphere proved by Kidambi & Newton [30] (see also [40]), the existence of quasiperiodic orbits on invariant tori for lattice vortex systems demonstrated by Lim [33], the proof by Bagrets & Bagrets [5] of the non-integrability (and resulting emergence of chaotic regimes) for general systems of four or more point vortices in a plane (see also Ziglin [53]), and several other related results such as in [10], [12], [15]-[17], [30], [33]-[35], [40], [50] and [52].

As a result of this impressive body of research, we now know, for example, that the equations of motion for point vortices moving in an ideal fluid in a plane have exactly three independent first integrals in involution. Hence the dynamics of more than three vortices is likely to be chaotic, while the dynamics of more than two point vortices in a half-plane appears to be non-integrable (see [31]). However, it is natural to ask if, in analogy with recent results on coaxial vortex rings ([10], [12]), there are some initial configurations that produce quasiperiodic motions on invariant tori and nearby periodic orbits? We provide an affirmative answer to this question here by proving that if the strengths of all the vortices have the same sign, such quasiperiodic and periodic orbits do exist for an ample set of initial configurations of vortices.

This paper is organized as follows: In Section 2.2 we briefly review the Hamiltonian concepts that are used in the sequel, including the KAM and Poincaré-Birkhoff theorems. Then, in Section 2.3, we review generalizations of the Poincaré-Birkhoff theorem, including a related conjecture of Arnold [3]. The nomenclature and theorems are then assembled to analyze the Hamiltonian equations of motion of four or more point vortices in a plane or half-plane, which we introduce in Section 2.4. We show how the Hamiltonian function for the point vortex equations can be decomposed as the sum of a certain integrable and non-integrable part in Section 2.5. In Section 2.6, we prove that the size of the integrable part can be controlled by the diameter of the initial configuration of vortices. This enables us to apply, in Section 2.7, KAM and generalized Poincaré-Birkhoff theory to prove the existence of invariant KAM tori and periodic trajectories for vortex systems that initially have certain configurations of sufficiently small diameter. Finally, we conclude in Section 2.8 with some remarks on the results obtained herein and possible related future research.

2.2 Symplectic preliminaries

In this section we present a very brief summary of the concepts and results of Hamiltonian dynamics we use in the sequel; additional information may be found in [1], [2], [4], [24], [37], [40], [38] and [43]. We assume, for convenience, in the sequel that all functions and fields are smooth, and we begin our discussion with continuous systems.

A (continuous) dynamical system in $\mathbb{R}^{2n}$ is said to be *Hamiltonian* (and to have *n-degrees of-freedom*) if it can be represented in the form

$$\dot{q}_k = \frac{\partial H}{\partial p_k}, \quad \dot{p}_k = -\frac{\partial H}{\partial q_k}, \qquad (1 \le k \le n) \tag{2.1}$$

where the dot denotes, as usual, differentiation with respect to t. H is called the *Hamiltonian function* of the system. There is a naturally associated two-form $\omega := \sum_{k=1}^{n} dq_k \wedge dp_k$ and a corresponding standard Poisson bracket defined as

$$\{f, g\} := \sum_{k=1}^{n} \left(\frac{\partial f}{\partial q_k}\frac{\partial g}{\partial p_k} - \frac{\partial f}{\partial p_k}\frac{\partial g}{\partial q_k} \right),$$

such that $(\mathbb{R}^{2n}, \omega)$ is a *symplectic manifold* pair and we also have the structure of a *Poisson manifold* $(\mathbb{R}^{2n}, \{\cdot,\cdot\})$. With this notation, (2.1) can be written in the more compact form

$$\dot{q}_k = \{q_k, H\}, \quad \dot{p}_k = \{p_k, H\}, \qquad (1 \le k \le n) \tag{2.2}$$

and the *first integrals* (*invariants* or *constants of motion*) f, of the system are defined by $\{f, H\} = 0$. Note that H is an integral, owing to the fact that $\{H, H\} = 0$.

If there exist n functionally independent integrals $\Phi_1 := H, ..., \Phi_n$ *in involution* (i.e. $\{\Phi_k, \Phi_m\} = 0$ for all $1 \le k, m \le n$) for (2.2), the system is said to be *Liouville-Arnold (LA)-integrable.* When the system is LA-integrable and has some recurrent orbits, a basic theorem of Liouville and Arnold gives a complete description of the motion as being quasiperiodic and confined to compact invariant manifolds diffeomorphic to the n-torus $\mathbb{T}^n$. Moreover, there exist *action-angle variables* $I := (I_1, ..., I_n), \theta := (\theta_1, ..., \theta_n)$ such that (2.2) can be transformed into the system

$$\dot{I}_k = \frac{\partial H}{\partial \theta_k} = 0, \quad \dot{\theta}_k = -\frac{\partial H(I)}{\partial I_k} \qquad (1 \le k \le n). \tag{2.3}$$

The KAM theorem applies to small perturbations of LA-integrable systems.

Theorem (KAM). *Suppose in* (2.2) *that* H *can be decomposed as*

$$H = H_0 + \epsilon H_1, \tag{2.4}$$

where H_0 satisfies (2.3) and $\epsilon > 0$. Then if the system satisfies the nondegeneracy condition

$$\det\left(\frac{\partial^2 H_0}{\partial I_j \partial I_k}\right) \neq 0$$

and ϵ is sufficiently small, the Hamiltonian system has a set $\mathcal{T} = \{T_\alpha\}$ of invariant manifolds (KAM tori) of positive (Lebesgue) measure, with all T_α diffeomorphic to $\mathbb{T}^n$.

Thus the KAM theorem guarantees the existence of quasiperiodic solutions for non-integrable systems that are small perturbations of LA-integrable ones. The Poincaré-Birkhoff theorem and its generalizations, on the other hand, address the existence of periodic trajectories.

It is convenient to treat the Poincaré-Birkhoff theorem in the context of a discrete dynamical system obtained from the surface of section map. Whence we obtain a diffeomorphism $f : \mathbb{T}^{n-1} \times \mathbb{B}^{n-1} \to \mathbb{T}^{n-1} \times \mathbb{B}^{n-1}$, where $\mathbb{B}^{n-1}$ is the closed unit ball in $\mathbb{R}^{n-1}$. The map is a *symplectomorphism*, i.e. it satisfies $f^*\omega = \omega$, where f^* is the standard pullback map on forms. We note that this correspondence between continuous Hamiltonian systems and symplectomorphisms actually goes both ways under fairly mild restrictions (see [7] and [24]), so most results for discrete systems have obvious discrete analogs and vice versa. For example, an LA-integrable map f can be defined in a rather manifest way, whereupon one obtains the discrete analog of the KAM theorem. In the same spirit, one obtains a continuous version of the following result; the proof of which was begun by Poincaré [42] and later completed by Birkhoff [8].

Theorem (P-B). *Let the symplectomorphism $f : \mathbb{S}^1 \times \mathbb{B}^1 \to \mathbb{S}^1 \times \mathbb{B}^1$ be a twist map, i.e. it moves the components of the boundary $\partial(\mathbb{S}^1 \times \mathbb{B}^1) = \mathbb{S}^1 \times \{\pm 1\}$ in opposite directions. Then f has at least two fixed points.*

2.3 Fixed point theorems

There have been many extensions of Theorem P-B; significant examples of which may be found in [3], [6], [7], [12], [19]-[24], [27], [37], [39] and [51]. Most of these generalizations include estimates for the number of fixed points for maps that are something like *twist maps*, which satisfy the property that $\psi : \mathbb{T}^m \times \mathbb{T}^m \to \mathbb{T}^m \times \mathbb{T}^m$ defined as $\psi(\xi, \eta) := (\xi, f_2(\xi, \eta))$

is an embedding, where $f = (f_1, f_2)$. Usually, the estimates are sharper when the fixed points are nondegenerate, which can always be achieved by an arbitrarily small C^1 perturbation (see [43]).

A number of the most sweeping generalizations of Theorem P-B have been achieved as a result of efforts to try to resolve the following conjecture.

Conjecture (**Arnold**). *Let* $f : \mathbb{T}^m \times \mathbb{B}^m \to \mathbb{T}^m \times \mathbb{B}^m$ *be a symplectomorphism that is exact* (*see* [3] and [24]), *homotopic to the identity, and links spheres on the boundary in the sense that for every* $\xi \in \mathbb{T}^m$, *the spheres* $\xi \times \partial\mathbb{B}^m$ *and* $f(\xi \times \partial\mathbb{B}^m)$ *are linked in* $\mathbb{R}^m \times \partial\mathbb{B}^m$. *Then* f *has at least* 2^m *fixed points* (*counting multiplicity*).

In fact, one of the most general extant versions of Theorem P-B is the following special case of Arnold's conjecture (cf. [24]).

Theorem (**Banyaga-Golé**). *Suppose that* f *is a symplectic twist map linking spheres in the manner of Arnold's conjecture. Then it has at least* 2^m *fixed points (counting multiplicity).*

The proof of the Banyaga-Golé theorem requires a strong admixture of both topological and variational techniques. In contrast, the next result of Blackmore *et al.* [12], which is all we shall need for the applications to point vortex dynamics in the sequel, employs essentially only some computations of Lefschetz numbers.

Theorem (**BCW**). *Every sufficiently small* C^1 *perturbation* f *of an LA-integrable symplectic twist map* $f_0 : \mathbb{T}^m \times \mathbb{B}^m \to \mathbb{T}^m \times \mathbb{B}^m$ *has at least* 2^m *fixed points (counting multiplicities).*

2.4 Hamiltonian formulation of point vortex dynamics

For our applications, we consider n point vortices in an ideal fluid in $\mathbb{R}^2$, or in the half-plane $\mathbb{H}^2 := \{(x, y) \in \mathbb{R}^2 : y > 0\}$, located at the points $(x_1, y_1), ..., (x_n, y_n)$ with the, respective, nonzero strengths $\Gamma_1, ..., \Gamma_n$, as shown in Fig. 2.1 and Fig. 2.2, respectively. The equations of motion of the configuration of point vortices in the plane can be expressed in the complex Hamiltonian form (cf. [1], [5], [13]-[17], [25], [28]-[36], [40], [41],

[45], [46] and [53]).

$$\kappa_j \dot{\bar{z}}_j = i \sum_{m=1, m\neq j}^{n} \kappa_j \kappa_m \left(z_m - z_j\right)^{-1} = 2i \frac{\partial H}{\partial z_j}, \qquad (1 \leq j \leq n) \tag{2.5}$$

which is equivalent to

$$\kappa_j \dot{x}_j = \frac{\partial H}{\partial y_j}, \quad \kappa_j \dot{y}_j = -\frac{\partial H}{\partial x_j}, \qquad (1 \leq j \leq n) \tag{2.6}$$

where

$$H = -\sum_{1\leq j<m\leq n} \kappa_j \kappa_m \log\left|z_m - z_j\right|, \tag{2.7}$$

$z_j := x_j + iy_j \in \mathbb{C}$ (the complex plane), the overbar denotes complex conjugation, and $\kappa_j := \Gamma_j/2\pi$, $1 \leq j \leq n$. Similarly, the equations of motion of the point vortices in the half-plane $\mathbb{H}^2$ are

$$\begin{aligned} &\kappa_j \dot{\bar{z}}_j = i \sum_{m=1, m\neq j}^{n} \kappa_j \kappa_m \left(z_m - z_j\right)^{-1} - i \sum_{m=1}^{n} \kappa_j \kappa_m \left(\bar{z}_m - z_j\right)^{-1} = 2i \frac{\partial \tilde{H}}{\partial z_j}, \\ &\iff \\ &\kappa_j \dot{x}_j = \frac{\partial \tilde{H}}{\partial y_j}, \quad \kappa_j \dot{y}_j = -\frac{\partial \tilde{H}}{\partial x_j}, \qquad (1 \leq j \leq n) \end{aligned} \tag{2.8}$$

where the Hamiltonian function is

$$\tilde{H} = -\sum_{1\leq j<m\leq n} \kappa_j \kappa_m \log\left|z_m - z_j\right| + \sum_{1\leq j\leq m\leq n} \kappa_j \kappa_m \log\left|\bar{z}_m - z_j\right|. \tag{2.9}$$

For the vortex dynamics applications that we shall consider in a Hamiltonian context, it is convenient to replace the standard Poisson bracket above with

$$\{f, g\} := \sum_{m=1}^{n} \kappa_m^{-1} \left(\frac{\partial f}{\partial q_m} \frac{\partial g}{\partial p_m} - \frac{\partial f}{\partial p_m} \frac{\partial g}{\partial q_m} \right), \tag{2.10}$$

where $q_m := x_m$, and $p_m := y_m$, $1 \leq m \leq n$, where our use here of the same notation as above should not cause any confusion in the sequel. We can now represent the equations of motion of point vortices in $\mathbb{R}^2$ in the

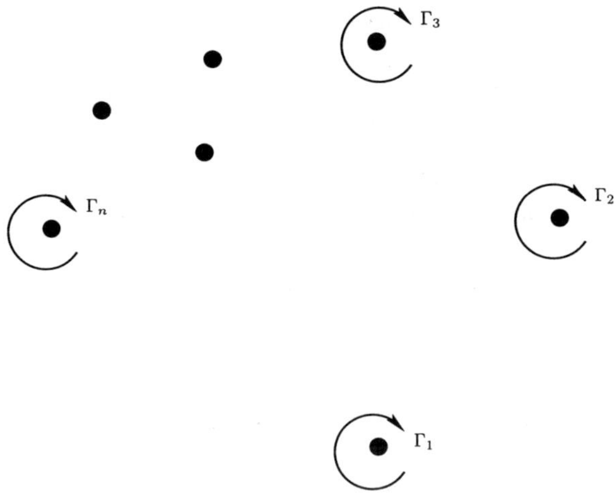

Fig. 2.1 Point vortices in the plane

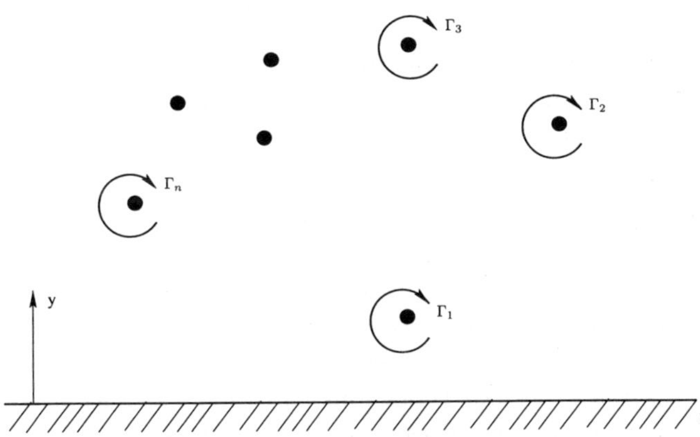

Fig. 2.2 Point vortices in a half-plane

very concise form

$$\dot{q}_m = \{q_m, H\} = \kappa_m^{-1} \frac{\partial H}{\partial p_m},$$
$$\dot{p}_m = \{p_m, H\} = -\kappa_m^{-1} \frac{\partial H}{\partial q_m}, \tag{2.11}$$

$(1 \le m \le n)$, where for convenience we repeat that

$$H = -\sum_{1\le j<m\le n} \kappa_j \kappa_m \log |z_m - z_j| , \tag{2.12}$$

and the equations for point vortex dynamics in $\mathbb{H}^2$ in the analogous Poissonian-Hamiltonian mode

$$\begin{aligned} \dot{q}_m &= \left\{q_m, \tilde{H}\right\} = \kappa_m^{-1} \frac{\partial \tilde{H}}{\partial p_m}, \\ \dot{p}_m &= \left\{p_m, \tilde{H}\right\} = -\kappa_m^{-1} \frac{\partial \tilde{H}}{\partial q_m}, \end{aligned} \tag{2.13}$$

$(1 \le m \le n)$, where we once again have

$$\tilde{H} = -\sum_{1\le j<m\le n} \kappa_j \kappa_m \log |z_m - z_j| + \sum_{1\le j\le m\le n} \kappa_j \kappa_m \log |\bar{z}_m - z_j| . \tag{2.14}$$

It will be enlightening to describe some of the invariants of the systems (2.11) and (2.13), and their consequences regarding integrability. Treating (2.11) from the point of view of classical mechanics, it is easy to see that the energy H, linear momentum $J := \kappa_1 z_1 + \cdots + \kappa_n z_n$, and angular momentum $K := \kappa_1 |z_1|^2 + \cdots + \kappa_n |z_n|^2$ are all conserved. Whence H, $\Re J$, $\Im J$, and K comprise a set of constants of motion of the system, where $\Re$ and $\Im$ denote, respectively, the real and imaginary parts of a complex number. One can also obtain these integrals taking the Noetherian approach of observing that H is invariant under the translation and rotation groups in the plane. From these four constants of motions we can construct the following three functionally independent integrals in involution: H, K and $L := (\Re J)^2 + (\Im J)^2$, satisfying $\{H, K\} = \{H, L\} = \{K, L\} = 0$. Consequently, the dynamics of three or fewer point vortices in the plane is LA-integrable - exhibiting periodic and quasiperidic motion on invariant tori. However, there cannot be more than three independent integrals in involution, so four or more point vortices in $\mathbb{R}^2$ is likely to exhibit chaotic dynamics (cf. [5] and [53]).

For the dynamics of point vortices moving in the half-plane, we observe that $\tilde{H}$ is invariant with respect to the one-dimensional Lie group of translations along the y-axis, so it follows that, in addition to $\tilde{H}$, (2.13) possesses the constant of motion $\Im J$. As $\tilde{H}$ and $\Im J$ are manifestly functionally independent, we infer that the equations of motion of one or two point vortices in $\mathbb{H}^2$ are LA-integrable. It appears from a variety of careful numerical experiments, such as those conducted by Knio *et al.* [31], that chaotic regimes occur for three or more vortices. Consequently, (2.13) should not, in general, have any additional independent integrals in involution, but it appears that this has not yet been proved.

We shall assume in the sequel that

$$(\mathcal{P}\mathbf{1})\quad \kappa_1 + \cdots + \kappa_n \neq 0.$$

This involves no real loss of generality, since we can always reduce the number of degrees-of-freedom of the system using the integrals in involution (as described say in [10], [16] and [33]) such that the sum of the strengths of the remaining point vortices is nonzero. It will be convenient to track the motion of (2.11) and (2.13) in terms of coordinates with respect to the *center of vorticity* of the vortex configuration, defined as

$$z_c := \frac{\sum_{m=1}^{n} \kappa_m z_m}{\sum_{m=1}^{n} \kappa_m} = (q_c, p_c) = (x_c, y_c)\,. \tag{2.15}$$

This amounts to making the change of variables $Z = (Z_1, ..., Z_n) = (X_1 + iY_1, ..., X_n + iY_n) := z - z_c$, where $z := (z_1, ..., z_n)$.

It follows from our discussion of integrals of (11) and (2.13) that for the motion of point vortices in $\mathbb{R}^2$, the coordinate system centered at z_c will be stationary, while for the dynamics described by (2.13), the coordinate system moves parallel to the x-axis. Consequently, the (Hamiltonian) equations of motion of (2.11) are unchanged by this transformation, and we may take $Z = z$ for this system. Note that we therefore have the additional property $\kappa_1 z_1 + \cdots + \kappa_n z_n = \kappa_1 Z_1 + \cdots + \kappa_n Z_n = 0$. On the other hand, the new coordinate system for (2.13) translates along the x-axis, and we readily find that the transformed equations of motion take the (non-Hamiltonian) form

$$\kappa_j \dot{X}_j = \Bigg\{ \sum_{m=1, m\neq j}^{n} \kappa_j \kappa_m D_{jm}^{-1} (Y_m - Y_j) + \sum_{m=1}^{n} \kappa_j \kappa_m \mathcal{D}_{jm}^{-1} \mathcal{Y}_{jm} - \left(\sum_{m=1}^{n} \kappa_m \right)^{-1} \sum_{l=1}^{n} \kappa_l \Bigg[\sum_{m=1, m\neq l}^{n} \kappa_j \kappa_m D_{lm}^{-1} (Y_m - Y_l) + \sum_{m=1}^{n} \kappa_j \kappa_m \mathcal{D}_{lm}^{-1} \mathcal{Y}_{lm} \Bigg] \Bigg\}, \tag{2.16}$$

$$\kappa_j \dot{Y}_j = - \sum_{m=1, m\neq j}^{n} \kappa_j \kappa_m D_{jm}^{-1} (X_m - X_j) + \sum_{m=1}^{n} \kappa_j \kappa_m \mathcal{D}_{jm}^{-1} (X_m - X_j),$$

where $D_{jm} := (X_m - X_j)^2 + (Y_m - Y_j)^2$, $\mathcal{Y}_{jm} := Y_m + Y_j + 2y_c(0)$, and $\mathcal{D}_{jm} := (X_m - X_j)^2 + \mathcal{Y}_{jm}^2$ for $1 \leq j, m \leq n$. Now just as in [10] and [12], we first use the identity $\kappa_1 Z_1 + \cdots + \kappa_n Z_n = 0$ to reduce this system by one degree-of-freedom, and then apply a Jacobi linear transformation to convert the reduced system to an equivalent new $(n-1)$-degree-of-freedom Hamiltonian form of the type

$$\dot{q}_m = \left\{ q_m, \tilde{\mathcal{H}} \right\} = \kappa_m^{-1} \frac{\partial \tilde{\mathcal{H}}}{\partial p_m}, \quad \dot{p}_m = \left\{ p_m, \tilde{\mathcal{H}} \right\} = -\kappa_m^{-1} \frac{\partial \tilde{\mathcal{H}}}{\partial q_m}, \tag{2.17}$$

$(1 \leq m \leq n-1)$, where the new Hamiltonian function $\tilde{\mathcal{H}}$ has essentially the same form as H in (2.11), except with two fewer variables. We leave the details of (2.17), which can be easily determined by following the procedure in [12], to the reader.

The dynamics of (2.11) with respect to z_c is not integrable if $n > 3$, and it appears that (2.13) with respect to a coordinate system moving with z_c (i.e. (2.17)) is not integrable when $n > 2$. However, in light of the work of Khanin [28] and some recent results on the Hamiltonian dynamics of coaxial vortex rings in an ideal fluid ([10], [12]), it is natural to speculate that some regular features of an integrable system might still persist for certain initial configurations of four or more point vortices in a plane or three or more point vortices in a half-plane. We shall verify these conjectures in what follows.

2.5 Decomposition of the Hamiltonian

As in [10] and [12], we first decompose the Hamiltonian H for point vortices in a plane, and also the Hamiltonian $\tilde{\mathcal{H}}$ for point vortices in a half-plane, into the sum of an integrable part and a non-integrable part whose size can be controlled if all of the vortices are initially sufficiently close together. The results that we shall prove all assume that the vortex strengths have the same sign. Accordingly as the equations of motion (2.11) and (2.13) are invariant under the change of variable and parameters $t \to -t$, $\kappa_m \to -\kappa_m$ $(1 \leq m \leq n)$, we assume without loss of generality that the vortex strengths are all positive; namely

> **($\mathcal{P}$2)** κ_m are nonzero real numbers of the same sign $\Leftrightarrow$ $\kappa_m > 0$ for all $1 \leq m \leq n$.

This implies that the center of vorticity z_c belongs to the convex hull of the vortex points (and the interior of the hull if the configuration is not collinear).

From now on, we also assume (by translation) that the origin of coordinates in the plane or half-plane is at the center of vorticity; to wit

> **($\mathcal{P}$3)** $z = (z_1, ..., z_n)$ for (11) ($= (z_1, ..., z_{n-1})$ for (2.17)) is with respect to z_c, so $\sum_{j=1}^{n} \kappa_j z_j = 0$.

Let $\delta(z)$ denote the diameter of the vortex configuration. A key result in estimating a non-integrable part of the Hamiltonian function for (2.11) and (2.17) is the following useful result, which shows that the diameter of the vortex configuration remains small for all time if it is initially sufficiently small.

Lemma 1. *Let* (2.11) *and* (2.17) *satisfy properties* $\mathcal{P}$1-$\mathcal{P}$3. *Then for every* $\epsilon > 0$ *in each of these dynamical systems, there exists a* $0 < \sigma < \epsilon$ *such that* $\delta(z(t)) < \epsilon$ *for all* $t \geq 0$ *whenever* $\delta(z(0)) < \sigma$.

Proof. For the system (2.11), the desired result follows immediately from $\mathcal{P}$2 and the invariance of $K = \sum_{j=1}^{n} \kappa_j |z_j|^2$. To obtain the analogous result for (2.17) is only slightly more difficult. In particular, we infer from $\mathcal{P}$2, $\mathcal{P}$3 and the invariance of the Hamiltonian function, in a manner completely analogous to that given in [10] and [12], that there exists a positive constant C such that $\delta(z(t)) \leq C\delta(z(0))$ for all $t \geq 0$. This completes the proof. □

We are now ready to describe the decomposition of the Hamiltonians of (2.11) and (2.17) into an integrable part plus a controllably small non-integrable part for sufficiently small diameter configurations of point vortices. For $n \times n$ symmetric matrices $A = (A_{jm})$ and $B = (B_{jm})$ with positive entries and positive constants c and ϵ, we define the *configuration space* $\mathcal{C} = \mathcal{C}(A, B, c, \epsilon)$ to be the set of points $z := (z_1, ..., z_n)$, or $z := (z_1, ..., z_{n-1})$ for the half-space system, satisfying the properties: (i) $\kappa_1 z_1 + \cdots + \kappa_n z_n = 0$; (ii) $z_j \neq z_m$ whenever $j \neq m$; (iii) $\delta(z) \leq \epsilon$; and (iv) the following inequalities obtain for all $1 \leq j, m \leq n$, or $1 \leq j, m \leq n-1$, respectively :

$$|\log|z_j - z_m| - A_{jm} \log|z_j| - B_{jm} \log|z_m||$$

$$+ ||z_j - z_m| - A_{jm}|z_j| - B_{jm}|z_m|| \leq c\,\delta(z).$$

It is easy to see (via a scaling) that there are choices of A, B and c such that there are subsets of positive measure in every neighborhood of $z_c = (q_c, p_c)$ admitting point vortex configurations in $\mathcal{C}$.

In light of the above definition of $\mathcal{C}$, first we decompose the Hamiltonian function (2.12) as

$$H = H_0 + H_1, \tag{2.18}$$

where

$$\begin{aligned} H_0 &:= - \sum_{1 \leq j < m \leq n} \kappa_j \kappa_m \left(A_{jk} \log|z_j| + B_{jk} \log|z_m|\right) \\ &= - \sum_{m=1}^{n} \beta_m \log|z_m|, \end{aligned} \tag{2.19}$$

with $\beta_1, ..., \beta_n$ positive constants defined in the obvious way, and $H_1 := H - H_0$. In an entirely analogous fashion, we write the Hamiltonian function for (2.17) in the form

$$\tilde{\mathcal{H}} = \mathcal{H}_0 + \mathcal{H}_1, \tag{2.20}$$

with

$$\begin{aligned} \mathcal{H}_0 &:= - \sum_{1 \leq j < m \leq n-1} \chi_{jm} \left(A_{jk} \log|z_j| + B_{jk} \log|z_m|\right) \\ &= - \sum_{m=1}^{n-1} \tilde{\beta}_m \log|z_m|, \end{aligned} \tag{2.21}$$

where χ_{jm} are the positive analogs of $\kappa_j\kappa_m$ for the reduced half-space system, and $\tilde{\alpha}, \tilde{\beta}_1, ..., \tilde{\beta}_{n-1}$ are positive constants. Then we define $\mathcal{H}_1 := \tilde{\mathcal{H}} - \mathcal{H}_0$. It is easy to see that both H_0 and $\mathcal{H}_0$ are LA-integrable, since they each have a separable Hamilton-Jacobi form. However, we still have to show that H_0 and $\mathcal{H}_0$ can be made arbitrarily large in comparison to H_1 and $\mathcal{H}_1$, respectively, for all times in order to apply the KAM and generalized Poincaré-Birkhoff theorems to the motion of the point vortex configuration in the plane and the half-space. The necessary estimates shall be derived in the next section.

2.6 Estimates of the non-integrable perturbations

We shall now show that the size of the non-integrable second term in (2.18) and (2.20) can be kept as small as we desire in comparison to the corresponding integrable first term if the initial configuration of vortices is chosen properly. Our two basic results for verifying this are expressed in the following lemmas, whose proofs we only sketch since they are essentially the same as those provided in [10] and [12]:

Lemma 2. *Let z represent the configuration of point vortices for the system* (2.11) *in the plane or the system* (17) *in the half-plane, and suppose that $\mathcal{P}1 - \mathcal{P}3$ are obtained. Then given any $c, \epsilon > 0$ sufficiently small, there exist matrices A and B and positive constants $d \geq c$ and $\nu \leq \epsilon$ such that if the initial configuration of vortices satisfies $z(0) \in \mathcal{C}(A, B, c, \nu)$, it follows that $z(t) \in \mathcal{C}(A, B, d, \epsilon)$ for all $t \geq 0$.*

Proof. Property (iii) in the definition of the configuration space follows immediately from Lemma 1 for both systems (2.11) and (2.17). As (i) is a direct consequence of $\mathcal{P}3$, it only remains to prove the part of the result associated with the inequalities (iv) in the definition of $\mathcal{C}$. To this end, we readily compute from the definitions of the Hamiltonian functions and the diameter that, owing to $\mathcal{P}2$, we have

$$-H \leq \sum_{1 \leq j < m \leq n} \kappa_j \kappa_m \log \delta(z) = C \log \delta,$$

where C is a positive constant. Moreover, a similar computation for (2.17) also yields the inequality $-\tilde{\mathcal{H}} \leq C \log \delta$ for sufficiently small diameters. Whence the desired results follows directly from the definitions and the fact that H and $\tilde{\mathcal{H}}$ are constants of motion of their respective dynamical systems, in a manner entirely analogous to that in the proofs of the corresponding property for coaxial vortex rings given in [10] and [12]. Thus the proof is complete. □

Lemma 3. *Let the systems* (2.11) *and* (2.17) *satisfy the same hypotheses as in Lemma 2. For any* $c, \epsilon > 0$ *sufficiently small, there exist matrices* A *and* B *and positive constants* c *and* $\nu \leq \epsilon$ *such that if the initial configuration of vortices satisfies* $z(0) \in \mathcal{C}(A, B, c, \nu)$, *then*

$$\left|\frac{H_1}{H_0}\right|, \left|\frac{H_1'}{H_0'}\right| \leq \epsilon,$$

and

$$\left|\frac{\mathcal{H}_1}{\mathcal{H}_0}\right|, \left|\frac{\mathcal{H}_1'}{\mathcal{H}_0'}\right| \leq \epsilon,$$

for all $t \geq 0$, *where the prime denotes the derivative.*

Proof. The desired estimates are direct consequences of Lemma 1, Lemma 2 and the definitions of H_0, H_1, $\mathcal{H}_0$, $\mathcal{H}_1$ and the configuration space $\mathcal{C}$. In particular, the inequalities (iv) in the description of $\mathcal{C} = \mathcal{C}(A, B, c, \epsilon)$ play a pivotal role in the computations (cf. [10] and [12]). Hence the proof is complete. □

2.7 Application of KAM and fixed point theorems

The stage is now set to apply Theorems KAM and BCW in order to prove the existence of quasiperiodic trajectories on invariant KAM tori and periodic orbits for certain initial configurations of vortices - in both the plane and half-plane - having sufficiently small diameters. We begin with the vortex dynamics in the plane. The action-angle variables (I, θ) defined as

$$I_k e^{i\theta_k} := q_k + ip_k, \qquad (1 \leq k \leq n) \tag{2.22}$$

enable us to express H_0 in the form

$$H_0 = H_0(I) = -\sum_{k=1}^{n} \beta_k \log I_k. \tag{2.23}$$

Accordingly the Hamiltonian dynamical system generated by H_0 assumes the particularly simple, uncoupled, obviously LA-integrable form

$$\dot{I}_k = \frac{\partial H_0}{\partial \theta_k} = 0, \quad \dot{\theta}_k = -\frac{\partial H_0}{\partial I_k} = \frac{\beta_k}{I_k} \qquad (1 \leq k \leq n). \tag{2.24}$$

Next, using the same form of action-angle variables as in (2.22), except with one less degree-of-freedom, we find for the integrable part of the Hamiltonian function of the vortex dynamics in the half-plane that

$$\mathcal{H}_0 = \mathcal{H}_0(I) = -\sum_{k=1}^{n-1} \tilde{\beta}_k \log I_k, \tag{2.25}$$

which generates the LA-integrable system

$$\dot{I}_k = \frac{\partial \mathcal{H}_0}{\partial \theta_k} = 0, \quad \dot{\theta}_k = -\frac{\partial \mathcal{H}_0}{\partial I_k} = \frac{\tilde{\beta}_k}{I_k} \qquad (1 \leq k \leq n-1). \tag{2.26}$$

With these preliminaries taken care of, we can now apply the KAM theorem to prove the existence of invariant tori.

Theorem 1. *Let both systems* (2.11) *and* (2.17) *satisfy* $\mathcal{P}1 - \mathcal{P}3$. *Then there exists an* $\epsilon > 0$ *for which there are initial configurations of the vortices in both the plane and half-plane satisfying* $0 < \delta(z(0)) < \epsilon$ *such that the following properties obtain:*

(a) *The dynamical system* (2.11) *has a set of invariant* n*-tori of positive measure in* $\mathbb{R}^{2n}$, *on each of which the flow is quasiperiodic.*

(b) *There is a set of invariant* $(n-1)$*-tori of positive measure in* $\mathbb{R}^{2(n-1)}$ *for* (2.17) *comprised of quasiperiodic orbits.*

Proof. It follows from Theorem KAM and Lemmas 1, 2 and 3 that it suffices to verify that the nondegeneracy condition holds uniformly as $I \to 0$. Now we compute from (2.23) that

$$\Delta := \det\left(\frac{\partial^2 H_0}{\partial I_j \partial I_k}\right) = (-1)^n \prod_{k=1}^{n} \beta_k I_k^{-2},$$

and from (2.25) that

$$\tilde{\Delta} := \det\left(\frac{\partial^2 \tilde{\mathcal{H}}_0}{\partial I_j \partial I_k}\right) = (-1)^n \prod_{k=1}^{n-1} \tilde{\beta}_k I_k^{-2},$$

so $|\Delta| \to \infty$ and $\left|\tilde{\Delta}\right| \to \infty$ as $I \to 0$. Moreover, it is easy to show that $|H_1/H_0|, |\mathcal{H}_1/\mathcal{H}_0|, |H_0/\Delta|, \left|\mathcal{H}_0/\tilde{\Delta}\right|, |H_1/\Delta|$ and $\left|\mathcal{H}_1/\tilde{\Delta}\right| \to 0$ as $I \to 0$, uniformly in θ on compact subsets (cf. [10] and [12]). Whence the nondegeneracy follows, and both parts (a) and (b) of the theorem are proved. □

We note that the existence of invariant tori for point vortex dynamics in the half-plane $\mathbb{H}^2$ as described by (b) of the above theorem with respect to the center of vorticity, corresponds to invariant sets diffeomorphic to $\mathbb{R} \times \mathbb{T}^{(n-1)}$ with respect to the original cartesian coordinates in (2.13).

Theorem 2. *For each $m \in \mathbb{N}$ there exists an $\epsilon_m > 0$ for which there are initial configurations of the vortices satisfying $0 < \delta(z(0)) < \epsilon_m$ such that the dynamical systems* (2.11) and (2.17) *have periodic orbits of at least m different periods, provided that the dynamical systems satisfy the same hypotheses as in Theorem 1.*

Proof. First, we deal with the dynamical system for point vortex motion in the plane. It is clear from the form of (2.24) that it has Poincaré surface of section maps whose m-fold compositions have fixed points when I is sufficiently small. The m-fold composition of the Poincaré maps of (2.11), which we denote as f_m, can be viewed, owing to Lemmas 1, 2 and 3, as arbitrarily small C^1 perturbations of an LA-integrable symplectic twist map f_0 corresponding to an m-fold composition of a Poincaré map of (2.11). Hence, if I is sufficiently small, it follows from the proof of Theorem 1 and (2.24) that the f_m are also symplectic twist maps. Consequently, we obtain the desired result for (11) from Theorem BCW (cf. [12]).

It remains to prove the existence of periodic orbits for the half-plane dynamics, but to this end the argument for vortex dynamics in a plane can be repeated almost verbatim. In an entirely analogous manner, we can view the Poincaré map for (2.17) as an arbitrarily small C^1 perturbation of the LA-integrable symplectic twist map obtained from the Poincaré map associated to (2.26). Thus we obtain the desired result for the half-plane vortex dynamics described by (2.17) by applying Theorem BCW just as in the case of (2.11), thereby completing the proof. □

2.8 Concluding remarks

We have shown how the KAM theorem and some generalized versions of the Poincaré-Birkhoff theorem can be used to prove the existence of KAM tori, quasiperiodic trajectories and periodic orbits for certain configurations of small diameter in non-integrable point vortex motion in the plane or the half-plane. As a system of two or three such vortices in the plane, or one or two vortices in the half-plane, respectively, is known to have LA-integrable dynamics, the following conjecture is quite plausible in view of our work in this paper:

> **Conjecture:** *Let a system of any finite number of point vortices in a plane or half-plane be configured as two or three separate groups, or two separate groups, respectively, each comprised of vortices of the same sign. Then if the diameters of each of these groups is sufficiently small, there exist initial configurations that produce motions with quasiperiodic flows on KAM tori and periodic orbits with countably many periods.*

In addition to trying to resolve this conjecture in our future research, we also plan to investigate some questions that arise naturally from our work in this paper. One such question concerns the possibility of finding concentrated configurations of point vortices having strengths of different signs in the plane or half-plane that lead to regular dynamics, which includes both quasiperiodic orbits on invariant KAM tori and nearby periodic trajectories. Another question is related to the manner in which such configurations can admit chaotic regimes as the diameter of the initial configuration is increased. In particular, just when and how does a transition from regular to chaotic motion occur as the initial configuration expands and its other geometric characteristics are varied? These investigations are apt to profit from the approach of considering a configuration of four (three) or more point vortices in the plane (half-plane) to be a perturbation of the integrable system of three (two) vortices in the plane (half-plane), along the lines of our studies in [14] and [48].

Acknowledgment

The first author is deeply indebted to Lu Ting for sharing, over the course of several years, his insights on point vortex dynamics, which proved to be indispensable in writing this paper. Preliminary versions of some of the results contained in this paper appeared in the Proceedings of the Conference on Dynamical Systems - Theory and Applications, Łodź, Poland, Dec. 2003.

Bibliography

[1] H. Aref, Integrable, chaotic, and turbulent motion in two-dimensional flows, *Ann. Rev. Fluid Mech.* **15** (1983), 345-389.

[2] V. Arnold, Small denominators and the problem of stability of motion in classical and celestial mechanics, *Russian Math. Surveys* **18** (1963), 85-91.

[3] V. Arnold, Sur une propriété topologique des applications globalment canoniques de la mechanique classique, *C. R. Acad. Sci. Paris Série I Math.* **261** (1965), 3719-3722.

[4] V. Arnold, *Mathematical Methods of Classical Mechanics*, Springer-Verlag, New York, 1978.

[5] A. Bagrets and D. Bagrets, Non-integrability of two problems in vortex dynamics, *Chaos* **7** (1997), 368-375.

[6] D. Bernstein and A. Katok, Periodic orbits for small perturbations on completely integrable Hamiltonian systems with convex Hamiltonians, *Invent. Math.* **88** (1987), 225-241.

[7] M. Bialy and L. Polterovitch, Hamiltonian systems, Lagrangian tori and Birkhoff's theorem, *Math. Ann.* **292** (1992), 619-627.

[8] G. Birkhoff, An extension of Poincaré's last geometric theorem, *Acta Math.* **47** (1925), 210-231.

[9] D. Blackmore, Hamiltonian analysis of vortex filaments, *Proc. 4th Int. Conf. on Nonlinear Dynamics*, Shanghai, 2002, 160-164.

[10] D. Blackmore and O. Knio, KAM theory analysis of the dynamics of three coaxial vortex rings, *Physica D* **140** (2000), 321-348.

[11] D. Blackmore and C. Wang, Morse index for autonomous linear Hamiltonian systems, *Int. J. Diff. Eqs. and Appl.* **7** (2003), 295-309.

[12] D. Blackmore, J. Champanerkar and C. Wang, A generalized Poincaré-Birkhoff theorem with applications to coaxial vortex ring motion, *Discrete and Contin. Dyn. Syst.-B* **5** (2005), 15-33.

[13] D. Blackmore and C.C. Lim, KAM tori and vortex clusters on the sphere (in preparation).

[14] D. Blackmore, O. Knio and L. Ting, Perturbed three point vortex problems, *Proc. 3rd MIT Conf. on Comp. Fluid and Solid Mech.*, June, 2005.

[15] V. Bogomolov, Two-dimensional fluid dynamics on a sphere, *Izv. Atmos. Oc. Phys.* **15** (1979), 18-22.

[16] A. Bolsinov, A. Borisov and I. Mamaev, Lie algebras in vortex dynamics and celestial mechanics - IV, *Reg. & Chaot. Dyn.* **4** (1999), 23-50.

[17] A. Borisov and V. Lebedev, Dynamics of three vortices on a plane and a sphere - II, *Reg. & Chaot. Dyn.* **3** (1998), 99-114.

[18] A. Calini, Recent developments in integrable curve dynamics, *Geometric Approaches to Differential Equations*, Lect. Notes of Australian Math. Soc.

15, Cambridge University Press, 2000, 56-99.

[19] A. Carter, An improvement of the Poincaré-Birkhoff fixed point theorem, *Trans. Amer. Math. Soc.* **269** (1982), 285-299.

[20] C. Conley and E. Zehnder, The Birkhoff-Lewis fixed point theorem and a conjecture of V. I. Arnold, *Invent. Math.***73** (1983), 207-253.

[21] W.Y. Ding, A generalization of the Poincaré-Birkhoff theorem, *Proc. Amer. Math. Soc.* **88** (1983), 341-346.

[22] A. Floer, Symplectic fixed points and holomorphic spheres, *Comm. Math. Phys.* **120** (1989), 575-611.

[23] J. Franks, Generalizations of the Poincaré-Birkhoff theorem, *Annals of Math.* **128** (1988), 139-151.

[24] C. Golé, *Symplectic Twist Maps*, World Scientific, Singapore, 2001.

[25] W. Gröbli, *Speciale Probleme über die Bewegung Geradliniger Paralleler Wirbelfäden*, Zürcher and Furrer, Zürich, 1877.

[26] H. Helmholtz, Über integrale hydrodynamischen gleichungen weiche den wirbelbewegungen entsprechen, *J. Reine Angew. Math.* **55** (1858), 25-225.

[27] F. Josellis, Lyusternik-Schnirelman theory for flows and periodic orbits for Hamiltonian systems on $T^n \times R^n$, *Proc. London Math. Soc.* **68** (1994), 641-672.

[28] K. Khanin, Quasi-periodic motion of vortex systems, *Physica D* **4** (1982), 261-269.

[29] G. Kirchhoff, *Vorlesungen über Mathematische Physik, Vol. 1*, Teubner, Leipzig, 1876.

[30] R. Kidambi and P. Newton, Motion of three point vortices on a sphere, *Physica D* **116** (1998), 143-175.

[31] O. Knio, L. Collerec and Juvé, Numerical study of sound emission by 2D regular and chaotic vortex configurations, *J. Comput. Phys.* **116** (1995), 226-246.

[32] H. Lamb, *Hydrodynamics,* Dover, New York, 1932.

[33] C.C. Lim, Existence of KAM tori in the phase-space of lattice vortex systems, *ZAMP* **41** (1990), 227-244.

[34] C.C. Lim, Relative equilibria of symmetric N-body problems on a sphere: inverse and direct results, *Comm. Pure Appl. Math.* **51** (1998), 341-371.

[35] C.C. Lim, J. Montaldi and M. Roberts, Relative equilibria of point vortices on the sphere, *Physica D* **148** (2001), 97-135.

[36] C.C. Lin, *On the Motion of Vortices in Two Dimensions*, Toronto University Press, Toronto, 1943.

[37] J. Mawhin and M. Willem, *Critical Point Theory and Hamiltonian Systems,* Springer-Verlag, New York, 1989.

[38] J. Moser, *Stable and Random Motions in Dynamical Systems*, Princeton Univ. Press, Princeton, 1973.

[39] J. Moser, Proof of a generalized form of the fixed point theorem due to G. D. Birkhoff, *Geometry and Topology*, Lecture Notes in Math., **597**, Springer-Verlag, New York, 1977.

[40] P. Newton, *The N-Vortex Problem: Analytical Techniques*, Springer-Verlag, New York, 2001.

[41] E. Novikov, Dynamics and statistics of a system of vortices, *Sov. Phys. JETP* **41** (1975), 937-943.

[42] H. Poincaré, Sur un thèoréme de géométrie, *Rend. del. Circ. Math. du Palermo* **33** (1912), 375-407.

[43] C. Robinson, Generic properties of conservative systems I; II, *Amer. J. Math.* **92** (1970), 562-603;897-906.

[44] E. Spanier, *Algebraic Topology*, McGraw-Hill, New York, 1966.

[45] J. Synge, On the motion of three vortices, *Can. J. Math.***1** (1949), 257-270.

[46] J. Tavantzis and L. Ting, The dynamics of three vortices revisited, *Phys. Fluids* **31** (1988),1392-1409.

[47] L. Ting and R. Klein, *Viscous Vortical Flows*, Lecture Notes in Physics, Vol. **374**, Springer-Verlag, Berlin, 1991.

[48] L. Ting and D. Blackmore, Bifurcation of motions of three vortices and applications, *Proc. 2004 ICTAM Conf.*, Warsaw, 2004.

[49] L. Ting, D. Blackmore and O. Knio, Perturbed three point vortex problems - the parabolic case. (in preparation)

[50] T. Tokieda, Tourbillons dansants, *C. R. Acad. Sci. Paris Série I* **333** (2001), 943-946.

[51] C. Wang and D. Blackmore, The Morse index and periodic orbits of Hamiltonian systems. (preprint).

[52] J. Weiss and J. McWilliams, Non-ergodicity of point vortices, *Phys. Fluids* **A3** (1991), 835-844.

[53] S. Ziglin, The non-integrability of the problem of motion of four vortices of finite strengths, *Physica D* **4** (1982), 268-269.

Chapter 3

Experiments on Heave/Pitch Limit-Cycle Oscillations of a Supercritical Airfoil Close to the Transonic Dip[1]

Guido Dietz, Günter Schewe, Holger Mai
Institute of Aeroelasticity
German Aerospace Center DLR, Bunsenstraße 10 37073 Göttingen, Germany

Dedicated to Lu Ting on his eightieth birthday

Abstract

Results from flutter experiments of the supercritical airfoil NLR 7301 at flow conditions near the transonic dip are presented. The airfoil was mounted with two degrees-of-freedom in an adaptive solid-wall wind tunnel, and boundary-layer transition was tripped. The results link the global aerodynamic force behavior to the observed LCOs and the identified transonic dip. The time lag of the lift response to the pitching motion of the airfoil appears to be responsible for the characteristic shape of the transonic dip. The amplitude limitation of the LCOs results from a slightly nonlinear dependency of lift and moment on the amplitude of the airfoil motion. LCOs can be controlled by relatively small forces, but amplitudes strongly depend on the damping of the aeroelastic system.

[1]This contribution is extracted from: Dietz, G., Schewe, G., and Mai, H. (2004). Experiments on heave/pitch limit-cycle oscillations of a supercritical airfoil close to the transonic dip, *J. Fl. Str.* **19**(1), pp. 1–16.
Schewe, Mai, and Dietz (2003) and the original paper together earned the "DLR Wissenschaftspreis 2004" award.

3.1 Introduction

The aeroelastic interaction of a typical transport-aircraft structure and the transonic flow exhibits two well-known phenomena: firstly, in the "transonic dip"(Tijdeman, 1977) the flutter speed shows a noticeable minimum between the critical Mach number where local supersonic regions occur in the flowfield and the Mach number where massive flow separation limits the operational flight regime; secondly, amplitude-limited oscillations of the structure, so-called limit-cycle oscillations (LCO), may occur (Edwards, 1996) instead of "classical" flutter where the amplitude of the structural oscillations grow exponentially at flight speeds above the critical flutter speed. It is assumed in the literature that both phenomena are linked to the inherently nonlinear aerodynamics at transonic speeds (Bendiksen, 1992). Since the relative flow speeds are close to the speed of sound, shock waves occur which may interact with the boundary layer eventually leading to flow separation. All of these mutual interactions are sources of nonlinearities. However, the aerodynamic phenomena, whose interactions are in particular responsible for the occurrence of the transonic dip as well as the amplitude limitation of the LCOs, have not yet completely been identified.

The physical understanding of the transonic dip and the LCO is of technical interest as a fundamental of an accurate prediction since the avoidance of transonic flutter is a key aircraft design problem (Edwards *et al.*, 1983). Furthermore, LCOs might be used to extend the operational flight regime, although they are usually seen as undesirable vibrations that limit the aircraft functionality (Meijer and Cunningham Jr., 1992). One could imagine flying in an unstable regime by using an active damping system, with the stable LCO as a fall-back position if the damping system fails. Obviously the amplitudes of the LCO have to be sufficiently small so that the structure does not fail by exceeding its ultimate load limit. Thus, LCOs could degrade the safety problem, as in the case of classical flutter, into a fatigue problem. In any case, reliable prediction methods for flutter at transonic speeds are needed, and the physics of transonic flutter must be understood, in order for the modeling to be robust and accurate. Thus, nonlinear aeroelasticity of large aspect-ratio wings is the subject of recent and current investigations. The present study concentrates on two spatial dimensions since a heaving and pitching airfoil can be seen as a representative section of a wing exhibiting bending and torsional oscillations.

Experimental investigations on nonclassical aeroelastic instability and response phenomena have been rare. Wind-tunnel test results of a

NACA 0012 airfoil single-side mounted with two degree-of-freedom were published within the Benchmark Models Program at NASA Langley by Rivera *et al.*, 1991, 1992. A conventional flutter boundary showing a transonic dip, a plunge instability region and stall flutter was found. Moreover, unsteady pressure measurements at the conventional flutter boundary were presented. One time series of a combined heave and pitch motion exhibiting an LCO behavior at transonic flow conditions was shown (Rivera *et al.*, 1991), but the amplitudes of the motion were not scaled. Edwards *et al.* (2001) measured flutter boundaries and LCO behavior of a semi-span business jet wing-fuselage model MAVRIC-I. The influence of the dynamic pressure and the angle of attack on the LCO behavior was reported by showing maps of the signal amplitude of wingtip accelerometers depending on the inflow Mach number and dynamic pressure. Furthermore, unsteady pressure distributions were measured at three span positions. The data set provides code validation test cases but for a more or less complex configuration. To the best of our knowledge, the first heave/pitch LCOs of a two-side mounted supercritical rigid airfoil in transonic flow were observed by Schewe and Deyhle (1996) in their exploratory experiment with an airfoil model shaped as NLR 7301. Similar data obtained in the same test set-up [Knipfer and Schewe (1999); corrected structural-dynamic parameters in Schewe *et al.* (2003)] are still the subject of numerical investigations.

Numerical studies of nonclassical flutter in transonic flow have been performed based on various numerical CFD and CSM models. Bendiksen (1992) studied the influence of shock dynamics on flutter and divergence in transonic flow by integrating the Euler equations whilst neglecting the structural damping. Nonlinear aeroelastic phenomena such as LCOs, weak divergence as well as flutter-divergence interactions were observed and described. However, similar to Schulze (1997) the reported simulated LCO amplitudes were quite high. Bendiksen (1992) partially discussed the transonic flutter behavior on the basis of the fluid-structure energy exchange: he observed that in transonic flow the phase difference between heave and pitch motion tends to adjust such that single degree-of-freedom flutter occurs. Bendiksen (1992) argued that the averaged power per cycle contributed by the lift due to the shock motion is maximized in this case. Morton and Beran (1999) investigated numerically the pitch-and-plunge flutter of a rigid NACA 64A006 airfoil in inviscid transonic flow. Although the predictions of flutter onset with three independent aeroelastic codes based on Euler equations were in agreement, the simulated LCO amplitudes differed noticeably. They speculated that differences in the numerical dissipation and the applied time steps caused the discrepancies. Weber *et al.* (2001) as well as Tang *et al.* (2003) computed LCOs of the NLR 7301 by integrating the thin-layer Navier-Stokes equations within their CFD model. These authors tried to simulate the experimental test case MP77 proposed by Knipfer and Schewe (1999). They tested different turbulence models, but the computed LCO amplitudes were an order of magnitude higher than the measured ones. Weber *et al.* (2001) suspected that, in part, the discrepancy was due to the neglected wind-tunnel interference. Thus, Castro *et al.* (2001) continued the work of Weber *et al.* (2001), taking into account the presence of the porous wind-tunnel walls. It was shown that modeling the wind-tunnel wall porosity significantly affects the LCO behavior. The simulated LCO amplitudes were also quite sensitive to the porosity parameter, whereas the flutter frequency as well as the phase difference between the pitch and the heave motion remained nearly constant. The numerical investigation indicated that wind-tunnel walls strongly influence the flutter characteristics compared with free-flight conditions. Nevertheless, with a wall porosity between 12% and 16% instead of 6% in the experiment, the predicted LCO amplitudes were comparable with the measured ones. The results confirmed the authors' previous decision to avoid this additional challenge in simulation by providing well-defined boundary conditions, using an adaptive test-section (Wedemeyer *et al.*, 1998) for the experiments reported here. Thomas *et al.* (2002) investigated numerically the LCO behavior of the NLR 7301 in the vicinity of MP77 using a harmonic balance approach. Assuming inviscid flow led to nearly linear aeroelastic behavior while the LCOs in viscous flow exhibited high amplitudes that are comparable with Weber *et al.* (2001) and Tang *et al.* (2003). However, the effects of viscosity play an important role in correct modeling of the LCO response phenomena (Thomas *et al.*, 2002). Besides the importance of damping in the accurate prediction of LCOs, Meijer and Cunningham Jr. (1992) showed by means of a semi-empirical LCO prediction method that the time lag of the aerodynamic response to the structural motion strongly affects the LCO eigenmode.

Here, we report results from experiments on heave/pitch limit-cycle oscillations of the supercritical airfoil NLR 7301 with tripped boundary-layer transition at flow conditions close to the transonic dip. The measurements were carried out in the adaptive test-section of the TWG Göttingen. The experimental results link the global aerodynamic force behavior to the observed LCOs and the detected transonic dip. It is shown which global aerodynamic force behavior is responsible for the transonic dip and for the amplitude limitation of the LCOs. Furthermore, the influence parameters on the amplitude of LCOs are discussed.

3.2 Test set-up

Only those details regarding the set-up and the test procedure that are important in order to understand the reported results are presented here. More detailed information may be found in Schewe *et al.* (2002).

The experiments were conducted in the Transonic Wind tunnel Göttingen (TWG). The TWG is a continuously working facility with a 1 m×1 m adaptive test-section. The ratio of the tunnel height to the chord of the investigated airfoil models is 3.333. Therefore the top and bottom walls were adapted to the stationary flow at the mean angle of attack of the airfoil, where the static aeroelastic equation is satisfied. The wall interference is minimized by a one-step method of wall adaptation based on a Cauchy type integral (Wedemeyer *et al.*, 1998). The displacement thickness of the turbulent wind-tunnel wall boundary layer is predicted by Head's method (Cebeci and Bradshaw, 1979) and is added to the wall shapes; top and bottom wall displacement thicknesses are obtained according to the measured pressure gradients at each wall while the gradient is neglected for the sidewalls (Jacobs, 2002). This adaptation to the mean steady position of the airfoil yields nearly a minimum residual wall interference for a moderately oscillating airfoil, as shown in pretests (Jacobs, 2002). The lowest value for the wind-tunnel resonance frequency can be estimated using the method of Voß(1998): the ratio of flutter frequency to wind-tunnel resonance frequency is smaller than 35% for the reported data.

Two set-ups were employed to obtain the presented results. One set-up allows free heave and pitch motions of the airfoil for flutter tests. A theoretical model representing the structural dynamics of this set-up is described in Section 3.3. The same set-up was also used for forced harmonic heave oscillations of the airfoil with a blocked pitch motion and hydraulic actuators in the heave direction. The other set-up forces harmonic pitch oscillations of the airfoil by means of hydraulic rotation cylinders. In both set-ups, the airfoil is mounted on each side to a piezoelectric balance of high stiffness (Schewe, 1991) in order to measure the steady and unsteady loads lift, drag and pitching moment. Two laser triangulators on each side of the wind tunnel measure the instantaneous heave and pitch of the model. For the flutter-test set-up a control system has been developed to introduce forces in heave direction which are measured piezoelectrically. A digital signal processor (DSP) device derives the heave velocity from the laser-triangulator signals such that a voltage proportional to this velocity can be passed into one electrodynamical exciter on each side. Small heave motions of the airfoil can then either be amplified or, by applying the inverted signal, damped. Tests revealed that in open loop state this flutter control system can be seen as nonintrusive.

Two airfoil models were shaped as the supercritical NLR 7301 airfoil with the contour given by Zwaan (1979) except that this geometry was cut off at $x/c = 1$. Thus, the trailing edges of the models are blunt with approximately 0.1% chord. In the present investigation the chords meet the leading and trailing edges of the models so that the given angle-of-attack α differs from the NLR results by $\alpha = \alpha_{\mathrm{NLR}} - 0.183^o$. Both models are made of a carbon-fiber composite structure having a chord of $c^* = 0.3$ m and a span of $b^* = 1$ m. They are light weight, very stiff and may be assumed to be rigid. In order to investigate the flow around a forced oscillating airfoil, one model is equipped with 64 miniature pressure transducers measuring both steady and unsteady pressure differences. These sensors are arranged beneath the model surface and the pressure taps are located in a zig-zag pattern of $\pm 0.5\%$ span around the middle section at 50% span. Laminar-turbulent boundary-layer transition has been tripped at 7% chord on the suction and 14% chord on the pressure side by zig-zag tape with a height of 0.0467% chord. The effectiveness of the transition tripping has been checked by infrared imaging. The tripping has been applied in the attempt of obtaining a better comparison with numerical simulations, since the accurate prediction of the transition region is still an unsolved problem, in particular for unsteady flows.

3.3 Equations of motion

The heave and pitch motions of the rigid airfoil in the flutter-test set-up are modeled by a two-degree-of-freedom system as shown in Fig. 3.1. The corresponding equations of motion can be written in nondimensional form as

$$v_\infty^2 \mathbf{M} \frac{\partial^2 \mathbf{u}(t)}{\partial t^2} + 2 v_\infty \mathbf{D} \frac{\partial \mathbf{u}(t)}{\partial t} + \mathbf{K}\, \mathbf{u}(t) = \frac{2}{\pi} \frac{v_\infty^2}{\mu} \mathbf{f}(t) , \tag{3.1}$$

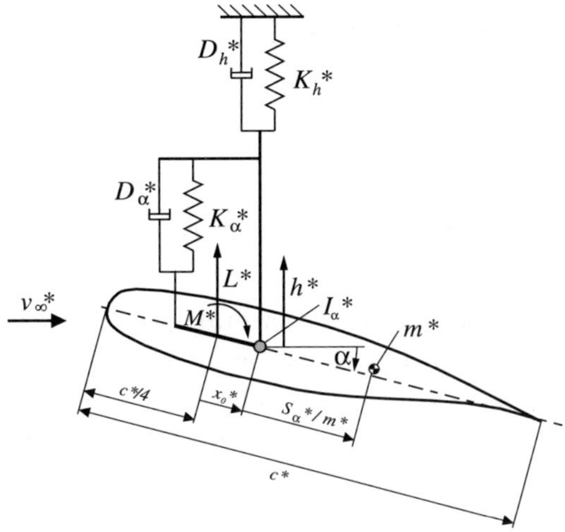

Fig. 3.1 Schematic of the structural model.

where

$$\mathbf{M} = \begin{bmatrix} 1 & -x_\alpha \\ -x_\alpha & r_\alpha^2 \end{bmatrix}, \quad \mathbf{D} = \begin{bmatrix} \delta_h \omega_h/\omega_\alpha & 0 \\ 0 & \delta_\alpha r_\alpha^2 \end{bmatrix}, \quad \mathbf{K} = \begin{bmatrix} (\omega_h/\omega_\alpha)^2 & 0 \\ 0 & r_\alpha^2 \end{bmatrix},$$

$$\mathbf{f}(t) = \begin{bmatrix} c_l(t) \\ c_m(t) + x_0 c_l(t) \end{bmatrix} \quad \text{and} \quad \mathbf{u}(t) = \begin{bmatrix} h(t) \\ \alpha(t) - \alpha_0 \end{bmatrix}.$$

Equation (3.1) describes the motion of the airfoil, its nondimensional heave $h(t) = h^*(t)/c^*$ at the elastic axis as well as the difference between its angle of attack $\alpha(t)$ and its off-wind value α_0 in the nondimensional time $t = t^* v_\infty^*/c^*$. With the dimensional quantities chord length c^*, mass m^*, static mass moment S_α^*, mass moment of inertia I_α^*, spring constants $K_{\alpha/h}^*$ and damping coefficients $D_{\alpha/h}^*$ as well as the free-stream speed v_∞^* and density ρ_∞^* the structural-dynamic parameters are made nondimensional as follows:

$$\begin{aligned}
&\text{elastic axis to 1/4-chord point distance:} && x_0 = \tfrac{x_0^*}{c^*} && = 0.0, \\
&\text{elastic axis to center of mass distance:} && x_\alpha = \tfrac{S_\alpha^*}{m^* c^*} && = 0.0420, \\
&\text{radius of gyration about the elastic axis:} && r_\alpha = \sqrt{\tfrac{I_\alpha^*}{m^* c^{*\,2}}} && = 0.1828, \\
&\text{Lehr's pitch-damping coefficient:} && \delta_\alpha = \frac{D_\alpha^*}{2\sqrt{K_\alpha^* I_\alpha^*}} && = 0.15\%, \\
&\text{Lehr's plunge-damping coefficient:} && \delta_h = \frac{D_h^*}{2\sqrt{K_h^* m^*}} && = 0.43\%, \\
&\text{uncoupled natural frequency ratio:} && \omega_h/\omega_\alpha = \frac{\sqrt{K_h^*/m^*}}{\sqrt{K_\alpha^*/I_\alpha^*}} && = 0.6991.
\end{aligned} \tag{3.2}$$

The aerodynamic loads lift $L^*(t)$ and pitching moment $M^*(t)$ that act on the airfoil with span b^* through the 1/4-chord point can be nondimensionalized as

$$\begin{aligned} &\text{lift coefficient:} && c_l(t) = L^*(t)/(\tfrac{\rho^*_\infty}{2}\, v^{*\,2}_\infty c^* b^*)\,, \\ &\text{moment coefficient:} && c_m(t) = M^*(t)/(\tfrac{\rho^*_\infty}{2}\, v^{*\,2}_\infty c^{*\,2} b^*)\,. \end{aligned} \tag{3.3}$$

The reciprocal of the reduced natural pitching frequency $k_\alpha = \sqrt{K^*_\alpha / I^*_\alpha} c^* / v^*_\infty = 86.95\,\text{m/s}\,/v^*_\infty$ can be interpreted as a nondimensional free-stream speed $v_\infty = 1/k_\alpha$ and is called the reduced speed. The mass ratio $\mu = (m^*/(\pi/4\; c^{*\,2} b^*))/\rho^*_\infty = 371.62\,\text{kg/m}^3\,/\rho^*_\infty$ describes the ratio of the structural inertial forces to the aerodynamic loads, whereas the flutter index $\text{fi} = 2v_\infty/\sqrt{\mu}$ characterizes the ratio of the aerodynamic loads to the elastic forces.

3.3.1 *Structural-dynamic parameters*

Equations (3.2) present the structural-dynamic parameters of the flutter-test set-up. The parameters, chord length, mass and torsional-spring constant, were measured directly. The mass moment of inertia, the static mass moment and the bending-spring constant were measured and then corrected by the results of a ground vibration test (GVT). The correction procedure minimizes the discrepancies between the theoretically obtained eigenfrequencies and maximizes the collinearity of the eigenvectors to the values that were measured. The relative difference between the eigenfrequencies of the structural model without aerodynamic loads to those of the ground vibration test are smaller than 0.3%, the collinearity of the eigenvectors is greater than 99.7%. The relative error of the measured eigenfrequencies is very small whereas the damping constants may be obtained from the GVT with a relative error of $\mathcal{O}\,(20\%)$. However, the structural-dynamic parameters are detected quite accurately, as the energy balance discussed in Section 3.4.3 indicates.

3.3.2 *Method of flutter calculations*

The stability of Eq. (3.1) is analyzed in flutter calculations. These flutter calculations are performed according to a k-method in order to calculate the flutter-stability limit from the measured aerodynamic characteristics of the airfoil. Therefore the motion of the structure is assumed to be harmonic,

$$\mathbf{u}(t) = \bar{\mathbf{u}} + \hat{\mathbf{u}}\mathrm{e}^{\imath kt} = \begin{bmatrix} \bar{h} \\ \bar{\alpha} - \alpha_0 \end{bmatrix} + \begin{bmatrix} \hat{h}\mathrm{e}^{\imath\phi_h} \\ \hat{\alpha}\mathrm{e}^{\imath\phi_\alpha} \end{bmatrix} \mathrm{e}^{\imath kt}\,, \tag{3.4}$$

with a complex $\hat{\mathbf{u}}$ and a real reduced frequency $k = \omega^* c^* / v_\infty^*$. The aerodynamic loads are also assumed to be harmonic with complex aerodynamic derivatives $c_{l/m,h/\alpha}$,

$$\begin{aligned} \mathbf{f}(t) &= \bar{\mathbf{f}} + \mathbf{A}\hat{\mathbf{u}}\mathrm{e}^{\imath kt} \qquad (3.5)\\ &= \begin{bmatrix} \bar{c}_l \\ \bar{c}_m + x_0 \bar{c}_l \end{bmatrix} + \begin{bmatrix} c_{l,h} & c_{l,\alpha} + x_0 c_{l,h} \\ c_{m,h} + x_0 c_{l,h} & \begin{matrix} c_{m,\alpha} + x_0 c_{l,\alpha} \\ + x_0 (c_{m,h} + x_0 c_{l,h}) \end{matrix} \end{bmatrix} \begin{bmatrix} \hat{h}\mathrm{e}^{\imath\phi_h} \\ \hat{\alpha}\mathrm{e}^{\imath\phi_\alpha} \end{bmatrix} \mathrm{e}^{\imath kt}. \end{aligned}$$

For simplicity of analysis, the small damping of the system is approximated by structural damping that can be seen as an imaginary spring stiffness instead of the viscous damping,

$$2 v_\infty \mathbf{D}\, \frac{\partial \mathbf{u}(t)}{\partial t} \approx \imath\, 2 \begin{bmatrix} \omega_h/\omega_\alpha & 0 \\ 0 & 1 \end{bmatrix} \mathbf{D}(\mathbf{u}(t) - \bar{\mathbf{u}})\,. \tag{3.6}$$

Introducing Eqs. (3.4)-(3.6) into Eq. (3.1) yields

$$\begin{aligned} \left\{ -v_\infty^2 k^2 \mathbf{M} + \left(\mathbf{K} + \imath\, 2 \begin{bmatrix} \omega_h/\omega_\alpha & 0 \\ 0 & 1 \end{bmatrix} \mathbf{D}\right) - \frac{2}{\pi} \frac{v_\infty^2}{\mu} \mathbf{A} \right\} \hat{\mathbf{u}}\mathrm{e}^{\imath kt} \qquad (3.7)\\ + \left(\mathbf{K}\bar{\mathbf{u}} - \frac{2}{\pi} \frac{v_\infty^2}{\mu} \bar{\mathbf{f}} \right) = 0\,. \end{aligned}$$

The first term in parentheses constitutes an eigenvalue problem with $\lambda = 1/(v_\infty k)^2$,

$$\left(\mathbf{K} + \imath\, 2 \begin{bmatrix} \omega_h/\omega_\alpha & 0 \\ 0 & 1 \end{bmatrix} \mathbf{D} \right)^{-1} \left(\mathbf{M} + \frac{2}{\pi \mu k^2} \mathbf{A} \right) \hat{\mathbf{u}} = \lambda \hat{\mathbf{u}}\,, \tag{3.8}$$

while the last term in parentheses in Eq. (3.7) represents the static aeroelastic equation. Since k is real and positive, the real eigenvalues λ describe the flutter-stability limit.

The structural-dynamic parameters are fixed in the experiments as given in Eqs. (3.2). The aerodynamic derivatives $\mathbf{A} = \mathbf{A}(k, \mathrm{Ma}_\infty, \mathrm{Re}_\infty)$ are functions of the reduced frequency, Mach number and the Reynolds number. They were measured in forced harmonic oscillation experiments at different frequencies and Mach numbers while the Reynolds number has been kept approximately constant at $\mathrm{Re}_\infty \approx 2 \times 10^6$. In order to determine the flutter-stability limit, μ and k are varied systematically to find real positive eigenvalues λ that satisfy Eq. (3.8). One has to keep in mind that a variation of μ at constant Mach and Reynolds numbers for a given gas and fixed structural-dynamic parameters is associated with a variation of v_∞ since μ depends on ρ^*_∞.

In the results a measured flutter boundary and the flutter-stability limit analyzed on base of measured aerodynamic derivatives are compared with the stability limit of an analytical reference case: the stability limit for an infinitesimally thin flat plate in inviscid subsonic compressible flow is predicted assuming the same structural-dynamic parameters as presented in Eq. (3.2). The aerodynamic derivatives for the flat plate can be determined by assuming that the unsteady flow due to harmonic oscillation can be described by the Possio (1938) integral equation. The method of Carstens (1973) is used to solve this equation numerically and to obtain the required derivatives.

3.4 Results and discussion

Experimental data were obtained from two kinds of experiments. In the first kind, time series were measured while the airfoil exhibits self-sustained oscillations in the flutter-test set-up. The energy exchange between the fluid and the structure can be calculated from these data. The examination of this energy balance links the global aerodynamic force behavior to the observed LCOs. Moreover, the flutter tests yield flutter boundaries. These flutter boundaries are compared with the results of a stability analysis of the governing Eq. (3.1). This comparison suggests, which global aerodynamic force behavior may be responsible for the transonic dip. For the stability analysis the aerodynamic response to the pitch as well as the heave motion of the airfoil is needed. Thus, this has been measured in the second kind of experiments, forced harmonic oscillation experiments. In the following these results are presented first. Afterward, the energy exchange between the fluid and the structure is discussed.

3.4.1 *Transonic dip*

Figure 3.2 shows the measured flutter and buffeting boundary in the flutter index fi/Mach number Ma_∞ plane revealing the slope of a transonic dip. The squares mark conditions where the Mach number Ma_∞ has been varied

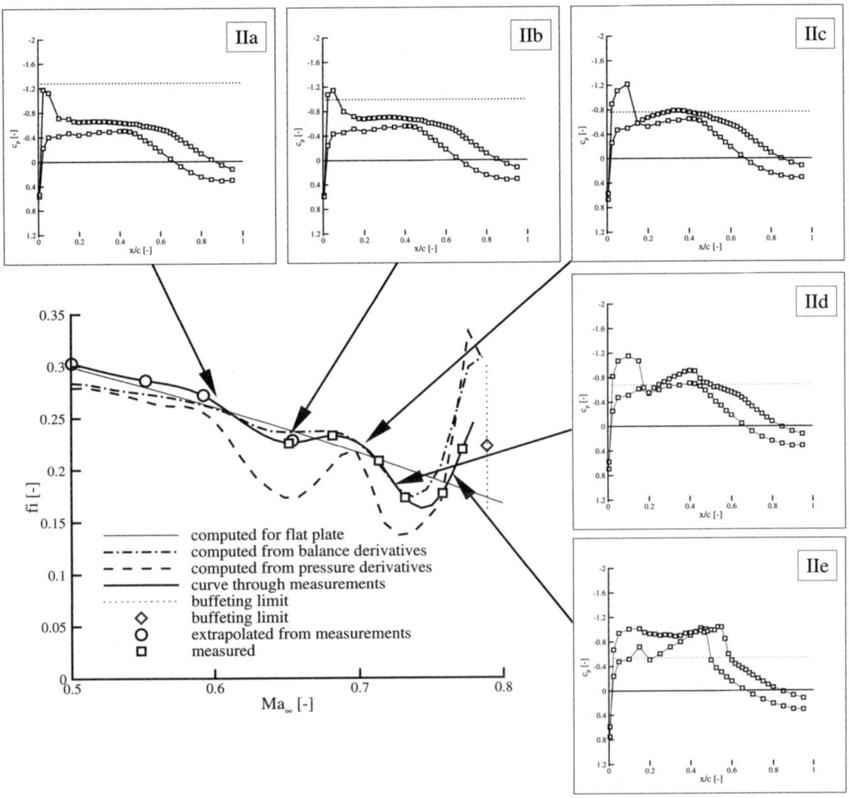

Fig. 3.2 Comparison of a measured/extrapolated (Zimmerman and Weissenburger, 1964) flutter and buffeting boundary at $\alpha = 0.46^o \pm 0.13^o$, $\text{Su}_0 = 110.4\text{K}/T_0^* = 0.365 \pm 1.5\%$ with flutter-stability limits analyzed from derivatives measured by a balance and by pressure transducers at $\alpha = 0.33^o \pm 0.05^o$, $\text{Re}_\infty = 2.21 \times 10^6 \pm 9.3\%$ as well as to a flutter-stability limit analyzed from theoretically obtained derivatives for a flat plate. Corresponding measured stationary pressure distributions with lines marking $c_{p\ \text{crit}}$.

at a constant stagnation pressure until oscillations of the airfoil are neither amplified nor damped. The flutter-control system is used in the procedure to find such a condition (Schewe *et al.*, 2003). The circles mark conditions that are extrapolated from detected values of the pitch and heave eigenfrequencies at several Mach numbers Ma_∞ for different stagnation pressure values according to Zimmermann and Weissenburger's 1964 criterion. The mean angle of attack $\bar{\alpha}$ decreases slightly with Mach number since $\bar{c}_m$ which remains nearly constant is always < 0; its off-wind value α_0 is fixed and the dynamic pressure $\rho_\infty^* {v_\infty^*}^2/2$ increases with Mach number at a constant stagnation pressure. Although α_0 has been adapted to each stagnation pressure, the angle of attack of the measured flutter boundary varies about $\pm 0.13^o$. Furthermore, the variation of Mach number Ma_∞ and stagnation pressure corresponds to a variation of the Reynolds number $\mathrm{Re}_\infty/10^6 \approx 2.09 \pm 0.75$ since the chord length of the airfoil is fixed and the stagnation temperature and thus the Sutherland constant $\mathrm{Su}_0 = 110.4\mathrm{K}/T_0^*$ remains almost constant. At $\mathrm{Ma}_\infty = 0.789$, buffeting is assumed to occur since the observed oscillations changed from combined pitch/heave oscillations with a pronounced peak in the power spectrum of the motion data at a typical reduced frequency $k \approx 0.3$ to pitch dominated oscillations of the structure at about $k = 0.482$ showing a broader peak in the power spectrum. This frequency doesn't match any structural eigenfrequency and is thus determined by the flow. So, self-excited shock-induced oscillations seem to occur which interact with the structure as buffeting.

In Fig. 3.2 the measured flutter boundary is compared with the flutter-stability limit analyzed based on the aerodynamic derivatives of an infinitesimal thin flat plate in inviscid subsonic compressible flow (Section 3.3.2). The corresponding stationary pressure distributions at several Mach numbers in Fig. 3.2 reveal that discrepancies between the stability boundary of the flat plate and the NLR 7301 correspond to the appearance of shock waves in the flowfield. The measured flutter boundary is also compared with flutter-stability limits analyzed from derivatives measured by the balance and by the pressure transducers described in Section 3.2. The slope of both curves with the two local minima matches qualitatively well to the measured one. But the quantitative agreement between the pressure-based data and the measured boundary is moderate. This is probably caused by a small torsion of the model. Although its carbon-fiber composite structure is very stiff and the model is mounted at both sides, the midsection rotates against the roots. Static deformation tests have shown that this torsion may amount up to 0.05^o at common loads. The pitching amplitudes of the midsection may differ from those of the roots by up to 4%, with a phase lag smaller than 1.5^o at typical flutter frequencies as accelerometer responses in forced pitch oscillation experiments revealed. This small torsion of the model may influence the pressure-based derivatives measured in the midsection. Thus, the agreement between the balance-based stability limits considering the global forces and the measured flutter boundary is much better.

In order to clarify which global force behavior is responsible for the observed transonic dip, the sensitivity of the stability limit to the four com-

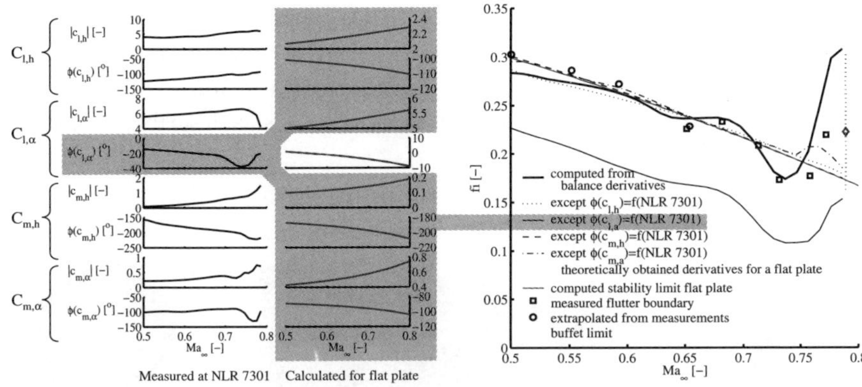

Fig. 3.3 Left: Magnitude and phase distributions of measured and theoretically obtained derivatives versus Mach number for forced harmonic pitch and heave oscillations at reduced frequencies corresponding to the stability limit of Fig. 3.2; Right: Comparison of a measured flutter and buffeting boundary at $\alpha = 0.46^o \pm 0.13^o$, $\mathrm{Su_0} = 110.4\mathrm{K}/T_0^* = 0.365 \pm 1.5\%$ with flutter-stability limits analyzed from measured balance derivatives at $\alpha = 0.33^o \pm 0.05^o$, $\mathrm{Re}_\infty = 2.21 \times 10^6 \pm 9.3\%$ combined with theoretically obtained derivatives for a flat plate.

plex derivatives is surveyed. Figure 3.3 shows on the left the magnitude and phase distributions of balance-based data measured from the NLR 7301 and theoretically obtained derivatives for the flat plate versus Mach number for forced harmonic pitch and heave oscillations at reduced frequencies corresponding to the stability limit of Fig. 3.2. On the right side of Fig. 3.3, the measured flutter and buffeting boundary is compared with flutter-stability limits analyzed from the measured balance derivatives combined with the theoretically obtained derivatives. For the combinations, in general, the theoretical values are applied, except for either one measured magnitude or one measured phase distribution. Replacing the magnitude distributions did not reveal slopes that differ essentially from the flat plate stability limit. Only the stability limit where the measured phase distribution of the $c_{l,\alpha}$ is applied (gray shaded combination in Fig. 3.3) matches the measured curve qualitatively. This suggests that the time lag of the lift response to the pitch motion of the airfoil may be responsible for the characteristic shape of the transonic dip. This proposition will be discussed in Section 3.4.3. Next, the behavior of the fluid-structure system in the unstable region close to the transonic dip will be discussed.

3.4.2 *Limit-cycle oscillations*

In the present study various types of LCOs with small amplitudes, even multiple nested ones, have been observed in the unstable region close to the transonic dip. In principle, the experiments in the adaptive test-section of the TWG revealed similar manifestations of LCOs as described by Schewe *et al.* (2003).

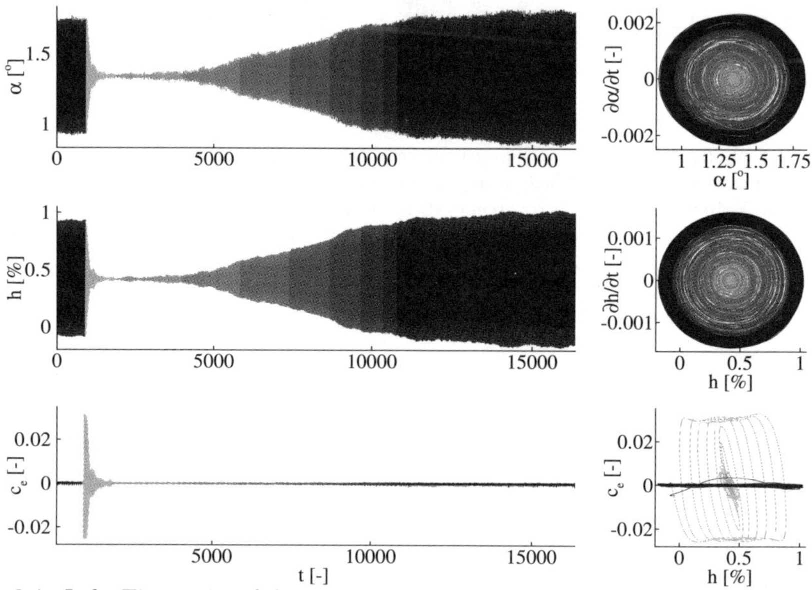

Fig. 3.4 Left: Time series of the pitch α and the heave h motion as well as the exciter-force coefficient $c_e(t) = F_e^*(t)/(\frac{1}{2}\rho_\infty^* v^{*\,2} c^* b^*)$. Right: Corresponding phase-space representation. $\mathrm{Ma}_\infty = 0.691$, $\mathrm{Re}_\infty = 1.46 \times 10^6$, $\mathrm{Su}_0 = 0.365$, $\mu = 1013.2$, $v_\infty = 2.65$, $\alpha_0 = 1.78^o \pm 0.04^o$.

Figure 3.4 shows on the left the time history of the pitch $\alpha(t)$ and the heave $h(t)$ motion at flow conditions in the unstable region of the transonic dip. At the beginning of the time series, the airfoil exhibits combined pitch and heave motions. By use of the flutter-control system these oscillations are damped out between $900 < t < 1800$. The forces applied by the electrodynamic exciters in heave direction c_e are shown in the lower part of Fig. 3.4 and are nondimensionalized in the same manner as the lift coefficient. The maximum amplitude of the forces that are applied to control the LCO $\hat{c}_e \approx 0.03$ are remarkably small compared with typical lift coefficients. At $t \approx 1800$ the flutter-control system is switched off into open-loop state and the amplitudes of the oscillations grow to the LCO state without any artificial excitation. On the right of Fig. 3.4, the corresponding phase-space representation of the observed oscillations are shown. Due to the appropriate scaling, the trajectories in the LCO state are shaped approximately circular. This indicates that the pitch and heave motion of the airfoil behave almost harmonically. A flutter-divergence interaction [Bendiksen (1992)] could not be observed, since the mean angle of attack for the LCOs is the same as the one for the damped state.

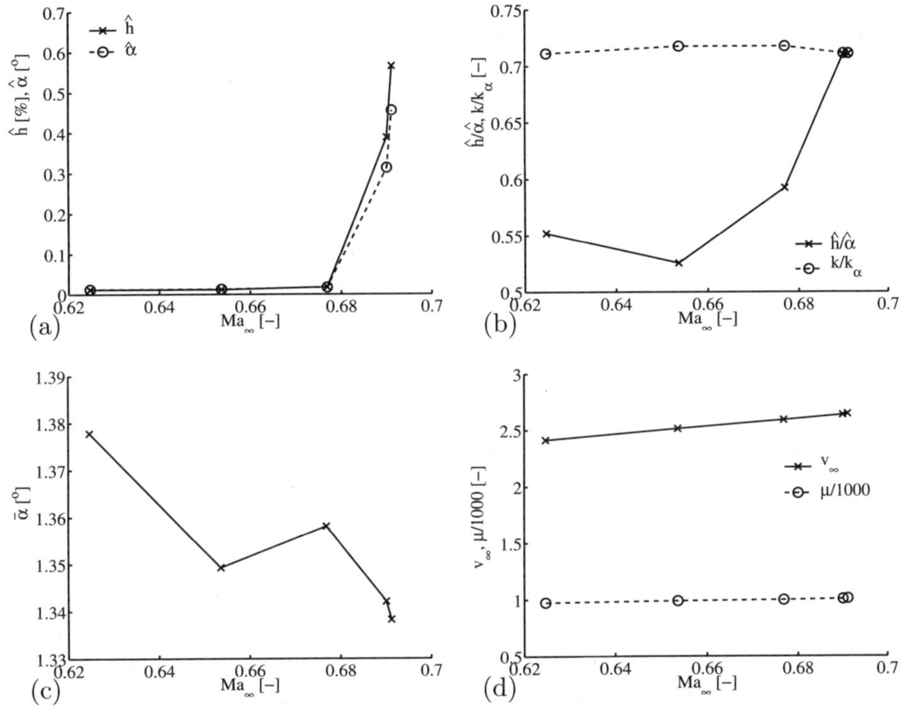

Fig. 3.5 Mach-number dependent behavior of LCO relevant properties in the vicinity of the flow parameters described in Fig. 3.4: (a) $\hat{h}$ heave and $\hat{\alpha}$ pitch amplitude, (b) $\hat{h}/\hat{\alpha}$ heave/pitch amplitude ratio and reduced flutter frequency k as well as (c) $\bar{\alpha}$ mean angle of attack, (d) Corresponding Mach-number dependent changes of the parameters reduced speed v_∞ and mass ratio μ.

Figure 3.5 shows several LCO property trends dependent on the Mach number Ma_∞ in the vicinity of the flow parameters described in Fig. 3.4. The Mach number is varied, while the stagnation pressure and the stagnation temperature of the wind tunnel flow is kept constant. Thus, the dimensional freestream velocity v_∞^* increases and the dimensional freestream density ρ_∞^* decreases with increasing Mach number. This yields the increasing reduced velocity v_∞ and the slightly increasing mass ratio μ that can be seen in Fig. 3.5(d). The mean angle of attack $\bar{\alpha}$ decreases with increasing Mach number since $\bar{c}_m \approx -0.05$ and is nearly constant and the v_∞^2/μ increases with Mach number (Fig. 3.5(c), last term in Eq. (3.7)). Figure 3.5(a) should not be interpreted as a bifurcation diagram in order to distinguish between a sub- and a supercritical bifurcation [Schewe *et al.* (2002), Schewe, Mai, and Dietz (2003)] since the Mach number resolution is poor and at least two stability parameters Ma_∞ and $\bar{\alpha}$ are changed simultaneously. But, the oscillation-amplitude trend presented in Fig. 3.5(a) indicate that the LCO behavior in this case is more or less "benign". Figure 3.5(a) shows finite amplitudes of motion even at stable flow conditions. Although the turbulence level of the wind tunnel is small with 0.1% at $\mathrm{Ma}_\infty = 0.5$ up to 0.25% at $\mathrm{Ma}_\infty = 0.8$ [Jacobs (2002)], these small disturbances provoke small oscillations of the airfoil. These disturbances also initiate the growing amplitudes in unstable regions as shown in Fig. 3.4. Figure 3.5(b) shows only slightly varying frequency ratios k/k_α. In the present investigation generally a phase difference $\phi_\alpha - \phi_h$ of nearly 180° is observed between the heave and the pitch motion. Thus, these LCOs can be considered as a single degree-of-freedom (SDOF) motion. The amplitude ratio $\hat{h}/\hat{\alpha}$ at $\phi_\alpha - \phi_h = 180^o$ can be interpreted as the distance per chord between a center of rotation upstream of the airfoil and the elastic axis. Around this center the chord of the airfoil rotates approximately while performing the identified LCOs. This supports Bendiksen's 1992 observation that in transonic flow the phase difference between heave and pitch motion tends to adjust such that SDOF flutter occurs. Figure 3.5(b) shows that the amplitude ratio $\hat{h}/\hat{\alpha}$ increases with increasing Mach number. Thus, the center of rotation of the chord of the airfoil moves upstream with increasing Mach number. However, compared with the results of Schewe *et al.* 1996, 2002, 2003, the reported LCO amplitudes are considerably smaller than those shown in the numerical results which are mentioned in the introduction. Nevertheless, there might be other coexisting or nested LCO attractors of higher amplitude around the presented ones that were not found. At least one additional "stall" attractor and one repeller between both attractors should exist theoretically. However, at other flow conditions coexisting nested LCOs were observed with a similar behavior as described by Schewe *et al.* (2003). So-called subcritical bifurcations (Schewe *et al.*, 2003) were also observed. In this case a repeller exists between the stable fixed point and the first stable LCO attractor and a traditional flutter-stability analysis would predict safe flight conditions although a gust could initiate limit-cycle oscillations.

3.4.3 *Energy exchange in limit-cycle oscillations*

The law of conservation of energy states that the sum of the kinetic energy in the heave and the pitch motion E_{kin} together with the potential energy in the springs E_{pot} must be equal to the constant initial energy $E_0 = E_0^*/(\rho_\infty^*/2\ v_\infty^{*\,2} c^{*\,2} b^*)$ in the structural-dynamic system plus the sum of work of the nonconservative forces ΣW, while energy is dissipated due to the structural damping. The power dissipated by the structural damping is $P_v = P_v^*/(\rho_\infty^*/2\ v_\infty^{*\,3} c^* b^*)$. The lift and the pitching moment perform the energy exchange between flow and structure at $P_l + P_m$ per unit time. Besides this the flutter-control system puts the power ($P_e > 0$) in or takes the power ($P_e < 0$) out of the structural-dynamic system. So,

$$E_0 = E_{\mathrm{kin}} + E_{\mathrm{pot}} - W = E_{\mathrm{kin}} + E_{\mathrm{pot}} - \int (P_v + P_l + P_m + P_e)\,\mathrm{d}t \quad (3.9)$$

represents an integral of the equation of motion (3.1) and thus describes the energy balance for the two degree-of-freedom system around its steady position $\bar{u}$ with

$$E_{\mathrm{kin}} = \tfrac{\pi\mu}{4} \tfrac{\partial u}{\partial t}^{\mathrm{T}} \mathbf{M} \tfrac{\partial u}{\partial t}, \quad E_{\mathrm{pot}} = \tfrac{\pi\mu}{4 v_\infty^2} (u - \bar{u})^{\mathrm{T}} \mathbf{K} (u - \bar{u}), \quad P_v = -\tfrac{\pi\mu}{v_\infty} \tfrac{\partial u}{\partial t}^{\mathrm{T}} \mathbf{D} \tfrac{\partial u}{\partial t},$$
$$P_m = (c_m - \bar{c}_m) \tfrac{\partial \alpha}{\partial t}, \quad P_l = (c_l - \bar{c}_l) \left(\tfrac{\partial h}{\partial t} + x_0 \tfrac{\partial \alpha}{\partial t} \right), \quad P_e = c_e \tfrac{\partial h}{\partial t}. \quad (3.10)$$

Figure 3.6 shows the time history of the total energy $E = E_{\mathrm{kin}} + E_{\mathrm{pot}}$ compared with the work of the nonconservative forces for a measured time series where the system exhibits an LCO that is damped at time $t = 1800$ until $t = 4800$ and is excited at time $t = 8500$. The total energy and the work agree satisfactorily. That indicates that the structural-dynamic parameters of Eqs. (3.2) and the time series of the motion and the forces were measured with sufficient accuracy, since many possible inaccuracies, e.g., improper damping coefficients, would lead to an accumulating discrepancy of the integral with respect to time of the power terms compared with the sum of energy (cf. Fig. 3.6, with Lehr's plunge-damping coefficient assumed to be 110% δ_h). The time traces and the amplitude dependency of the different power terms of this energy balance give some enlightenment as to which aerodynamic force behavior is responsible for the amplitude limita-

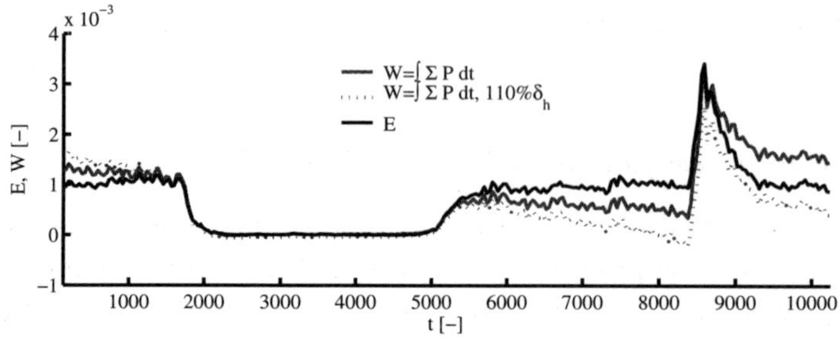

Fig. 3.6 Comparison of the total energy $E = E_{\text{kin}} + E_{\text{pot}}$ and the work applied by nonconservative forces $W = \int \Sigma P dt$ calculated according to Eq. (3.10) from a measured time series and the structural-dynamic parameters of Eq. (3.2). The flutter-control system is used at time 1800 to damp an LCO until time 4800 and at time 8500 to excite the system.

tion of the observed LCOs. Therefore the LCO amplitude at a given set of flow and structural parameters is varied by using the flutter-control system in order to extract energy out of or to add energy to the oscillating system.

Figure 3.7 demonstrates the results of this method: on the left several components of power averaged per oscillation cycle are plotted against the limit-cycle pitch amplitude squared. The LCO amplitude has been adjusted by the flutter-control system in the following way: at conditions near the transonic dip where the airfoil exhibits limit-cycle oscillations, the LCO has been damped out artificially. Then, the damping factor of the flutter-control system was reduced gradually. For different damping factors the LCO amplitudes were recorded and the energy exchange between the fluid and the structure was observed. When the damping factor has been reduced to zero, the curve $\Sigma\bar{P} - \bar{P}_e$ crosses the zero axis at the naturally occurring LCO amplitude, since no power is exchanged with the flutter-control system. Then the flutter-control system has been switched to excitation and the amplification factor was increased gradually until the LCO exhibited an amplitude greater than twice the unexcited one. Later the flutter-control system was gradually increased to damp the oscillations again in order to demonstrate that no hysteresis occurs. It is remarkable that for all LCO amplitudes larger than 10% of its maximum value the LCO properties $\hat{h}/\hat{\alpha} = 0.709 \pm 1.4\%$, $\phi_\alpha - \phi_h = 178.2^o \pm 0.7^o$ and $k = 0.270 \pm 0.8\%$ remain almost constant although the flutter-control system introduces forces in heave direction. Figure 3.7 demonstrates on the left that both the power $\bar{P}_v$ due to the structural damping and the power $\bar{P}_m$ due to the pitching moment withdraw energy from the oscillating system,

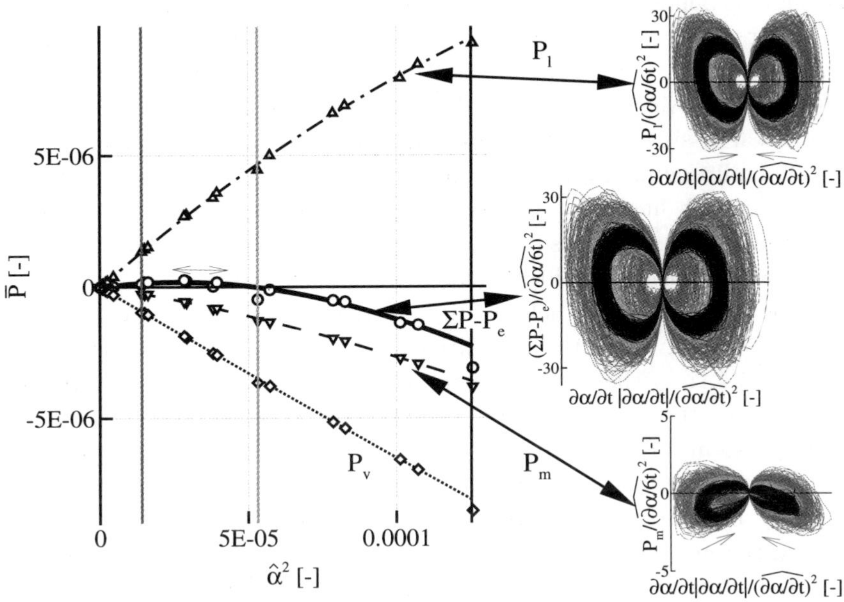

Fig. 3.7 Left: Components of power $\bar{P}_l, \bar{P}_m, \bar{P}_v, \Sigma\bar{P} - \bar{P}_e$ averaged per oscillation cycle versus limit-cycle pitch amplitude squared $\hat{\alpha}^2$. Right: Instantaneous components of power $P_l, P_m, \Sigma P - P_e$ versus pitching velocity times its absolute value $\partial\alpha/\partial t\,|\partial\alpha/\partial t|$. Ordinate and abscissa are normalized by the amplitude of the pitching velocity squared $\widehat{\partial\alpha/\partial t}^2$. $\mathrm{Ma}_\infty = 0.691 \pm 0.05\%$, $\mathrm{Re}_\infty = 1.47 \times 10^6 \pm 0.8\%$, $\mathrm{Su}_0 = 0.367 \pm 0.6\%$, $\mu = 1008.9 \pm 0.6\%$, $v_\infty = 2.64 \pm 0.3\%$, $\alpha_0 = 1.78^o \pm 0.04^o$.

because both components are negative (cf. sign convention in Eqs. (3.9), (3.10)). Since the abscissa is the limit-cycle pitch amplitude squared $\hat{\alpha}^2$, $\bar{P}_v$ appears as a straight line. In the case of "classical" flutter exhibiting exponentially growing oscillation amplitudes, the sum of power $\Sigma\bar{P} - \bar{P}_e$ would be a straight line in such a diagram, too. But, since the curves of $\bar{P}_m$ and $\bar{P}_l$ are slightly nonlinear, the sum of power $\Sigma\bar{P} - \bar{P}_e$ is also nonlinear and is able to cut the abscissa at a distinct amplitude $\hat{\alpha}^2$. This argument, while having in mind that the motion of the airfoil is almost harmonic, leads to the following proposition: the amplitude dependency of the imaginary part of the lift or of the pitching-moment response to the motion of the airfoil is necessary for the amplitude limitation of the observed LCOs. A slight variation of the structural damping would change the LCO amplitude significantly, since the curve $\Sigma\bar{P} - \bar{P}_e$ crosses the abscissa at a fine angle and would then rotate around the origin. Thus, the LCO amplitude strongly depends on the structural damping in the system. This statement is in agreement with Meijer and Cunningham Jr. (1992) who found that the damping is important for the correctly modeling LCO amplitude. It also supports Morton and Beran's 1999 speculation, that differences in the numerical dissipation and the applied time steps cause discrepancies in the LCO amplitude prediction. A stable LCO attractor requires that $\Sigma\bar{P} - \bar{P}_e$ decreases with increasing oscillation amplitude while cutting the abscissa. Conversely, if a repeller is present, the sum of power would have to increase with the oscillation amplitude close to $\Sigma\bar{P} - \bar{P}_e = 0$.

The diagram on the left of Fig. 3.7 reveals that the power due to the lift feeds energy into the oscillations. This fortifies the presumption of Section 3.4.1: in the vicinity of the transonic dip, the time lag of the lift response to the motion of the airfoil yields a power flux into the oscillating system. The significance of the lift supports the argumentation of Bendiksen (1992). He claims that the tendency to transonic SDOF flutter may be caused by a maximum work contribution of the lift variation due to the moving shock wave. Bendiksen (1992) argues on the basis of a shock-doublet approximation of Ashley (1980) that this work contribution is maximum if the shock motion is in phase with the heave velocity at the shock location and if the heave velocity at this location is maximized. The phase differences between heave and pitch motion observed in the present investigation agree with his argument. Moreover, the phase lag of the lift response to the pitching of the airfoil demonstrates a tendency to grow toward -90° in the vicinity of the transonic dip (cf. Fig. 3.3), such that the lift response tends to get in phase with the heave velocity. If this is caused by the shock motion being in phase with the local heave velocity, this cannot be determined since neither local forces nor flowfield information were measured in the present flutter experiments. The local aerodynamics and the local energy exchange between the fluid and structure will be the subject of future investigations.

The right side of Fig. 3.7 shows several instantaneous components of power versus pitching velocity times its absolute value ($\partial\alpha/\partial t\,|\partial\alpha/\partial t|$). The orientation of these trajectories is indicated by the arrows. The abscissa is chosen thus, in order to distinguish between the up- and downstroke of the airfoil. The trajectories are plotted for three different LCO amplitudes corresponding to the vertical gray lines in the left diagram. Abscissa and ordinate are normalized by the amplitude of the pitching velocity squared in order expose discrepancies in shapes. However, it is very difficult to recognize that the power due to the lift and the moment, averaged per cycle, decreases as shown on the left of Fig. 3.7. Moreover, there is no part of the oscillation cycle that one can identify in which an eminent power loss takes place. Thus, the nonlinearities leading to the amplitude limitation of the measured LCOs are clearly small. In the lower right part of Fig. 3.7, a viscous damper would occur as a wedge $\wedge$ with the kink in the origin. Therefore, it is shown that the pitching moment acts more or less like a damper; but, compared with the amplifying power due to the lift, the damping power due to the moment was small in all observed cases (cf. axis limits on Fig. 3.7, right). However, the nonlinear increase of the damping moment could clearly be observed, whereas the nonlinear decrease of the driving power to the lift could not be confirmed for any condition.

3.5 Concluding remarks

Results from experiments on heave/pitch flutter of the supercritical airfoil NLR 7301 with tripped laminar-turbulent transition at transonic flow conditions have been reported. The global aerodynamic force behavior was linked to the observed transonic dip phenomena and limit-cycle oscillations.

The most significant results from this investigation can be summarized as follows.

(i) The phase lag of the lift response to the airfoil pitch motion may be responsible for the reduction of the flutter stability at transonic speeds called transonic dip.
(ii) A slightly nonlinear oscillation-amplitude dependency of the lift and the pitching moment affects the amplitude limitation of the limit-cycle oscillations.
(iii) The LCO amplitude can be controlled by relatively small forces.
(iv) The amplitudes nevertheless strongly depend on the structural damping of the aeroelastic system.

Since these results are obtained using one airfoil shape with a limited range of structural-dynamic parameter variation, it is not claimed that the present results are generally transferable to the flutter behavior of large aspect-ratio wings. In particular, the mass ratio μ in the order of $\mathcal{O}(1000)$ is high compared with realistic configurations. Thus, for real aircraft, the aerodynamics are more dominant compared with the structural dynamics than in our experiments, according to Eq. (3.1). Therefore, the aerodynamic amplitude-limiting effects are stronger, but also the oscillation amplitudes in real flight will increase more rapidly, since the ratio of aerodynamic work to total energy scales with $1/\mu$ (cf. Eq. (3.10)). Nevertheless, the results indicate the following consequences.

(a) Although or just because LCOs are controllable with small forces, its amplitude strongly depends on the system damping. Thus, LCOs are, in the authors' opinion, only marginally usable in order to extend the operational flight regime of aircraft with large aspect-ratio wings (cf. Introduction). However, LCOs are technically very important in the context of flutter since traditional flutter-stability analysis underestimates the instability of the coupled fluid-structure system in the case of a subcritical bifurcation.

(b) Since manifestations of LCOs are rather sensitive to flow and structural parameters as well as to boundary conditions, the capability of a computational aeroelasticity code to reproduce the LCO amplitude is a telling test of the code.

(c) The reduction of the flutter stability, called transonic dip, is caused by the aerodynamic response to the motion of an airfoil with finite thickness. At transonic conditions, the unsteady flow field depends on the steady flowfield and therefore on the airfoil incidence and shape (Tijdeman and Seebass, 1980). Thus, arranging the steady flowfield appropriately may cause unsteady flow that does not lead to a pronounced dip. So, design criteria for airfoil shapes in order to minimize the transonic dip could be formulated from the aeroelastic point of view, if the aerodynamics that leads to the time lag of the lift response to the airfoil pitch motion is understood.

The present investigation links the global force behavior to the observed motions of an airfoil. For future work we intend to examine the local aerodynamics leading to the reported global force behavior. Furthermore, we intend to enlarge the range of structural-dynamic parameter variation for flutter testing and to investigate wing models in flutter experiments in order to ensure the transferability of the obtained results to the flutter behavior of real large aspect-ratio wings.

Acknowledgment

The first author is greatly indebted to Lu Ting for providing encouragement while working on his doctoral thesis advised by Egon Krause at the RWTH Aachen, Germany. This support was indispensable for him in order to assess entropy-layer instabilities in a supersonic flowfield by means of stability analysis.

We are grateful to Dr V. Carstens for providing the code to solve the Possio integral equation and wish to thank Dr K. Weinman, Dr W. Wegner, Dr R. Voß and Dr P. Petrie-Repar for fruitful discussions and suggestions. Thanks also to J. Nuhn for thoroughly acquiring the data as well as to J. Berold and D. Möhle being responsible for the models and the test set-ups. Finally, we would like to thank the staff of the DNW-TWG for their cooperation. This work was supported by the Hermann von Helmholtz-Gemeinschaft Deutscher Forschungszentren (HGF) within a project called "Aerostabil".

Bibliography

Ashley, H. (1980). Role of shocks in the "sub-transonic" flutter phenomenon, *J. Aircr.* **17**(3), pp. 187–197.

Bendiksen, O. O. (1992). Role of shock dynamics in transonic flutter, AIAA 92-2121, AIAA.

Carstens, V. (1973). Berechnung der instationären Druckverteilung an harmonisch schwingenden Gittern in ebener Unterschallströmung, IB 253-73 J06, DLR.

Cebeci, T. and Bradshaw, P. (1979). *Momentum Transfer in Boundary Layers*, McGraw-Hill, New York.

Edwards, J. W. (1996). Transonic shock oscillations and wing flutter calculated with an interactive boundary layer coupling method, in *EUROMECH-Colloquium 349, Simulation of Fluid-Structure Interaction in Aeronautics*, Göttingen, Germany.

Jacobs, M. (2002). Treatment of the wall boundary layer in the wall adaptation procedure and steady wall adaptation for dynamic tests, private communication, DNW BU GuK, Göttingen.

Knipfer, A. and Schewe, G. (1999). Investigation of an oscillating supercritical 2d wing section in transonic flow, AIAA 99-0653, AIAA.

Meijer, J. J. and Cunningham Jr., A. M. (1992). Development of a method to predict transonic limit cycle oscillation characteristics of fighter aircrafts, CP 507, AGARD.

Morton, S. A. and Beran, P. S. (1999). Hopf-bifurcation analysis of airfoil flutter at transonic speeds, *J. Aircr.* **36**(2), pp. 421–429.

Possio, C. (1938). L'azione aerodinamica sul profilo oscillante in un fluido compressibile a velocita iposonora, *L'Aerotecnica* **18**, pp. 441–458.

Schewe, G. (1991). Force measurements in aeroelasticity using piezoelectric multicomponent transducers, in *Proc. Int. Forum on Aeroelasticity and Structural Dynamics*, DGLR 91-06, Aachen, Germany.

Schewe, G. and Deyhle, H. (1996). Experiments on transonic flutter of a two-dimensional supercritical wing with emphasis on non-linear effects, in *Proc. RAeS Conference on "Unsteady Aerodynamics"*, London, UK.

Schulze, S. (1997). Transonic aeroelastic simulation of a flexible wing section, R 822, AGARD.

Tang, L., Bartels, R. E., Chen, P. C., and Liu, D. D. (2003). Numerical investigation of transonic limit cycle oscillations of a two-dimensional supercritical wing, *J. Fl. Str.* **17**(1), pp. 29–41.

Tijdeman, H. (1977). *Investigations of the Transonic Flow Around Oscillating Airfoils*, Diss., Technische Hogeschool Delft, The Netherlands, also NLR TR 77090.

Tijdeman, H. and Seebass, R. (1980). Transonic flow past oscillating airfoils, *Ann. Rev. Fluid Mech.* **12**, pp. 181–222.

Voß, R. (1998). Wall correction methods for dynamic tests, AG 336, AGARD.

Zimmermann, N. H. and Weissenburger, J. T. (1964). Prediction of flutter onset speed based on flight testing at subcritical speeds, *J. Aircr.* **1**(4), pp. 190–202.

Zwaan, R. J. (1979). Summary of data required for the AGARD SMP activity "Standard Aeroelastic Configurations" – Two-dimensional configurations, MP 79015 U, NLR.

Chapter 4

Vortices in Superconductors

Max Gunzburger and Janet Peterson
School of Computational Science
Florida State University, Tallahassee FL 32306-4120, USA

Dedicated to Lu Ting on the occasion of his eightieth birthday

Abstract

Fluid vortices have fascinated mankind from time immemorial, and have been the subject of interest since the dawn of scientific inquiry. Superconducting vortices are of much more recent vintage, but are an equally fascinating phenomenon. Vortices in superconductors are tubes of magnetic flux, or equivalently, cylindrical current loops, that penetrate into a material sample. Knowledge about the structure and dynamics of superconducting vortices are of importance both to the understanding of the basic physics of superconductors and to the design of devices. Vortices in superconductors can be modeled on several scales, from the microscopic to the macroscopic. Here, we concentrate on providing a brief review of mezoscopic-scale models for superconductivity, namely the Ginzburg-Landau equations and variants thereof. At this scale, superconducting vortices are analogous to viscous fluid models that consist of a collection of discrete vortices. The basic model holds for homogeneous, isotropic superconductors; the variants can account for inhomogenieties and anisotropies due to impurities, thickness variations, and thermal fluctuations. These all affect the vortex state, as do changes in the applied magnetic field and current strengths and directions. We use computational simulations to illustrate these effects.

4.1 Introduction

The two hallmarks of superconductivity are *perfect conductivity* and *perfect diagmagnetism*. The former alludes to the flow of current without resistance; in fact, the resistivity of the superconducting state is below the detection capability of any past or current measuring device. The latter refers to the expulsion of a magnetic field from a sample as it is cooled below the critical temperature at which it becomes superconducting. Perfect conductivity in metals at very low temperatures was discovered by Kamerlingh-Onnes in 1911 while perfect diamagnetism was first observed by Meissner and Ochsenfeld in 1934. This description of superconductivity applies to metal superconductors in the bulk, i.e., away from boundaries. The situation for other superconductors, including the high-critical temperature superconductors first developed by Bednorz and Müller in 1985, is more complicated. In particular, mixed states occur in which both the normal, i.e., nonsuperconducting, and superconducting phases co-exist. In this mixed state, the magnetic field can penetrate the sample in a tube-like configuration as depicted in the sketch in Fig. 4.1. These tubes of magnetic flux are referred to as (superconducting) *vortices*.

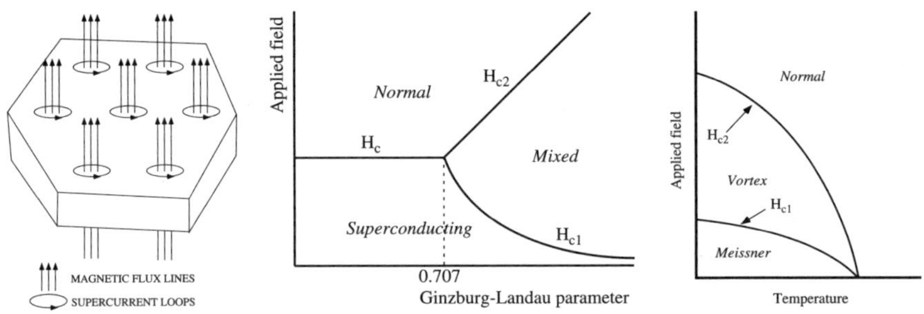

Fig. 4.1 Left: magnetic flux tubes, i.e., vortices, in a superconductor. Center: the response of a superconductor to an applied magnetic field for type-I ($\kappa < 1/\sqrt{2}$) and type II ($\kappa > 1/\sqrt{2}$) superconductors. Right: the phase diagram for an ideal type-II superconductor.

Superconductivity can be modeled at several scales; Fig. 4.1 depicts a mezoscale description of superconductivity. At the microscopic scale, the BCS theory [2] provided, in 1957, the first satisfactory theoretical explanation of superconductivity. Before that time, phenomenological continuum theories had been developed, e.g., the London [25] and the Ginzburg-Landau (GL) [21] models that date from 1935 and 1950, respectively. In

1959, the latter theory was connected [22] to the BCS theory. Also in 1959, the GL theory was used [1] to predict the vortex-like structures depicted in Fig. 9.1; this represents a remarkable triumph for the GL theory since, in the first place, that theory was not developed to cover that phenomena and, in the second place, even in 1959, these structures had not been readily observed in experiments. A more detailed account of the BCS, London, and GL theories than that given here can be found, e.g., in [26].

In the mezoscale description of superconductivity, vortices are individually resolved. In the GL model, the vortices have finite cores, while the London model can be viewed as a simplification of the GL model in which the isolated vortices are line singularities. Unfortunately, the distances between vortices is very small, e.g. a few hundred Angstroms, so that such descriptions cannot be used for the modeling and design of most devices. For that purpose, macroscale models are appropriate. Several such models of varying complexity have been developed; their range of validity depends on the values of certain material parameters. See [3] for a full review of mezoscopic and macroscopic models for superconductivity, and for how the various models can be derived in a systematic manner.

Perhaps of special interest to those familiar with fluid vortices is the connection between the London model and a macroscopic vortex density model for superconductivity. The former involves isolated line vortices, while the latter, which can be derived from the former through an averaging process, involves a transport equation for the evolution of a vortex density function. This transport equation bears a striking similarity to the vorticity transport equation for inviscid fluids; it also bears an important difference which accounts for the observation that fluid vortices are transported with the velocity field while superconducting vortices are transported perpendicular to the current field. Again, see [3] for more details.

Here, we focus on the mezoscale Ginzburg-Landau model for superconductivity and several enhanced versions of that model that have been developed to extend the physical range of its applicability.

4.2 The Ginzburg-Landau model

We now review the Ginzburg-Landau (GL) theory; a more detailed account can be found in, e.g., [26]. The GL theory of superconductivity applies to homogeneous and isotropic materials in the steady state. The primary dependent variables employed by the theory are the vector-valued magnetic potential $\mathbf{A}$ and a complex-valued order parameter ψ. Based on Landau's theory of second-order phase transitions, Ginzburg and Landau noted that the Gibbs free energy associated with a superconducting material is given by

$$\mathcal{G}(\psi, \mathbf{A}) = \int_\Omega \left(f_n + \alpha|\psi|^2 + \frac{\beta}{2}|\psi|^4 \right) d\Omega + \int_\Omega \frac{1}{m_s} \left| \left(i\hbar\nabla + \frac{e_s}{c}\mathbf{A} \right) \psi \right|^2 d\Omega$$
$$+ \int_\Omega \left(\frac{|\nabla \times \mathbf{A}|^2}{8\pi} - \frac{\nabla \times \mathbf{A} \cdot \mathbf{H}}{4\pi} \right) d\Omega \,.$$

The three terms on the right-hand side respectively correspond to the condensation, kinetic, and magnetic energies; here, $\mathbf{H}$ denotes a given applied field. The coefficient α is temperature dependent and changes sign (from positive to negative) as the temperature T decreases from above to below the critical temperature T_c. For $T > T_c$, the sample is in a normal (non-superconducting) state while for $T < T_c$, the sample is superconducting. The other coefficients are roughly temperature independent. In particular, m_s and e_s are the mass and charge of the superconducting charge carriers; these are often chosen to be twice the corresponding quantities for an electron. The basic thermodynamic postulate of the GL theory is that the superconducting sample is in a state such that the Gibbs free energy is a minimum.

Variables of physical interest may be recovered from ψ and $\mathbf{A}$ through the relations

$$\begin{aligned}
\mathbf{h} &= \nabla \times \mathbf{A} = \text{magnetic field} \\
\mathbf{j} &= \frac{e_s}{m_s}(\hbar\nabla\phi - \frac{e_s}{c}\mathbf{A})|\psi|^2 = \text{supercurrent} \\
N_s &= |\psi|^2 = \text{density of available of superconducting charge carriers,}
\end{aligned}$$

where ϕ denotes the phase of the order parameter, i.e., $\psi = |\psi| \exp^{i\phi}$.

There are two length scales associated with GL free energy, the penetration depth and coherence lengths given by

$$\lambda = \sqrt{-\frac{\beta m_s c^2}{4\pi\alpha e_s^2}} \qquad \text{and} \qquad \xi = \sqrt{-\frac{\hbar}{2m_s\alpha}} \,,$$

respectively; they provide the scale for variations in the magnetic field and order parameter, respectively. The ratio $\kappa = \lambda/\xi$ is referred to as the

Ginzburg-Landau parameter; λ, ξ, and κ are material parameters.

Superconductors can be divided into two classes according to the value of κ. For $\kappa < 1/\sqrt{2}$, the surface energy associated with phase transitions from the superconducting to the normal state is positive so that phase transitions are avoided. These types of superconductors, which includes metals, are referred to as Type-I superconductors; the description of superconductivity given at the beginning of the introduction is largely that of a type-I superconductor.

For type-II superconductors, $\kappa > 1/\sqrt{2}$ and the surface energy associated with phase transitions is positive so that phase transitions are favored. Most superconductors of practical use are of type-II; high-temperature superconductors are strongly type-II with, typically, $\kappa \geq 50$. A natural question follows: if the surface energy associated with phase transitions in type-II superconductors is negative, what stops phase transitions from occurring over arbitrarily small length scales? The answer is *fluxoid quantization*. Let $\partial\Sigma$ denote a closed curve lying in the material sample such that $|\psi| \neq 0$ everywhere on the curve, i.e., the curve $\partial\Sigma$ nowhere intersects a normal region; let Σ denote a surface bounded by the closed curve $\partial\Sigma$. We define the fluxoid to be

$$\Phi' = \int_\Sigma \mathbf{h} \cdot d\Sigma + \frac{m_s c}{e_s^2} \int_{\partial\Sigma} \frac{\mathbf{j}}{|\psi|^2} \cdot d(\partial\Sigma) = \frac{\hbar c}{e_s} \int_{\partial\Sigma} \nabla\phi \cdot d(\partial\Sigma)\,.$$

Then, it turns out that $\Phi' = (2\pi\hbar c/e_s)n$, where n is an integer, i.e., Φ' is quantized. Again, see [26] for details.

Fluxoid quantization and the negative surface energies associated with phase transitions results in the existence of a *mixed* normal-superconducting state in type-II superconductors. In this state, we have that $0 < |\psi| < \psi_0 = \sqrt{-\alpha/\beta}$, i.e., $|\psi|$ varies between the normal and superconducting values, and $\mathbf{h} \neq \mathbf{0}$, i.e., the field penetrates the sample. This penetration takes the form of tubes of flux, i.e., superconducting vortices, as depicted in Fig. 4.1; for pure, low-temperature superconductors, the vortices arrange themselves in a regular, hexagonal lattice. The dependence of both type-I and type-II superconductors on the applied magnetic is also sketched in Fig. 4.1 as is the phase diagram for an ideal type-II superconductor. At least in the bulk, for low enough values of the fields and temperatures, we have a Meissner region in which the field is excluded from the sample and $|\psi|$ attains its full superconducting value ψ_0. For high enough values of the fields or temperatures, superconductivity is destroyed. For type-II superconductors, for intermediate values of the magnetic field and temperature, one also has the mixed, vortex state; so, in this case, one has the two critical fields $H_{c1}(T)$ and $H_{c2}(T)$ which separate the three regions in the phase diagram.

If we nondimensionalize lengths by λ, ψ by ψ_0, $\mathbf{H}$ by $\sqrt{2}H_c = \sqrt{8\pi\alpha^2/\beta}$, and $\mathbf{A}$ by λH_c, the Gibbs free energy takes the form

$$\mathcal{G}(\psi, \mathbf{A}) = \int_\Omega \left(f_n - |\psi|^2 + \frac{1}{2}|\psi|^4 \right) d\Omega + \int_\Omega \left| \left(\frac{i}{\kappa}\nabla + \mathbf{A} \right) \psi \right|^2 d\Omega + \int_\Omega \left(|\nabla \times \mathbf{A}|^2 - 2\nabla \times \mathbf{A} \cdot \mathbf{H} \right) d\Omega \,.$$

The importance of the Ginzburg-Landau parameter κ is now clear; indeed, it is the sole parameter appearing in the nondimensionalized Ginzburg-Landau model. Also, one now has that $H_{c1} \approx \ln\kappa/(2\kappa)$ and $H_{c2} = \kappa$.

The Euler-Lagrange equations that correspond to the minimization of the Gibbs free energy are the celebrated *Ginzburg-Landau equations.* We do not list them here since they are merely the time-independent versions of (4.1).

The vortex state in a two-dimensional square-shaped sample of side 100λ is depicted in Fig. 4.2 for $\kappa = 5$ and the (nondimensional) applied field has magnitude $H = 0.5\kappa$ and is directed perpendicular to the sample cross-section. The level curves of the magnitude of the nondimensionalized order parameter and a vector plot of the supercurrent are provided; for the sake of clarity, only the contours for $|\psi| \leq 0.5$ are provided. The simulations were carried out using a finite element code [15; 16; 17; 18]. Note the hexagonal lattice of vortices near the center of the sample. The flux tubes sketched in Fig. 4.1 are perpendicular to the paper in Fig. 4.2. At the center, or core, of the vortices, the order parameter goes to zero. Note the supercurrent circulating around the core of each vortex and around the boundary of the sample. The latter acts as a shielding current that prevents the magnetic field from penetrating into the sample, except as flux tubes through the vortex cores.

4.2.1 *The time-dependent Ginzburg-Landau equations*

The motion of vortices, i.e., of the magnetic flux tubes, induces a Lorentz force, hence a resistance, and thus a loss of perfect conductivity. To account for such motions, time-dependent versions of the Ginzburg-Landau model have been developed; see, e.g., [23]. Such models can be used to study the equilibrium structure of the vortex state and how that structure is achieved and can also account for effects due to applied currents.

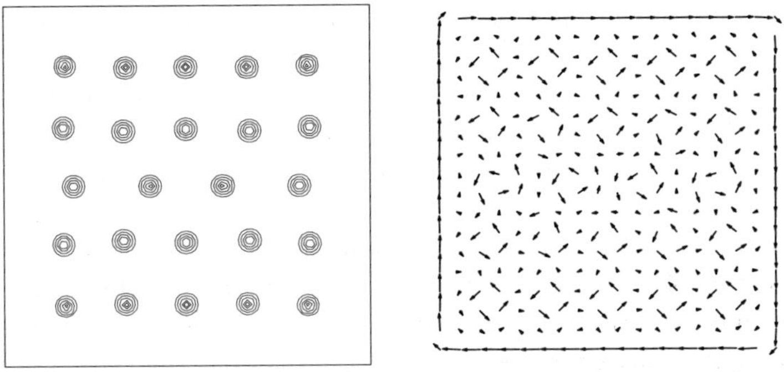

Fig. 4.2 Vortex lattice in a two-dimensional, type-II superconductor. Left: level curves of the magnitude of the order parameter. Right: the supercurrent distribution.

The time-dependent Ginzburg-Landau (TDGL) equations in the presence of an applied field $\mathbf{H}$ are given, in nondimensional form, by

$$\begin{cases} -\left(\dfrac{\partial \psi}{\partial t} + i\kappa\Phi\psi\right) = \left(\dfrac{i}{\kappa}\nabla + \mathbf{A}\right)^2 \psi - \psi + |\psi|^2\psi \\ -\sigma\left(\dfrac{\partial \mathbf{A}}{\partial t} + \nabla\,\Phi\right) = \nabla \times (\nabla \times \mathbf{A} - \mathbf{H}) + |\psi|^2\mathbf{A} + \dfrac{i}{2\kappa}(\psi^*\nabla\psi - \psi\nabla\,\psi^*) \end{cases} \tag{4.1}$$

Here, the scalar potential Φ is an additional primary variable. The normal "conductance" σ is a second parameter appearing in the TDGL model. Just as κ is the ratio of the relaxation length scales for $\mathbf{A}$ and ψ, σ is the (inverse) ratio of the relaxation time scales for the two dependent variables. The system (4.1) holds within a superconducting sample; exterior to that sample, the Maxwell equations hold.

The physically interesting variables N_s and $\mathbf{h}$ are given as before; we also have that

$$\mathbf{E} = -\frac{\partial \mathbf{A}}{\partial t} - \nabla\,\Phi = \text{electric field}$$

$$\mathbf{j} = \left(\frac{1}{\kappa}\nabla\phi - \mathbf{A}\right)|\psi|^2 - \sigma\left(\frac{\partial \mathbf{A}}{\partial t} + \nabla\,\Phi\right) = \text{current.}$$

Now, the current not only has a resistanceless, supercurrent component, but also has a resistive, Ohmic, normal current component, e.g., the second term in the above equation for the current. In general, one should append to the TDGL equations another equation for Φ. However, in many situations, Φ may be determined as a gauge choice; see, e.g., [14]. Additionally, initial conditions on ψ and $\mathbf{A}$ must be imposed, along with boundary and interface conditions imposed on the boundary of the superconducting sample. These include the continuity of the tangential components of the magnetic field, i.e., of $\nabla \times \mathbf{A} \times \mathbf{n}$ across the boundary of the sample, and, if a vacuum exists exterior to the sample, the vanishing of the normal component of current at the boundary.

In the steady-state Ginzburg-Landau model, a gauge may always be chosen so that $\Phi = 0$. Thus, the Ginzburg-Landau equations can be obtained by merely setting the left-hand side of the TDGL equations to zero.

If an applied current is also applied to the sample, then the system (4.1) must be slightly amended. For example, if a constant applied current J is applied in the y-direction and a constant applied field H is applied in the z-direction, we can choose the electric potential $\Phi = -\frac{1}{\sigma}Jy$ and, since, the total applied field = applied field + field due to the current, we have that $\mathbf{H}$ is in the z-direction and $|\mathbf{H}| = H - Jx$.

Many analytical results have been derived for both the GL and TDGL models; see, e.g., [8; 14; 15; 16; 19]. For example, for the TDGL model, the global existence and uniqueness of appropriately defined weak solutions and the continuous dependence of the solution on the initial data have been proved. An invariant integral has been identified and the convergence results for semi-discrete and fully-discrete Galerkin finite-dimensional approximations have been obtained.

In Fig. 4.3, finite element simulations of the TDGL equations are used to show, through a series of snapshots in increasing time, the nucleation of vortices at the boundaries of a sample and the development of the steady-state vortex lattice of Fig. 4.2. Figure 4.3 also illustrates the motion of vortices induced by an applied current. The applied field is perpendicular to the paper; the applied current is in the vertical direction and the vortices move from left to right.

4.2.2 *A simplified GL model valid for high values of κ*

As has already been mentioned, for most practical superconductors, the value of the Ginzburg-Landau parameter κ is "large," e.g., 50 or more. One can take advantage of this to derive simplified versions of the GL or TDGL equations [7]. The exact nature of these equations depends on the size of the applied magnetic field. The $\kappa \to \infty$ limit with $\mathbf{H} = O(\kappa)$ is of greatest interest for technological applications since one is interested in high fields. In this setting, to leading order, the steady-state GL equations

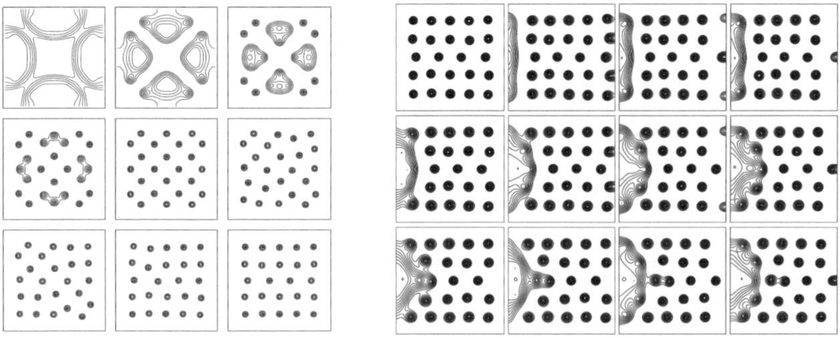

Fig. 4.3 Left: the nucleation of vortices and the subsequent formation of the vortex lattice in a two-dimensional, type-II superconductor. Right: vortex motion in the presence of a constant applied current in the vertical direction.

simplify to

$$\nabla \times \mathbf{A}_0 = \mathbf{H} \qquad \text{and} \qquad (i\nabla + \mathbf{A}_0)^2 \psi - \psi + |\psi|^2 \psi = 0 \,. \tag{4.2}$$

Note that $\mathbf{A}_0$ is simply the magnetic potential due to the applied field, ignoring the superconductor. The important simplification effected is that the magnetic potential uncouples from the order parameter, i.e., one may solve the Maxwell equations for $\mathbf{A}_0$ and then solve (4.2) for ψ. Equations for corrections to the magnetic field have also been developed; see [7] for details.

Figure 4.4 shows the vortex state in a $10\lambda \times 10\lambda$; the full GL model is used to determine that state for several values of κ and may be compared to the vortex state obtained using the simplified model (4.2). We see that, even for moderate values of κ, the results for the full GL model are virtually indistinguishable from those for the simplified model.

4.3 The vortex state in non-ideal superconductors

We now explore some variants of the basic Ginzburg-Landau model that take into account several important effects that the GL model is not equipped to handle. We also provide the results of numerical simulations that serve to illustrate some of the resulting changes that occur in vortex structure. Details may be found in, e.g., [4; 5; 6; 9; 10; 20].

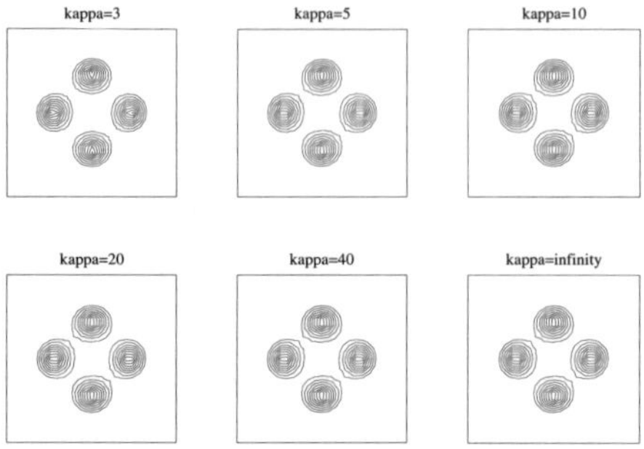

Fig. 4.4 Superconducting vortices in a 10×10 penetration length sample for $H = \frac{1}{2}\kappa$ for the full GL model for several values of κ and for the simplified GL model (4.2) valid in the limit $\kappa \to \infty$.

4.3.1 *Pinning of vortices*

As was already mentioned, the motion of vortices cause resistance. Further, in a clean superconductor, any applied current will cause vortices to move. Thus, the identification of mechanisms that can "pin" vortices, i.e., stop them from moving, are of great interest. Here, we discuss three such mechanisms and how they can be effectively modeled through variations in the GL equations.

Vortices can be pinned by normal impurities within the superconductor. The appearance of such inclusions can be modeled by allowing α and m_s to vary across the sample. In regions occupied by the superconducting material, $\alpha < 0$ and $m_s =$ twice the electron mass. In regions occupied by the normal, nonsuperconducting material, we choose $\alpha > 0$ and allow the mass to be a second material parameter. In nondimensional form, the parameters that define the model are $|\alpha_n/\alpha_s|$ and m_s/m_n, where the subscripts s and n refer to the superconducting and normal materials, respectively. It was shown in [4] that this model correctly accounts for physical behaviors at superconducting/normal and superconducting/vacuum interfaces.

Another pinning mechanism is variations in thickness in thin films. Such materials can be modeled by the full, three-dimensional Ginzburg-Landau model; however, if one takes advantage of the thinness of the film, one can derive simplified models that are much less costly to compute with. A variable thickness, thin film model is derived in [6], where it is shown that, to leading order, the magnetic field is unaffected by the superconductor. Thus, if the applied field is a constant, to leading order, the magnetic field is constant throughout the sample. As a result, the vector potential can be determined independently of the order parameter. The latter can be determined, to leading order, from a variable coefficient version of the first Ginzburg-Landau equation using the known vector potential. In the steady-state, this equation is given by

$$\left(\frac{i}{\kappa}\nabla + \mathbf{A}_0\right) \cdot d \left(\frac{i}{\kappa}\nabla + \mathbf{A}_0\right) \psi + d(|\psi|^2 - 1)\psi = 0\,,$$

where $d(x, y)$ denotes a function describing the thickness variation of the film and $\mathbf{A}_0$ the known vector potential. Higher-order corrections can also be determined. In addition, it has been shown, using this simplified model, that all superconducting materials, whether type-I or type-II in bulk, behave as type-II superconductors when made into sufficiently thin films. Furthermore, the rigorous connection between solutions of the thin film model and solutions of the three-dimensional GL equations has been established.

A third pinning mechanism is grain boundaries in anisotropic superconductors. Anisotropies can be accounted for within the Ginzburg-Landau formalism by assuming that the mass in the Gibbs free energy is a tensor, i.e., we replace m_s with a tensor M_s. If the lattice is aligned with the coordinate axes, M_s is diagonal, i.e., $M_s = \text{diag}(m_1, m_2, m_3)$. For details, see, e.g., [5]. This model accounts for anisotropic but homogeneous materials. It can be combined with the other models discussed in this subsection to treat anisotropic, inhomogeneous materials as well as anisotropic thin films.

We now use some numerical simulations to see, in Fig. 4.5, how the vortex lattice is affected by the three pinning mechanisms. The parameters in these models are as for Fig. 4.2, so that that figure gives the lattice structure in the absence of any of the pinning mechanisms. In the left of Fig. 4.5, we have pinning by the inclusion of some normal impurities within a two-dimensional superconductor. These impurities are located at the small squares evident in the figure; in these squares we have chosen $\alpha_n = -\alpha_s$ and $m_n = m_s$. Note that one vortex is attracted to each of the 10 small impurities. In the center figure, we have pinning by thin regions in a thin film. The thin regions are again located at the small squares evident in the figure; in these squares, the thickness of the film is half of the thickness of the rest of the film. Again, note that one vortex is attracted to each of the 10 small thin regions. On the right of Fig. 4.5, we have pinning by a grain boundary in a two-dimensional superconductor. The position of that boundary is indicated by the diagonal line in the figure. Above that line, we have that $m_1/m_2 = 0.5$ while below we have the reverse, i.e., $m_1/m_2 = 2$. Note that vortices are attracted to the grain boundary. Thus, the simulations of Fig. 4.5 show how each of the three pinning mechanisms distort the vortex lattice of Fig. 4.2 by attracting vortices toward the pinning sites.

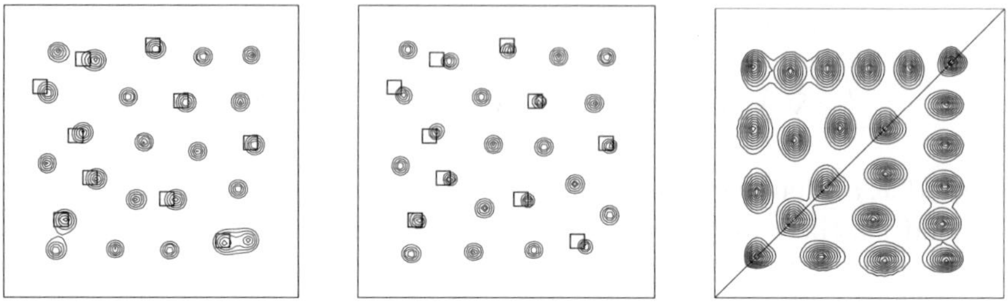

Fig. 4.5 The pinning of vortices by normal inclusions (left), by thinner regions in a variable thickness thin-film (center), and by a grain boundary in an anisotropic superconductor (right).

4.3.2 *Josephson junctions*

A simple Josephson junction consists of a thin slice of normal, nonsuperconducting material sandwiched between slabs of superconducting material. An important property of such junctions is that, if the normal slice is thin enough, a resistanceless supercurrent tunnels across the junction; this is known as the *DC-Josephson effect.* For junctions that can be modeled as one-dimensional objects, e.g., the normal slice is of finite thickness but of infinite extent in the other two directions and the superconducting slabs are of semi-infinite extent, De Gennes developed a theory relating properties in the superconductor on one side of the junction to the corresponding properties in the superconductor on the other side; see, e.g., [11]. If we let $(\cdot)^+$ and $(\cdot)^-$ denote evaluation, at the interfaces with the normal material, of a quantity in the superconductors on the two sides of the simple one-dimensional junction, the De Gennes relation is given by

$$\begin{pmatrix} \psi \\ \left(\frac{1}{\kappa}\nabla - i\mathbf{A}\right)\psi \end{pmatrix}^+ \begin{pmatrix} M_{11} & M_{12} \\ M_{21} & M_{22} \end{pmatrix} \begin{pmatrix} \psi \\ \left(\frac{1}{\kappa}\nabla - i\mathbf{A}\right)\psi \end{pmatrix}^- ,$$

where the M_{ij} are real and are determined by the particular junction and depend on its thickness, the type of material, etc. De Gennes also gave the expression

$$J = \frac{1}{M_{12}} |\psi^+ \psi^-| \sin(\phi^+ - \phi^-)$$

for the tunneling supercurrent flowing across the one-dimensional junction in terms of quantities in the superconductor on each side of the junction. The De Gennes theory does not give any detailed information about the vortex state near or at the junction.

The model used in §4.3.1 for treating normal impurities can also be used for Josephson junctions. In fact, it was shown in [4] that all of the De Gennes theory for Josephson junctions can be recovered from the variable α, variable m_s model described in §4.3.1. For example, if d denotes the thickness of the normal slice, one finds that the above De Gennes relations hold with

$$M_{11} = \cosh(2\kappa d\sqrt{\alpha}) \qquad M_{12} = \frac{m_n}{\sqrt{\alpha}}\sinh(2\kappa d\sqrt{\alpha})$$

$$M_{21} = \frac{\sqrt{\alpha}}{m_n}\sinh(2\kappa d\sqrt{\alpha}) \qquad M_{22} = \cosh(2\kappa d\sqrt{\alpha})\,.$$

Here, we have assumed that the same superconducting material is used on both sides of the junction; the model of [4; 20] can account for different superconductors on each side of the simple junction and remains in complete agreement with the De Gennes theory for that case as well. If one examines the expressions for the tunneling current J and for M_{12}, one sees that the model of [4; 20] predicts that, as expected, J decreases exponentially as d increases and as α increases. Furthermore, J also decreases as m_n increases. Since large m_n corresponds to a highly insulating material, this is in agreement with the experimental observation that junctions made from insulating materials need to be thinner than junctions made from metals in order to obtain the same tunneling current.

The model of [4; 20] not only recovers the De Gennes theory for simple Josephson junctions, but can also provide detailed information about the vortex state in and near the junction. Of course, it can also treat bounded samples in multi-dimensions. Here, in Fig. 4.6, we give the structure of the vortex state in three junctions having differing thicknesses. We now nondimensionalize lengths by ξ so that sample size is now 20ξ by 20ξ. For the thinnest junction of thickness one coherence length, we see that vortices are attracted to the normal material in the junction; the corresponding current plot shows the supercurrent crossing the junction, both as a shielding current near the boundary and as current loops circulating around the vortices in the junction. For the junction of thickness 4ξ, the vortices have become much "weaker" and much less supercurrent is crossing the junction. For the 8ξ junction, the normal material is already thick enough so that no supercurrent is tunneling across the junction; the two superconductors on each side of the normal slice have become essentially uncoupled. Note that there is some supercurrent flowing in the normal material parallel and near to the interfaces with the superconducting materials; this effect, known as the *proximity effect*, is also effectively treated by the model of [4; 20].

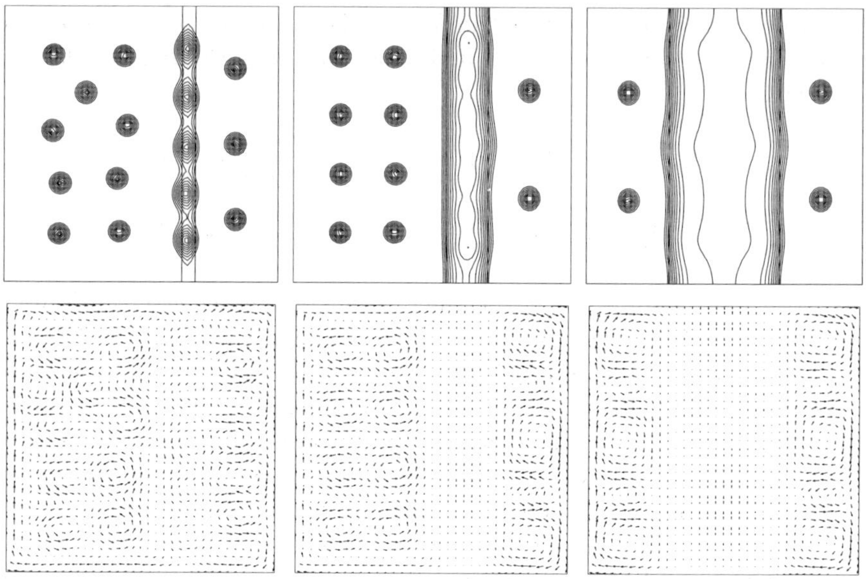

Fig. 4.6 Magnitude of the order parameter (top) and supercurrent (bottom) in a simple Josephson junction of thickness ξ, 3ξ, and 8ξ (left to right).

4.3.3 *Surface superconductivity in high fields*

The phase diagram of Fig. 4.1 holds in the idealized case of pure, homogeneous, isotropic superconductors of infinite extent. For samples that have interfaces with normal materials, e.g. bounded samples, the sharp transition from the vortex state for fields below H_{c2} to the nonsuperconducting, normal state for fields above H_{c2} does not occur as indicated by that phase diagram. In fact, superconductivity can persist for fields higher than H_{c2}. For semi-inifinite superconductors, it can be shown that superconductivity persists up to a field $H_{c3} \approx 1.69 H_{c2}$; see, e.g., [26]. Moreover, near H_{c3}, it is thought that the bulk of the superconductor is in the normal state and the superconducting state exists only near the surface of the sample which, in the semi-inifinite case, is just the planar interface between the superconductor and the normal material. However, these simple one-dimensional analytical theories cannot account for bounded samples in predicting the structure of the superconducting state nor can they account for how, e.g. corners, can affect the value of the highest applied field for which superconductivity persists. These theories cannot say much, if anything, about the structure of the superconducting state for fields between H_{c2} and H_{c3} and not near either.

The Ginzburg-Landau model can be used, in principle, to examine the structure of the superconducting state for any value of the applied field and for bounded samples. Thus, one can examine this state as one increases the field from a value below H_{c2} at which the superconductivity appears as a vortex lattice all the way up to a high enough field so that superconductivity is destroyed. In Fig. 4.7, we present some computational results of such a study. For a sample of size 5ξ by 5ξ, we see that the vortex state (in this case a single vortex) persists relatively unchanged as one crosses the value $H_{c2} = \kappa$, which is the value at which an infinitely large superconductor would lose superconductivity. At a value of the field above 1.4κ, the vortex state changes and four vortices appear; this state persists for fields above the value $H_{c3} = \kappa$ at which one-dimensional theories predict that superconductivity is totally destroyed. In fact, superconductivity persists for fields up to 2.09κ. This indicates that corners can aid in preserving superconductivity at higher fields.

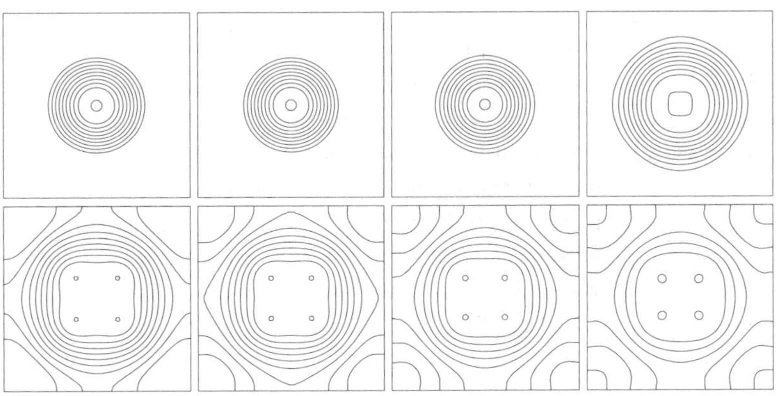

Fig. 4.7 Vortex state in increasing fields; left to right and top to bottom: magnetic field strength $= 0.9\kappa$, 1.0κ, 1.2κ, 1.4κ, 1.6κ, 1.7κ, 1.8κ, and 1.9κ.

Additional studies of the surface superconductivity state in anisotropic superconductors can be found in [10].

4.3.4 *Thermal fluctuations*

The idealized phase diagram of Fig. 4.1 is further complicated by thermal fluctuations, i.e., the motion of vortices due to the vibrations of molecules in the superconductor. This effect becomes especially pronounced as one approaches the critical temperature T_c above which an ideal superconductor loses its superconductivity. Indeed, thermal fluctuations can induce a (local) loss of superconductivity at temperatures below T_c and, conversely, induce (local) superconductivity at temperatures above T_c. Fluctuations can also cause depinning, especially near T_c. Indeed, should the vibratory motion of vortices due to thermal fluctuations be of sufficiently large amplitude, vortices may wander far enough away from a pinning site so that they become permanently dislodged and, for example, continue moving in a resistive fashion due to the application of a current.

Thermal fluctuations can be modeled (see [24]) by introducing a random Langevin forcing term in the first TDGL equation, i.e.,

$$\left(\frac{\partial\psi}{\partial t} + i\kappa\Phi\psi\right) + \left(\frac{i}{\kappa}\nabla + \mathbf{A}\right)^2 \psi - \psi + |\psi|^2\psi = \eta\,.$$

Here, η is a Gaussian random variable in time and space having zero mean. The variance increases as the temperature approaches T_c, i.e., as T approaches T_c, the probability of having larger fluctuations increases. See [12; 1] for more details.

We now present some computational results for thermal fluctuations. First, in Fig. 4.8, we show the nucleation of vortices and the development of the vortex lattice for the same configuration as in Fig. 4.2, but now we add a Langevin forcing term having variance 2. Note that, although things look somewhat chaotic at the beginning, eventually we arrive at a lattice as in Fig. 4.2, but now distorted by the effects of the thermal fluctuations. Of course, a true steady-state is never achieved, but instead the individual vortices eventually reach a quasi-steady-state in which they vibrate about some fixed location. In Fig. 4.9, we show the effects of increasing the temperature, which is modeled through increasing the variance in the random forcing term. We give snapshots, taken at the same time, of computational simulations for different values of the variance. The time chosen is large enough so that the quasi-steady-state described above has been reached.

We see that as the variance, i.e., the temperature, is increased, the effects of thermal fluctuations become more pronounced, and that for large enough variances, the vortex lattice becomes unrecognizable. This may model the transition to the *vortex liquid state*, i.e., the destruction of the Abrikosov lattice at fields below H_{c2}.

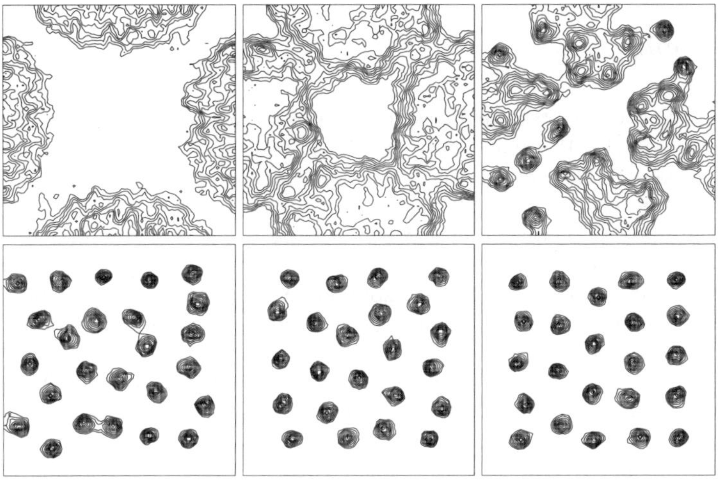

Fig. 4.8 Nucleation of vortices and development of the vortex lattice in the presence of thermal fluctuations having variance 2.

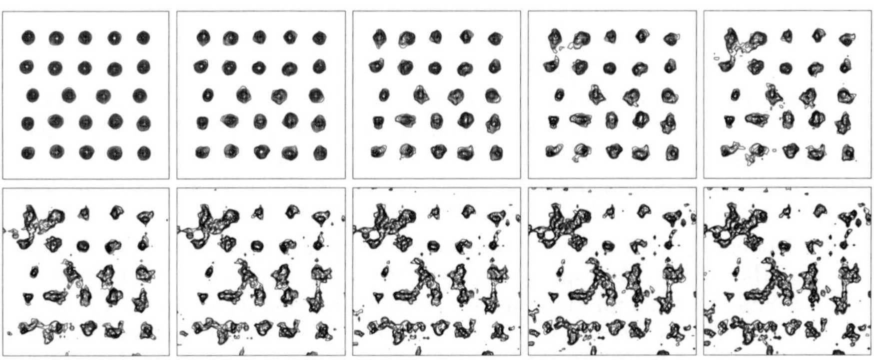

Fig. 4.9 The effect of increasing temperature, modeled through the variance of the random thermal fluctuations, on the vortex lattice; the variance ranges from 1 to 10 (left to right and top to bottom).

Acknowledgment

The first author acknowledges the deep and lasting influence that Lu Ting has had on his career and thanks him for all he has done to teach him how to do science properly and responsibly.

Bibliography

[1] A. Abrikosov; On the magnetic properties of supercondutors of the second group; *S*oviet Phys. JETP **5**, (1957), 1174-1182.

[2] J. Bardeen, L. Cooper, and J. Schrieffer; Theory of superconductivity, *Phys. Rev.* **1**08, (1957), 1175-1204.

[3] S. Chapman; A hierarchy of models for type-II superconductors, *SIAM Rev.* **4**2, (2000), 555-598.

[4] S. Chapman, Q. Du, and M. Gunzburger; A Ginzburg-Landau type model of superconducting/normal junctions including Josephson junctions, *Europ. J. Appl. Math.* **6**, (1995), 97-114.

[5] S. Chapman, Q. Du, and M. Gunzburger; On the Lawrence-Doniach and anisotropic Ginzburg-Landau models for layered superconductors, *SIAM J. Appl. Math.* **5**5, (1995), 156-174.

[6] S. Chapman, Q. Du, and M. Gunzburger; A model for variable thickness superconducting thin films, *J. Appl. Math. Phys.* **4**7, (1996), 410-431.

[7] S. Chapman, Q. Du, M. Gunzburger, and J. Peterson; Simplified Ginzburg-Landau type models of superconductivity in the high kappa, high field limit, *A*dv. Math. Sciences Appl. **5**, (1995), 193-218.

[8] S. Chapman, S. Howison, and J. Ockendon; Macroscopic models of superconductivity, *SIAM Review* **3**4, (1992), 529-560.

[9] S. Chapman and G. Richardson; Vortex pinning by inhomogeneities in type-II superconductors, *Phys. D* **1**08, (1997), 397-407.

[10] J. Clem, V. Kogan, J. Deang, and M. Gunzburger; Nucleation of superconductivity in finite anisotropic superconductors and the evolution of surface superconductivity toward the bulk mixed state; *Phys. Rev. B* **65**, (2002), 094514.

[11] P. de Gennes; *Superconductivity in Metals and Alloys*, Benjamin, New York, (1966).

[12] J. Deang, Q. Du, and M. Gunzburger; Stochastic dynamics of Ginzburg-Landau vortices in superconductors, *P*hys. Rev. B **6**4, (2001), 052506.

[13] J. Deang, Q. Du, and M. Gunzburger; Modeling and computation of random thermal fluctuations and material defects in the Ginzburg-Landau model for superconductivity, *J.* Comput. Phys. **1**81, (2002), 45-67.

[14] Q. Du; Global existence and uniqueness of solutions of the time-dependent Ginzburg-Landau model for superconductivity, *Applicable Anal.* **5**2, (1994) 1-17.

[15] Q. Du; Finite element methods for the time-dependent Ginzburg-Landau model of superconductivity, *Comp. Math. Appl.* **27**, (1994), 119–133.

[16] Q. Du, M. Gunzburger, and J. Peterson; Analysis and approximation of the Ginzburg-Landau model of superconductivity, *SIAM Review* **34**, (1992), 54-81.

[17] Q. Du, M. Gunzburger, and J. Peterson; Solving the Ginzburg-Landau equations by finite element methods, *Phys. Rev. B* **46**, (1992), 9027-9034.

[18] Q. Du, M. Gunzburger, and J. Peterson; Finite element approximation of a periodic Ginzburg-Landau model for type-II superconductors, *Numer. Math.* **64**, (1993), 85-114.

[19] Q. Du, M. Gunzburger, and J. Peterson; Modeling and analysis of a periodic Ginzburg-Landau model for type-II superconductors, *S*IAM J. Applied Math. **53**, (1993), 689-717.

[20] Q. Du, M. Gunzburger, and J. Peterson; Computational simulation of type-II superconductivity including pinning phenomena, *Phys. Rev. B* **51**, (1995), 16194-16203.

[21] V. Ginzburg and L. Landau; On the theory of superconductivity, *Soviet Phys. JETP* **50**, (1950), 1064-1082.

[22] L. Gor'kov; Microscopic derivation of the Ginzburg-Landau equations in the theory of superconductivity, *Soviet Phys. JETP* **9**, (1959), 1364-1367.

[23] Gor'kov and G. Éliashberg; Generalization of the Ginzburg-Landau equations for non-stationary problems in the case of alloys with paramagnetic impurities, *Soviet Phys. JETP* **27**, (1968), 328-334.

[24] P. Hohenberg and B. Halperin; Theory of dynamic critical phenomena, *Rev. Mod. Phys.* **49**, (1977), 435-479.

[25] F. London and H. London; The electromagnetic equations of the supraconductor, *Proc. Roy. Soc. London Ser. A* **14**, (1935), 71.

[26] M. Tinkham; *Introduction to Superconductivity*, McGraw-Hill, New York, 1975.

Chapter 5

Accurate Numerical Simulation of Three-dimensional Lid-driven Cavity Flows with Different Span Lengths

Katsuya Ishii
Information Technical Center
Nagoya University, Nagoya 464-6801, Japan

Tomonori Nihei
Dept. of Computational Science and Engineering
Nagoya University, Nagoya 464-6803, Japan

Shizuko Adachi
RIKEN, Wako, 351-0198, Japan

Dedicated to Lu Ting on his eightieth birthday

Abstract

The steady incompressible vortical flows at the Reynolds number $Re = 850$ in three-dimensional lid-driven rectangular cavities with square cross sections and different span lengths are studied using a combined compact finite difference (CCD) scheme with high accuracy and spectral-like resolution. The simulations have been done at the span wise aspect ratios $\Lambda = 1, 3, 4, 5$ and 6.55. The results of $\Lambda = 1, 3, 4$ shows that the streamlines in the cavities are chaotic and a streamline occupies the whole flow region. On the other hand, the tori of streamlines are found near the symmetric plane in the cavities with longer spans at $\Lambda = 5$ and 6.55, though the global chaotic streamlines are also observed along the central axis of the cavity and around the tori.

5.1 Introduction

The vortical flow in the lid-driven cavity has been extensively studied as a fundamental model for the vortex dynamics in closed systems with simple geometry as well as for a simplified version of many manufacturing devices [1],[2],[5],[7]. There are two geometric non-dimensional parameters: the aspect ratio $\Gamma = d/h$ and the spanwise aspect ratio $\Lambda = l/h$ where d, h and l are the width, the height and the span length of the cavity respectively. We consider the cavity with a square cross section ($\Gamma = 1$) with different spanwise aspect ratio Λ's, and study the side boundary effect on the overall flow structure and the Taylor-Görtler-like vortices in the secondary flow. Albensoeder *et al.* [2] recently studied the three dimensional flow in lid-driven cavities with different aspect ratios by a linear stability analysis of two dimensional flows in the cavity with infinite span length and experiments. They reported that the critical mode is stationary in the square cavity with a short wave, and in the experiment the cell structures near the symmetric plane are observed in the cavity flow with $\Gamma = 1$ and $\Lambda = 6.55$ at the Reynolds number $Re = Ud/\nu = 850$, where U is the speed of the sliding upper wall and ν is the kinematic viscosity. However, they observed only the flow pattern near the wall. The whole flow structure is not clear. In this paper, we focus on the global flow structure in the cavity. Since numerical simulations of these vortical flows with different spatial scales need high accuracy and high resolution, we use a spectral-like combined compact finite difference (CCD) scheme proposed by the authors [6].

5.2 Formulation and numerical method

5.2.1 *Governing equations*

We consider the flows in a three dimensional cavity with the width $d = 1$, the height $h = 1$, and the spanwise length Λ shown in Fig. 5.1.

The upper wall at $y = 1$ is sliding with the velocity $U = 1$ in the x-direction. The dimensionless incompressible Navier-Stokes equations are given by

$$\frac{\partial \mathbf{u}}{\partial t} + \mathbf{u} \cdot \nabla \mathbf{u} = -\frac{1}{\rho} \nabla p + \frac{1}{Re} \nabla^2 \mathbf{u} \tag{5.1}$$

$$\nabla \cdot \mathbf{u} = 0 \tag{5.2}$$

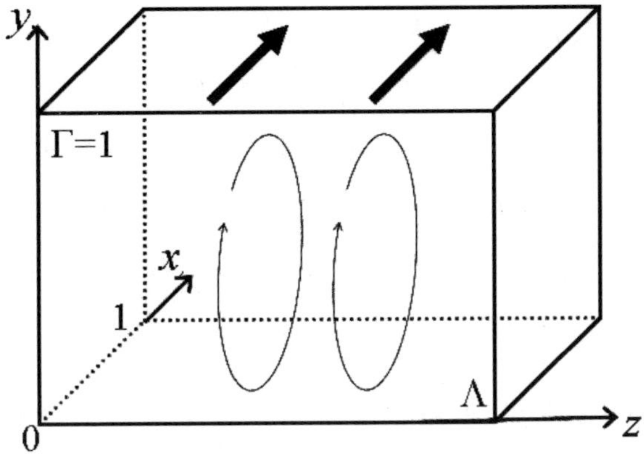

Fig. 5.1 coordinate system

where length, time, fluid velocity $\mathbf{u}$, pressure p have been made dimensionless using the scale h, h/U, U, ρU^2, respectively, and Reynolds number is $Re = \rho h U/\mu$, ρ is the fluid density and μ is viscosity. The no-slip boundary conditions are

$$\mathbf{u} = (1,0,0) \quad on \quad y = 1 \tag{5.3}$$

$$\mathbf{u} = 0 \quad on \quad y = 0 \tag{5.4}$$

$$\mathbf{u} = 0 \quad on \quad x = 0 \quad \text{and} \quad 1 \tag{5.5}$$

$$\mathbf{u} = 0 \quad on \quad z = 0 \quad \text{and} \quad \Lambda. \tag{5.6}$$

5.2.2 *Numerical method*

The governing equations are numerically solved by the Marker-and-Cell (MAC) method [3] and the fourth order Runge-Kutta method for the time integration. The spatial derivatives in the equation are evaluated by the spectral-like combined compact finite difference method. First we explain the spectral-like combined compact finite difference briefly. Consider a function $f(x)$ defined on the interval where N grid points are located with a uniform spacing h. Let f_i, f_i', f_i'' and f_i''' be the values of the function and its first, second and third derivatives at i-th grid point x_i respectively. The CCD scheme evaluates these derivatives implicitly from the following relations:

$$f_i' = \frac{a_1}{h}(f_{i+1} - f_{i-1}) + a_2(f_{i+1}' + f_{i-1}') + a_3 h(f_{i+1}'' - f_{i-1}'') + a_4 h^2 (f_{i+1}''' + f_{i-1}''') \tag{5.7}$$

$$f_i'' = \frac{b_1}{h^2}(f_{i+1} + f_{i-1} - 2f_i) + \frac{b_2}{h}(f_{i+1}' - f_{i-1}') + b_3(f_{i+1}'' + f_{i-1}'') + b_4 h(f_{i+1}''' - f_{i-1}''') \tag{5.8}$$

$$f_i''' = \frac{c_1}{h^3}(f_{i+1} - f_{i-1}) + \frac{c_2}{h^2}(f_{i+1}' + f_{i-1}') + \frac{c_3}{h}(f_{i+1}'' - f_{i-1}'') + c_4(f_{i+1}''' + f_{i-1}'''). \tag{5.9}$$

The accuracy of the scheme is determined by the parameters a_j, b_j and c_j. By relaxing the requirement for the highest accuracy, we have free parameters. The values of the free parameters are chosen to minimize errors in the wave number space for most waves except very short waves [6] as follows:

$$a_1 = \frac{8c_1 + 315}{240}, a_2 = -\frac{8c_1 + 195}{240}, a_3 = \frac{16c_1 + 255}{1200}, a_4 = -\frac{4c_1 + 45}{1800},$$
$$b_2 = -\frac{11b_1 - 15}{16}, b_3 = \frac{3b_1 - 7}{16}, b_4 = -\frac{b_1 - 3}{48},$$
$$c_2 = -c_1, c_3 = \frac{8c_1 + 15}{20}, c_4 = -\frac{4c_1 + 15}{60},$$

where the free parameters b_1 and c_1 are given by $b_1 = -0.86774$ and $c_1 = -2.1721$.

This CCD scheme has eighth-order accuracy and spectral-like resolution for the first order derivative. The equations (5.7-5.8) must be modified at the boundary point. At the boundary point $i = 0$, we use

$$f_0' = \frac{\alpha_1}{h}(f_1 - f_0) + \frac{\alpha_2}{h}(f_2 - f_0) + \alpha_3 f_1' + \alpha_4 h f_1'' + \alpha_5 h^2 f_1''' \tag{5.10}$$

$$f_0'' = \frac{\beta_1}{h^2}(f_1 - f_0) + \frac{\beta_2}{h^2}(f_2 - f_0) + \frac{\beta_3}{h} f_1' + \beta_4 f_1'' + \beta_5 h f_1''' \tag{5.11}$$

$$f_0''' = \frac{\gamma_1}{h^3}(f_1 - f_0) + \frac{\gamma_2}{h^3}(f_2 - f_0) + \frac{\gamma_3}{h^2} f_1' + \frac{\gamma_4}{h} f_1'' + \gamma_5 h^2 f_1'''. \tag{5.12}$$

These equations have sixth order accuracy for the first and second derivatives. If the value of the first derivative as well as that of the function are given, i.e.

$$\mathbf{u}_n = 0 \quad and \quad \partial_n \mathbf{u}_n = 0,$$

the finite difference scheme of seven order accuracy can be obtained for the second derivative.

In the MAC method, the Poisson equation for pressure needs to be solved in place of the continuity equation for the velocity $\mathbf{u}^n$;

$$\nabla^2 p = -\frac{div\mathbf{u}^n}{\Delta t} + \nabla\nabla : \mathbf{u}^n\mathbf{u}^n + \frac{1}{Re}\nabla^2 \mathrm{div}\mathbf{u}^n (= Q) \tag{5.13}$$

and the boundary condition is

$$\nabla p = \frac{1}{Re}\nabla^2 \mathbf{u} \qquad \text{on the wall.}$$

It generally takes a large portion of the entire computation time to solve the Poisson equation numerically. Since CCD schemes implicitly evaluate the derivatives, it is hard to apply a simple iterative method such as SOR to the CCD Poisson equation solver. We developed an alternating direction implicit (ADI) method for solving the Poisson equation. This algorithm can be parallelized easily. The interation m to $m+1$ can be separated into three parts:

(a) iteration along the x-axis to obtain "intermediate variables" $P^*_{i,j,k}, (\delta_x P)^*_{i,j,k}, (\delta_x^2 P)^*_{i,j,k}, (\delta_x^3 P)^*_{i,j,k}$,

$$P^*_{i,j,k} - \omega(\delta_x^2 P)^*_{i,j,k} = \omega(\nabla^2 p^m_{i,j,k} - Q_{i,j,k}) \tag{5.14}$$

$$\begin{aligned}(\delta_x P)^*_{i,j,k} &- \frac{a_1}{\Delta x}(P^*_{i+1,j,k} - P^*_{i-1,j,k}) \\ &- a_2\{(\delta_x P)^*_{i+1,j,k} + (\delta_x P)^*_{i-1,j,k}\} + a_3\Delta x\{(\delta_x^2 P)^*_{i+1,j,k} - (\delta_x^2 P)_{i-1,j,k}\} \\ &- a_4\Delta x^2\{(\delta_x^3 P)^*_{i+1,j,k} - (\delta_x^3 P)^*_{i-1,j,k}\} = 0\end{aligned} \tag{5.15}$$

$$\begin{aligned}(\delta_x^2 P)^*_{i,j,k} &- \frac{b_1}{\Delta x^2}(P^*_{i+1,j,k} + P^*_{i-1,j,k} - 2P^*_{i,j,k}) \\ &- \frac{b_2}{\Delta x}\{(\delta_x P)^*_{i+1,j,k} - (\delta_x P)^*_{i-1,j,k}\} - b_3\{(\delta_x^2 P)^*_{i+1,j,k} + (\delta_x^2 P)_{i-1,j,k}\} \\ &- b_4\Delta x\{(\delta_x^3 P)^*_{i+1,j,k} - (\delta_x^3 P)^*_{i-1,j,k}\} = 0\end{aligned} \tag{5.16}$$

$$(\delta_x^3 P)^*_{i,j,k} - \frac{c_1}{\Delta x^3}(P^*_{i+1,j,k} - P^*_{i-1,j,k})$$
$$-\frac{c_2}{\Delta x^2}\{(\delta_x P)^*_{i+1,j,k} + (\delta_x P)^*_{i-1,j,k}\} - \frac{c_3}{\Delta x}\{(\delta_x^2 P)^*_{i+1,j,k} - (\delta_x^2 P)_{i-1,j,k}\}$$
$$-c_4\{(\delta_x^3 P)^*_{i+1,j,k} + (\delta_x^3 P)^*_{i-1,j,k}\} = 0. \tag{5.17}$$

b) iteration along the y-axis to obtain "intermediate variables" $P^{**}_{i,j,k}, (\delta_y P)^{**}_{i,j,k}, (\delta_y^2 P)^{**}_{i,j,k}, (\delta_y^3 P)^{**}_{i,j,k},$

$$P^{**}_{i,j,k} - \omega(\delta_y^2 P)^{**}_{i,j,k} = P^*_{i,j,k} \tag{5.18}$$
$$(\delta_y P)^{**}_{i,j,k} - \frac{a_1}{\Delta y}(P^{**}_{i,j+1,k} - P^* *_{i,j-1,k})$$
$$-a_2\{(\delta_y P)^{**}_{i,j+1,k} + (\delta_y P)^{**}_{i,j-1,k}\} - a_3\Delta y\{(\delta_y^2 P)^{**}_{i,j+1,k} - (\delta_y^2 P)^{**}_{i,j-1,k}\}$$
$$-a_4\Delta y^2\{(\delta_y^3 P)^{**}_{i,j+1,k} - (\delta_y^3 P)^{**}_{i,j-1,k}\} = 0 \tag{5.19}$$
$$(\delta_y^2 P)^{**}_{i,j,k} - \frac{b_1}{\Delta y^2}(P^{**}_{i,j+1,k} + P^{**}_{i,j-1,k} - 2P^{**}_{i,j,k})$$
$$-\frac{b_2}{\Delta y}\{(\delta_y P)^{**}_{i,j+1,k} - (\delta_y P)^{**}_{i,j-1,k}\} - b_3\{(\delta_y^2 P)^{**}_{i,j+1,k} + (\delta_y^2 P)^{**}_{i,j-1,k}\}$$
$$-b_4\Delta y\{(\delta_y^3 P)^{**}_{i,j+1,k} - (\delta_y^3 P)^{**}_{i,j-1,k}\} = 0 \tag{5.20}$$
$$(\delta_y^3 P)^{**}_{i,j,k} - \frac{c_1}{\Delta y^3}(P^{**}_{i,j+1,k} - P^{**}_{i,j-1,k})$$
$$-\frac{c_2}{\Delta y^2}\{(\delta_y P)^{**}_{i,j+1,k} + (\delta_y P)^{**}_{i,j-1,k}\} - \frac{c_3}{\Delta y}\{(\delta_y^2 P)^{**}_{i,j+1,k} - (\delta_y^2 P)^{**}_{i,j-1,k}\}$$
$$-c_4\{(\delta_y^3 P)^{**}_{i,j+1,k} + (\delta_y^3 P)^{**}_{i,j-1,k}\} = 0. \tag{5.21}$$

c) iteration along the z-axis to obtain $p^{k+1}_{i,j,k}$ and "intermediate variables", $P^{***}_{i,j,k}, (\delta_z P)^{***}_{i,j,k}, (\delta_z^2 P)^{***}_{i,j,k}, (\delta_z^3 P)^{***}_{i,j,k},$

$$P^{***}_{i,j,k} - \omega(\delta_z^2 P)^{***}_{i,j,k} = P^{**}_{i,j,k} \tag{5.22}$$
$$(\delta_z P)^{***}_{i,j,k} - \frac{a_1}{\Delta z}(P^{***}_{i,j,k+1} - P^{***}_{i,j,k-1})$$
$$-a_2\{(\delta_z P)^{***}_{i,j,k+1} + (\delta_z P)^{***}_{i,j,k-1}\} - a_3\Delta z\{(\delta_z^2 P)^{***}_{i,j,k+1} - (\delta_z^2 P)^{***}_{i,j,k-1}\}$$
$$-a_4\Delta z^2\{(\delta_z^3 P)^{***}_{i,j,k+1} - (\delta_z^3 P)^{***}_{i,j,k-1}\} = 0 \tag{5.23}$$
$$(\delta_z^2 P)^{***}_{i,j,k} - \frac{b_1}{\Delta z^2}(P^{***}_{i,j,k+1} + P^{***}_{i,j,k-1} - 2P^{***}_{i,j,k})$$
$$-\frac{b_2}{\Delta z}\{(\delta_z P)^{***}_{i,j,k+1} - (\delta_z P)^{***}_{i,j,k-1}\} - b_3\{(\delta_z^2 P)^{***}_{i,j,k+1} + (\delta_z^2 P)^{***}_{i,j,k-1}\}$$
$$-b_4\Delta z\{(\delta_z^3 P)^{***}_{i,j,k+1} - (\delta_z^3 P)^{***}_{i,j,k-1}\} = 0 \tag{5.24}$$

$$
\begin{aligned}
(\delta_z^3 P)^{***}_{i,j,k} - \frac{c_1}{\Delta z^3}(P^{***}_{i,j,k+1} - P^{***}_{i,j,k-1}) \\
- \frac{c_2}{\Delta z^2}\{(\delta_y P)^{***}_{i,j,k+1} + (\delta_z P)^{***}_{i,j,k-1}\} - \frac{c_3}{\Delta z}\{(\delta_z^2 P)^{***}_{i,j,k+1} - (\delta_z^2 P)^{***}_{i,j,k-1}\} \\
- c_4\{(\delta_z^3 P)^{***}_{i,j,k+1} + (\delta_z^3 P)^{***}_{i,j,k-1}\} = 0 \qquad (5.25)
\end{aligned}
$$

$$
p^{m+1}_{i,j,k} = P^{***}_{i,j,k} + p^{**}_{i,j,k}. \qquad (5.26)
$$

The iterative process stops when the correction at the iterative m,

$$
\mathrm{corr}^{m+1} = \frac{\Sigma_{i,j,k}|\nabla^2 p_{i,j,k} - Q_{i,j,k}|}{\Sigma_{i,j,k}|Q_{i,j,k}|} \qquad (5.27)
$$

is smaller than 10^{-9}.

In the case of the cubic cavity ($f = 1$) at Re=100 to 400, the numerical results using this new scheme with a small uniform grid system ($33\times33\times33$) are in good agreement with the previous results [4]. In this study the $100 \times 100 \times 100\ f$ uniform grid system is used in the calculation, in which f is a spanwise aspect ratio.

5.3 Results

We consider the flows in the cavities with an aspect ratio $\Gamma = 1$ and the spanwise aspect ratios $\Lambda = 1, 3, 4, 5$ and 6.55 at the Reynolds number $Re = 850$. In all cases, we get steady state solutions. In the previous study of the cavity flow with $\Gamma = 1$ [5], we also got a steady flow structure at $400 < Re < 1000$, which consists of a main flow and secondary flows near the walls, and we observed chaotic streamlines in the whole flow region. In Fig. 5.2, we show the streamlines drawn from the projected velocity on the vertical left half plane (a) near the upstream end wall $x = 0.01$ and (b) of $x = 0.5$ in the case of $\Lambda = 1$.

We observe main upper flow and secondary flow along the bottom and side walls in Fig. 5.2(a). In Fig. 5.2(b), a main down flow region is found as well as a Taylor-Görtler-like streamwise vortex near the bottom and another streamwise vortex of the upper corner region. Three parts of streamlines in this flow are shown in Fig. 5.3(a,b,c), respectively. The start points of three streamlines are chosen arbitrarily but all streamlines looks chaotic and fill up the whole region of cavity. The flow structures of $\Lambda = 3$ and 4 are similar to that of =1, but that of $\Lambda = 5$ is different. In Fig. 5.4, we

(a) (b)

Fig. 5.2 Streamlines using the projected velocities on the vertical left half plane of (a) $x = 0.01$ (b) $x = 0.5$ in the cavity with $\Lambda = 1$.

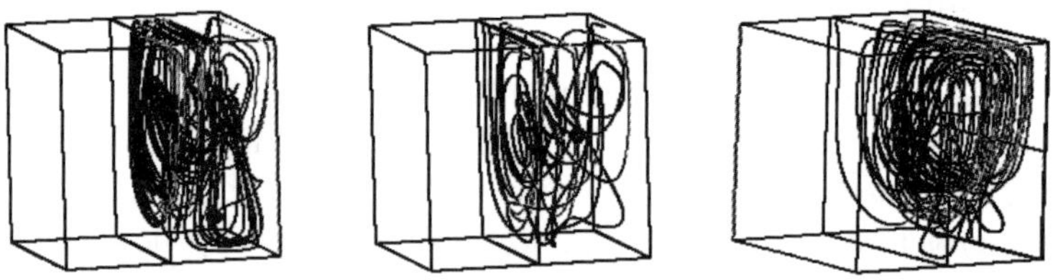

Fig. 5.3 Streamlines in the cavity with $\Lambda = 1$.

show the streamlines using the projected velocity on the vertical left half plane (with $1/2\ f$) of $x = 0.01$ in the cavity with $\Gamma = 4$ and 5.

Though the main upper flow region becomes wider and the secondary flow region is narrow in Fig. 5.4(a), the flow pattern on the cross section of $x = 0.01$ is similar to Fig. 5.2(a) of $f = 1$. On the other hand, we observe that there are at least two cells in the upper flow region and the secondary flow near the bottom in Fig. 5.4(b); one is the upper region near the side

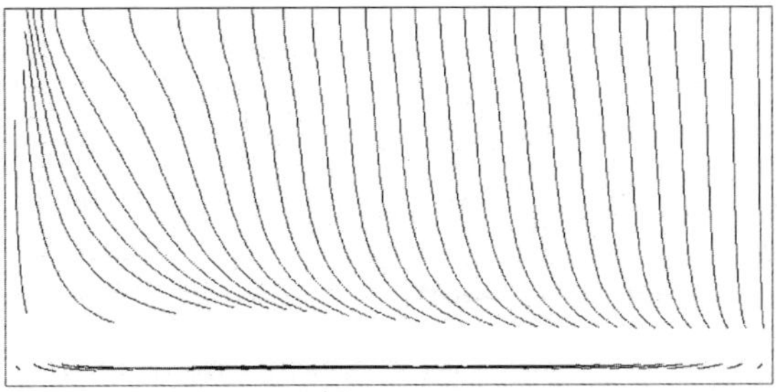
(a)

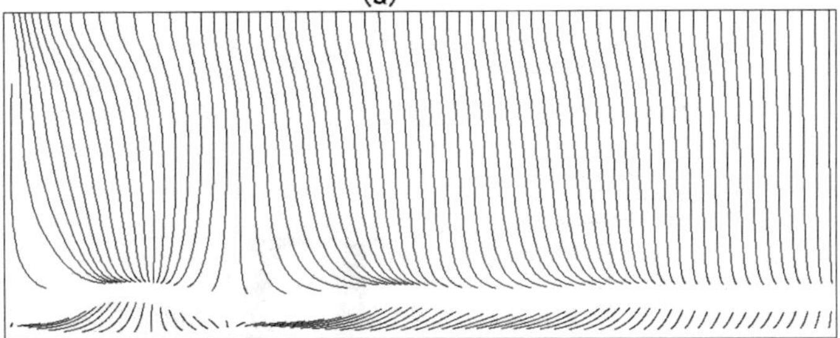
(b)

Fig. 5.4 Streamlines using the projected velocities on the vertical left half plane of x=0.01 in the cavity with $\Lambda = 6.55$.

wall with the width of 1/6 f and the other is located in the center. The bounded line between the main and secondary flow region along the bottom is also wavy. These characteristics is clearly found in the streamlines drawn from the projected velocity on the vertical left half plane of $x = 0.01$ in the cavity with $\Gamma = 6.55$ in Fig. 5.5. We find 2.5 cell structures with the width of $1/9\Lambda$ in the central region and the other cell structure near the side wall in Fig. 5.5.

These results agree with the experimental result [5]. These patterns of long span cavities are strongly related to the secondary flow of the overall flow structure. We investigate the pattern of streamlines for understanding the overall flow field.

We show parts of three streamlines at $\Lambda = 6.55$ in Figs. 5.6(a,b) and 5.7. The streamline in Fig. 5.6(a) and (b) is localized and makes two tori. The streamline in Fig. 5.6(a) is located in the cell near the symmetric plane of Fig. 5.5 and the streamline in Fig. 5.6(b) is located in the other cell. They implies that there is a closed streamline [4] inside each torus. On the other hand, the streamline in Fig. 5.7 is global and chaotic. It shows that

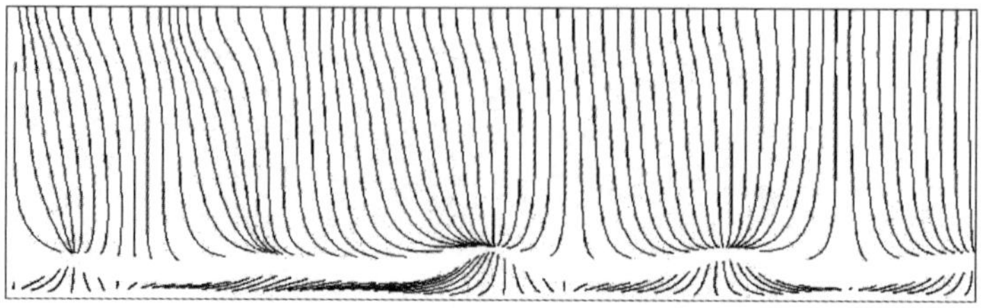

Fig. 5.5 Streamlines using the projected velocities on the vertical left half plane of x=0.01 in the cavity with $\Lambda = 6.55$.

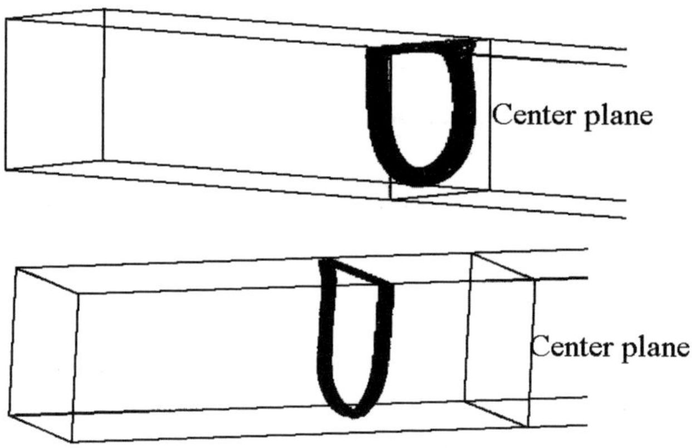

Fig. 5.6 Two typical localized streamlines in the cavity with of $\Lambda = 6.55$.

there is a strong secondary flow from the side wall to the symmetric plane.

5.4 Concluding remarks

The results show that the spectral-like CCD scheme is effective and useful for vortical incompressible three dimensional flow calculations. The lid-driven cavity flow with the spam aspect ratio $\Lambda \leq 4$ has chaotic streamlines in the whole region. On the other hand, the cell structures related to closed streamlines are clearly obtained in the numerical simuralion of the lid-driven

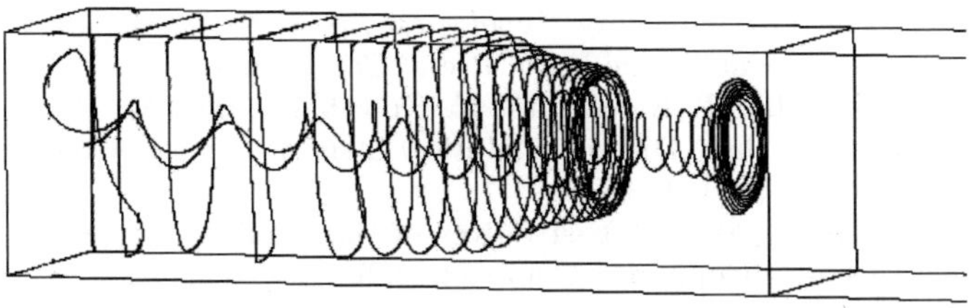

Fig. 5.7 A part of a typical chaotic streamline in the cavity with $\Lambda = 6.55$.

cavity flow over $\Lambda \sim 5$ at $Re = 850$, which agree with the expriments. In these flow fields, global chaotic streamlines are also observed along the central axis of the cavities. Therefore, the side wall effects can not be ignored even in the three-dimensional cavity flows with long span length.

Acknowledgment

The first author is deeply indebted to Lu Ting for sharing, over the course of many conversations, his insights on accurate computational method and vortex flow dynamics. These interactions proved indispensable in the writing of this paper.

This work was partially supported by the Grant-in-Aid for the 21st Century COE "Frontiers of Computational Science" and the Grant-in-Aid for Scientific Research from the Ministry of Education, Science and Culture No.16605004.

Bibliography

[1] C.K. Aidun, H.G. Triantafillopoulos, and J.D. Benson (1991) Global Stability of a Lid-Driven Cavity with Throughflow: Flow Visualization Studies, *Phys. Fluids* **A3**, 2081-2091.

[2] S. Albensoeder,H.C. Kuhlmann, H.J. Rath (2001) Three-dimensional Centrifugal-flow instabilities in the Lid-driven-cavity Problem, *Phys. Fluids* **13**, 121-135.

[3] F.H. Harlow and J.E. Welch (1965), Numerical Calculation of time-dependent Viscous Incomplessible Flow of Fluid with Free Surface, *Phys. Fluids* **8**, 2182.

[4] K. Ishii and R. Iwatsu (1990) Numerical Simulation of the Lagrangean flow structure in a driven-cavity, *Topological Fluid Mechanics*(eds. H.K.Moffatt, A. Tsinobar), Cambridge Univ. Press, Cambridge,54-63.

[5] R. Iwatsu, K. Ishii, T. Kawamura, K. Kuwahara and J.M. Hyun, Numerical Simulation of Three Dimensional Flow Structure in a Driven Cavity, *Fluid Dyn. Res.* **5** (1989), 173-189.

[6] T. Nihei,K. Ishii (2003) A Fast Solver of the Shallow Water Equations on a Sphere using a Combined Compact Difference Scheme, *J. Comput. Phys.* **187**, 639-659.

[7] P.N.Shankar and M.D. Deshpande (2000) Fluid mechanics in the Driven Cavity.*Annu. Rev. Fluid Mech.* **32**, 93-136.

Chapter 6

Geometric, Stochastic and Algebraic Vortices

Nicholas Kevlahan
Department of Mathematics & Statistics
McMaster University, Hamilton, Ontario, Canada

Dedicated to Lu Ting on his eightieth birthday

Abstract

Mathematical concepts can often be understood in one of three complementary ways: geometric, algebraic or stochastic. In this chapter I show how vortices may be interpreted in each of these three ways. Each point of view is appropriate for certain applications, and each sheds new light on the role played by vortices in fluid dynamics. This review is necessarily brief, and makes no attempt to be exhaustive, but I hope it gives an interesting overview of vortices in all their forms.

6.1 Introduction

Before examining the different ways of representing the vortices, it is natural to say a few words about why vortices are interesting and important to study. A vortex may be defined as a region of flow characterized by significant coherent vorticity $\boldsymbol{\omega}$, where vorticity is the curl of velocity,

$$\boldsymbol{\omega} = \nabla \times \boldsymbol{u}. \tag{6.1}$$

Coherence here means both coherence in space (e.g. a connected identifiable shape, usually tube-like), and coherence in time (e.g. the shape can be followed for a significant length of time). This definition is deliberately vague:

the precise meaning of the word vortex in fluid dynamics has been the subject of long and intense debate (see for example [4; 16] for a comparison of different definitions). Even the vague definition stated above is not unbiased; it has a definite geometric or topological flavour. A purely stochastic or statistical definition is also possible. Farge, Schneider & Kevlahan [10] proposed to define the vortices of two-dimensional turbulence as the non-Gaussian part of the vorticity field, i.e. that part which is left over when the vorticity is de-noised. Moreover, an algebraic understanding of vortices is used in numerical vortex methods. In vortex methods the solution of the vorticity equation (a partial differential equation) is reduced to a set of algebraic equations for the positions and strengths of vortex elements [7]. These vortex elements have very little geometry: in most cases they are nothing more than points. Even the definition of the word vortex depends on whether we take a geometric, stochastic or algebraic point of view.

Why have people put so much effort into defining and understanding vortices? First, we are drawn to vortices because they are beautiful and ubiquitous features of fluid flows. People have been fascinated by their persistence (e.g. leap-frogging smoke rings) and by their structure (e.g. the von Karman vortex street behind a bluff body). In turbulent flows the vortices are the only obvious organized structures in a sea of random fluctuations. Vorticity is also attractive mathematically since the vorticity equation is simpler than the primitive form of the Navier–Stokes equations, especially in two dimensions. For example, in two-dimensional inviscid flows all functionals of vorticity are conserved, and in three-dimensional inviscid flows the integral of helicity is conserved (which means the linking number of vortex rings is invariant). Unfortunately, the vorticity field is actually quite difficult to measure experimentally.

Vortices are also important because they control the dynamics of many useful flows. An indication of the prominence of vortices in fluid dynamics is given by the fact that the word 'vortex' or 'vortices' occurs in the titles of ten articles in the *Annual Review of Fluid Dynamics* in the last ten years (an average of one article on vortices per issue). Perhaps the best-known example is the starting vortex around an airfoil (which generates lift as a consequence of Kelvin's circulation theorem). Other examples include the use of vortices to improving mixing, the transport of pollutants in the atmosphere and oceans by vortices [35], and the generation of vorticity (and hence turbulence) by curved shocks [20]. We now consider two important vortex-dominated flows in more detail.

A good example of the key role of vortices is fluid–structure interaction at moderate Reynolds numbers (say 100 < Re < 300). Well-defined coherent vortices are generated via the no-slip boundary condition at the surface of a bluff body. These vortices exert a force on the body which causes it to move or vibrate. The aeolian noise generate by the wind blowing through telephone wires is the most familiar example of this. Swimming fish use controlled vortex-induced vibration to swim. More dramatic examples include the destruction of buildings (e.g. the Tacoma narrows bridge in 1940 and the Ferrybridge nuclear station cooling towers in 1965) and machines (e.g. heat exchanger tube bundles can shake themselves apart). Vortex-induced vibration is a major area of engineering research, and controlling it is vital for the construction of bridges, large buildings, airplanes and off-shore drilling platforms.

The precise role of vortices in fluid–structure interaction was recently clarified in a paper by Noca *et al.* [32; 33]. They showed that the force on a body exerted by a vortical flow is given by

$$\begin{aligned}\boldsymbol{F}(t) = &-\frac{1}{\mathcal{N}-1}\frac{d}{dt}\int_{V_\infty} \mathbf{x}\times\boldsymbol{\omega}\,\mathrm{d}V \\ &-\frac{1}{\mathcal{N}-1}\frac{d}{dt}\oint_{S_b(t)} \mathbf{x}\times(\boldsymbol{n}\times\boldsymbol{u})\,\mathrm{d}S - \oint_{S_b(t)} \boldsymbol{n}\cdot(\boldsymbol{u}-\boldsymbol{u}_S)\,\mathrm{d}S, \quad (6.2)\end{aligned}$$

where $\boldsymbol{n}$ is the inward-pointing normal vector of the body surface, $\mathcal{N}$ is the dimension of the space, V_∞ is the (infinite) fluid volume, $S_b(t)$ is the body surface and $\boldsymbol{u}_S$ is the velocity of the surface. If we assume a no-slip boundary condition at the surface of the body, and that the body is stationary and non-rotating, then

$$\boldsymbol{F}(t) = -\frac{1}{\mathcal{N}-1}\frac{d}{dt}\int_{V_\infty} \mathbf{x}\times\boldsymbol{\omega}\,\mathrm{d}V. \quad (6.3)$$

Note that (6.3) says that the force on the body is equal and opposite to the change in the fluid impulse. Equation (6.3) shows very clearly that the fluid forces on a body are due entirely to the movement and changing shape of the vortices.

Sound or noise is another phenomenon which may be related directly to vortex structure and dynamics. Kambe *et al.* [18] (following Lighthill [28]) showed that vortices generate a far-field sound wave with pressure given by

$$p_F = -\frac{\varepsilon^{(1)}(t_r)}{15\pi c^2 r} - \frac{\rho_0}{c^2}Q^{(3)}_{ij}(t_r)\frac{x_i x_j}{r^3} + \frac{\rho_0}{c^3}Q^{(4)}_{ijk}(t_r)\frac{x_i x_j x_k}{r^4} + \cdots \quad (6.4)$$

where the superscript (n) indicates the n-th time derivative, $t_r = t - r/c$ is the retarded time, $\varepsilon(t)$ is the rate of dissipation of energy, c is the speed of sound, $r = |\mathbf{x}|$ is the distance from the vortex source, and

$$Q_{ij}(t) = -\frac{1}{12\pi}\int (\boldsymbol{y}\times\boldsymbol{\omega})_i\, y_j\, \mathrm{d}^3\boldsymbol{y},$$

$$Q_{ijk}(t) = \frac{1}{32\pi}\int (\boldsymbol{y}\times\boldsymbol{\omega})_i\, y_j y_k\, \mathrm{d}^3\boldsymbol{y},$$

are moments of the vorticity distribution. Equation (6.4) clearly shows how changes in the configuration or shape of vortices directly generates sound. In the same paper Kambe *et al.* present results from experiments which show how colliding vortex rings create a sound burst. This is a nice example of how changes in vortex topology can have a practically important effect. Reducing aircraft noise may depend on a better understanding of how vortices interact!

As mentioned earlier, vortices are also considered to be key to understanding turbulence. The principal obstacle in dealing with turbulence, both numerically and analytically, is the fact that high Reynolds number turbulence contains a huge range of active length and time scales. A naive estimate (which ignores intermittency) suggests that the number of space–time degrees of freedom in a turbulent flow increases like Re^3. In atmospheric turbulence the active length scales range from millimetres to tens or hundreds of kilometres! This range of scales is also continuous, which means the usual separation of scales techniques (e.g. WKB) are not applicable. A different way of reducing the dimensionality of the problem must be found. The fact that turbulence contains coherent vortices, and that vortices may control much of the dynamics of the flow, has suggested a new way of tackling the problem [10]. One can divide turbulence into the coherent vortices (which are calculated exactly), and the rest of the flow (defined to be noise, whose effect can be modelled statistically). In this approach the problem of turbulence is reduced to vortex dynamics. Because turbulence is intermittent, the coherent vortices are conjectured to take up an increasingly small part of the flow as the Reynolds number increases, resulting in a significant reduction in the dimensionality of the problem. The simplest example of this approach is large eddy simulation (LES) where the coherent vortices are taken to be the largest Fourier modes of the flow [27]. More sophisticated methods using adaptive wavelet filtering to include all scales of the coherent vortices are also being attempted [13; 14; 19; 38].

The transition to turbulence is also characterized by changes in vortex structure and dynamics. One of the clearest examples is the flow past a cylinder as Reynolds number increases. Frisch [11] shows a beautiful sequence of visualizations of the vorticity as a function of Reynolds number. He comments that the spatial and temporal symmetries of the vorticity field are progressively lost as Reynolds number gradually increases, only to be recovered statistically in fully developed turbulence at high Reynolds number. In this case, as for all bounded flows, much of the turbulence originates in the wall vortices of the boundary layer. An understanding of vorticity production in the boundary layer is also thought to be essential for calculating the mean velocity profile of in channel or pipe flow [31].

We have seen that vortices dominate useful flows, as well as being attractive mathematically and visually beautiful in their own right. The rest of this paper sketches how vortices may be represented in three complementary ways: geometric, stochastic and algebraic. The geometric (or topological) point of view is described in Section 6.2, with a particular emphasis on the coherent vortices of turbulence. Section 6.3 considers two versions of the stochastic approach: a probabilistic representation of the vorticity equation, and a wavelet de-noising definition of coherent vortices. The algebraic vortex is represented in Section 6.4 by two numerical schemes: the vortex method, and the adaptive wavelet collocation method for the vorticity equation. Finally, we make some concluding comments in Section 6.5.

6.2 Geometric vortices

6.2.1 *Conservation laws*

The most obvious way of characterizing vortices is based on their geometry and topology. Geometric vortices are defined in terms of their shape, and how this shape changes in time through interaction with other vortices or self advection. Let us recall the three-dimensional incompressible vorticity equation,

$$D_t\boldsymbol{\omega} = (\boldsymbol{\omega}\cdot\nabla)\boldsymbol{u} + \nu\Delta\boldsymbol{\omega}, \tag{6.5}$$

where the D_t is the material derivative, and the velocity $\boldsymbol{u}$ is related to the vorticity via the Biot–Savart law (which makes the equation nonlinear). The first term on the right hand side of (6.5) is the change in vorticity due to vortex stretching (which is identically zero in two dimensions), and the second term is the diffusion of vorticity (which spreads vorticity out, and is essential for vortex reconnection). The geometrical and topological constraints on this equation are much simpler (and exact) if we set $\nu = 0$, which gives the inviscid equations. In this case one can prove a number of interesting conservation laws [30].

The best known conservation law is Kelvin's circulation theorem

$$D_t \left(\int_{S(t)} \boldsymbol{\omega} \cdot \mathrm{d}\boldsymbol{S} \right) = 0 \tag{6.6}$$

which states that the vorticity flux through a material surface $S(t)$ is conserved. Helicity is also conserved

$$D_t \left(\int_{\mathbb{R}^3} \boldsymbol{u} \cdot \boldsymbol{\omega} \, \mathrm{d}\mathbf{x} \right) = 0, \tag{6.7}$$

which means that the linking number of vortex rings is conserved (they cannot be unlinked). The fluid impulse and the moment of the fluid impulse are also conserved. In two dimensions, because the vortex stretching term is absent, all functionals of vorticity are conserved (including all L^p norms).

6.2.2 *Slender vortices*

The simplest geometrical vortices are vortex lines (i.e. the circulation is concentrated on a curve in $\mathbb{R}^3$). These curves move with the fluid, and must end at solid boundaries or form loops. Although a vortex line is an elegant and natural idealization of a vortex, it has the disadvantage that any non-zero curvature causes it to move infinitely fast! The concept of a *slender* vortex has therefore been developed to correct this problem [21; 22; 30; 37]. The simplest version of the slender vortex approximation [40] is the self-induction equation for an isolated filament

$$\frac{\partial \boldsymbol{X}}{\partial t} = \kappa \boldsymbol{b}, \tag{6.8}$$

where $\boldsymbol{X}(s,t)$ is the centreline of the vortex, $\kappa(s,t)$ is its curvature and $\boldsymbol{b}(s,t)$ is the binormal to the curve. Note that in the self-induction approximation the thickness δ of the vortex filament does not appear explicitly (although it has been absorbed into the time via the re-scaling $t = \log(1/\delta)\, t'$). The effect of other filaments may be included via the Biot–Savart law, provided the separation between filaments remains much greater than δ.

Slender vortex models have the advantage that the conservation laws of inviscid flow stated above remain valid. The disadvantage is that slender vortex models are only valid as long as the vortices remain well-separated. This means they cannot treat vortex reconnection or topology change. In addition, Zhou [44] has shown that these models can produce unphysical results because the core width is uniform along the vortex and does not change in time. Despite these shortcomings, the simplicity of slender vortices make them attractive mathematically. They have been used to examine the early stages of vortex interaction [23], and to understand some aspects of turbulence [37]. Interestingly, slender vortex models appear to be almost exact when applied to the quantized vortices of superfluids.

6.2.3 *Diffusive vortices*

Less mathematical, but more practically useful, geometric definitions of vortices have been developed to extract (or *educe*) vortices from experimental and numerical data. Bonnet *et al.* [4] made a comprehensive comparison of the most popular techniques. Because real vortices are diffusive, their definition must also be a bit fuzzy. The simplest approach is to threshold on the L^1 or L^2 norm of vorticity. Figure 6.1 shows isosurfaces of vorticity from the $4\,096^2$ direct numerical simulation (DNS) of Yokokawa *et al.* [43]. The isosurfaces reveal that the vortices are tube-like and fill much of the turbulence (although they are not distributed completely homogeneously). The effect of these vortices on flow dynamics and kinematics can be evaluated by conditional averaging on the thresholded vorticity. More sophisticated approaches attempt to include the temporal, as well as spatial, coherence of the vorticity.

The goal of vortex eduction is to determine if characteristics of the flow as a whole can be explained by the vortices alone. Such an approach has been quite successful when the flow includes large coherent vortices with relatively simply dynamics (such as in mixing layers, jets, and in the fluid–structure interaction at moderate Reynolds numbers). Vortex eduction has also been used extensively to investigate near-wall turbulence (which led to the discovery of hair-pin vortices), and the transition to turbulence on a flat plate. Another major application of vortex eduction has been the attempt to determine whether turbulence is contains coherent vortices and, if so, whether these vortices control the dynamics of the flow. It is now clear that the answer to the first question is yes: all turbulent flows are full of coherent vortices which are tube-like and persist for a significant time [17]. The answer to the second question is not so clear: vortices appear to be an essential part of turbulence, but they may not always control its dynamics.

The geometric approach includes the mathematically attractive slender vortex model, and many types of vortex eduction for real flows. However,

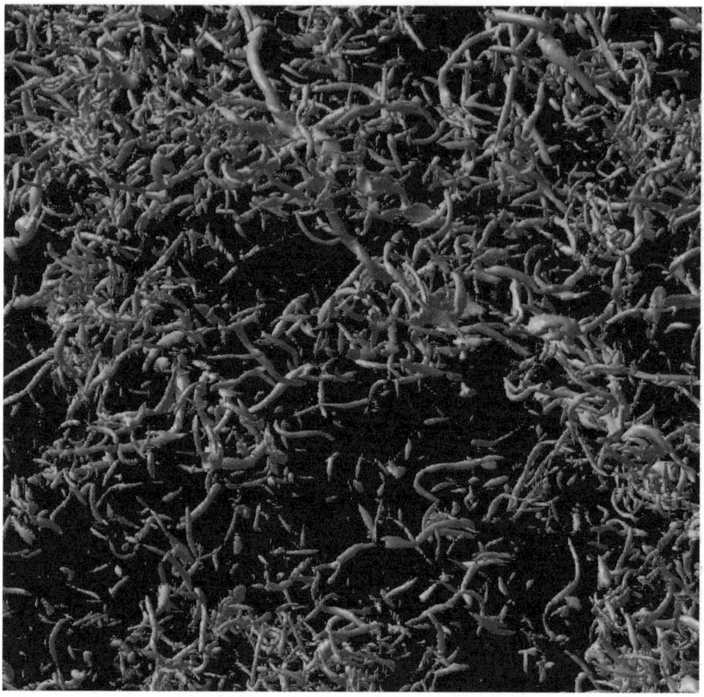

Fig. 6.1 Yokokawa *et al.* [42] DNS of homogeneous isotropic turbulence at $\mathrm{Re}_\lambda = 1\,217$. The figure shows isosurfaces of vorticity of at the level $\bar{\omega} + 4\sigma$, where $\bar{\omega}$ is the mean vorticity and σ is its standard deviation. The region shown is $(748^2 \times 1\,496)\eta$, where η is the Kolmogorov length.

shape and topology are not always the most useful ways of representing vortices. In the following section we will see how a stochastic point of view can provide insight not available from geometry alone. We will also see how the vorticity equations themselves may be cast in an entirely probabilistic form.

6.3 Stochastic vortices

6.3.1 *Proper orthogonal decomposition*

Perhaps the oldest stochastic (or statistical) representation of vortices is the proper orthogonal decomposition (POD). The POD was introduced by Lumley [29], and has since become one of the standard tools for analyzing vortical flows [3]. POD (which is also known as Karhunen–Loève decomposition, or principal components analysis) decomposes a turbulent velocity field into an ordered set of orthogonal eigenfunctions $\phi^{(n)}(\mathbf{x})$ ($n = 1, 2, 3, \ldots, N$), which are solutions of the linear eigenvalue problem

$$\int R_{ij}(\mathbf{x}, \mathbf{x}')\, \phi_j^{(n)}(\mathbf{x}')\, \mathrm{d}\mathbf{x}' = \lambda^{(n)} \phi_i^{(n)}(\mathbf{x}),$$

where $R_{ij}(\mathbf{x}, \mathbf{x}') = \langle \boldsymbol{u}(\mathbf{x}) \boldsymbol{u}^*(\mathbf{x}') \rangle$ is the two-point correlation tensor. The first N eigenfunctions provide an optimal N-term representation of the flow in terms of the L^2, or energy, norm. In addition, if the eigenvalues $\lambda^{(n)}$ decrease rapidly with n, and the flow is periodic or statistically stationary in time, the first eigenmode is a good representation of a typical coherent vortex. This is true for most mixing layers and jets. POD may also be interpreted as a generalized Fourier transform since, if the flow is homogeneous in space, the eigenfunctions are Fourier modes (the 'eddies' of classical homogeneous isotropic turbulence).

POD can be used to construct a reduced model of a turbulent flow by retaining only the first few eigenfunctions. This model captures the most important structure of the flow (as measured by energy), and may be used to clarify its dynamics. One must be careful, however, in interpreting POD eigenfunctions as geometric vortices. In fact, the POD eigenfunctions are only statistically representative of the average kinetic energy of the flow: there is no guarantee that they resemble the vortices in any particular flow realization. This is especially obvious in the case of homogeneous turbulence, where the most energetic Fourier mode (at the integral scale) does not resemble a real vortex! Indeed, most kinetic energy is at the integral scale $O(1)$, while most enstrophy is at the much smaller Taylor scale $O(\mathrm{Re}^{-1/2})$. POD vortices are therefore not (in general) the same as the geometric vortices of the previous section. Bonnet *et al.* [4] compares the POD statistical vortices with other, more geometric, eduction techniques.

6.3.2 *A de-noising definition of coherent vortices*

Instead of trying to find the optimal N-term representation for the kinetic energy of a flow, one can define vortices statistically in terms of their coherence. Farge *et al.* [10] have proposed defining the vortices of turbulence as that part of the flow which is not Gaussian noise. The basic idea is to use nonlinear wavelet filtering of the vorticity field to remove additive Gaussian noise from the vorticity. The wavelet modes associated with the noise are discarded, and the vortices are reconstructed from the remaining wavelet modes.

This approach is based on Donoho's [9] optimal de-noising theorem for additive Gaussian white noise. The theorem states that the optimal way to de-noise a signal ω sampled on N points and perturbed by an additive Gaussian white noise of variance σ is to take its orthonormal wavelet transform, and then retain only those wavelet coefficients with absolute value larger than the threshold $\epsilon_D = (2\sigma \log N)^{1/2}$ before reconstructing the de-noised signal $\omega_>$. In the case of turbulence, $\omega_>$ is then *defined* to be the coherent vortices of the flow.

Unfortunately, Donoho's theorem cannot be applied directly to a turbulent vorticity field. First, because we do not know in advance that the noise is Gaussian, and secondly because we do not know its variance. Moreover, the statistical theory of homogeneous turbulence suggests that the noise may be coloured rather than white (i.e. the noise may depend on the length scale). In order to address these difficulties Farge et al. proposed an iterative de-noising algorithm. In the first iteration the variance of the noise is taken to be equal to the variance of the vorticity field, and the noise is extracted. Because this is an overestimation, some coherence will remain in the noise, and so the filtering is repeated on the noise alone. The coherent part of the noise is then combined with the coherent vorticity extracted in the first iteration. This procedure is repeated until the noise is Gaussian to within the desired tolerance. In practice, for two-dimensional turbulence one iteration is often sufficient.

Farge *et al.* applied this algorithm to extract the coherent vortices from a 256^2 DNS of two-dimensional turbulence and found that the coherent vortices represent just 0.7% of the wavelet modes, but contain 94% of the enstrophy. Furthermore, the coherent vortices closely resemble the vortices defined using geometrical techniques, such the Q-criterion [15; 16].

By applying this wavelet de-noising at each time step, and modelling the effect of the neglected modes statistically, one obtains coherent vortex simulation (CVS). CVS assumes that the coherent vortices (defined in the specific way described here) control the dynamics of the flow. If some noise is retained along with the coherent vortices, one obtains wavelet direct numerical simulation (WNDS), where the effect of the neglected modes is not included. WDNS will be described in detail in Section 6.4.

In both POD and CVS the vortices are defined statistically, but this definition is still based on the usual deterministic vorticity equation (6.5). We now consider a probabilistic representation of the vorticity equation itself, where the vortices (defined as particles) are reduced to random variables.

6.3.3 *A probabilistic version of the vorticity equation*

As mentioned in the introduction, most mathematical concepts may be expressed in either geometrical, stochastic or algebraic form. In particular, many partial differential equations (PDEs) can be associated to a stochastic differential equation (SDE). Or rather, a Langevin type SDE has an associated PDE (called the Fokker–Planck equation) which describes the time evolution of the probability density function (PDF) of its solution (which is a random variable).

For example, one can associate to the three-dimensional linear advection–diffusion equation

$$\frac{\partial \boldsymbol{f}}{\partial t} = \boldsymbol{v} \cdot \nabla \boldsymbol{f} + \nu \Delta \boldsymbol{f} \tag{6.9}$$

an SDE for the stochastic process $\boldsymbol{X}(t; \boldsymbol{X}_0) \in \mathbb{R}^3$

$$d\boldsymbol{X}(t) = \boldsymbol{v}(t, \boldsymbol{X}(t)) + \sqrt{2\nu}\, \mathrm{d}\boldsymbol{W}_t, \tag{6.10}$$

where $\boldsymbol{W}_t$ is a three-dimensional Brownian motion. The PDF of the random variable $\boldsymbol{X}(t)$ is the solution $\boldsymbol{f}(t)$ of the advection–diffusion equation.

Comparing equation (6.9) to the two-dimensional vorticity equation

$$\frac{\partial \boldsymbol{\omega}}{\partial t} = \boldsymbol{u} \cdot \nabla \boldsymbol{\omega} + \nu \Delta \boldsymbol{\omega} \tag{6.11}$$

suggests that we should be able to find its associated Langevin equation (where $\boldsymbol{X}(t)$ becomes the Lagrangian trajectory of a stochastic point vortex). The problem is that in equation (6.11) the advecting velocity is not known in advance; it depends on the vorticity $\boldsymbol{\omega} = \nabla \times \boldsymbol{u}$. This means we require an additional SDE for the velocity, and we need to solve a coupled system of nonlinear SDEs.

In a recent paper Busnello *et al.* [5] derived just such an SDE for the three-dimensional vorticity equation. The vorticity and velocity are given by

$$\boldsymbol{\omega}(\mathbf{x}, t) = \boldsymbol{E}[U_t^{x,t} \boldsymbol{\omega}_0(X_t^{x,t})], \tag{6.12}$$

$$\boldsymbol{u}(\mathbf{x}, t) = \frac{1}{2} \int_0^\infty \frac{1}{s} \boldsymbol{E}[\boldsymbol{\omega}(\mathbf{x} + \boldsymbol{W}_s, t) \times \boldsymbol{W}_s]\, \mathrm{d}s, \tag{6.13}$$

where $\boldsymbol{E}[\cdot]$ is the expectation. Equation (6.13) is the stochastic form of the Biot–Savart law. The vorticity is thus a weighted average over the Lagrangian paths $(\boldsymbol{X}_s^{x,t})_{0 \le s \le t}$ of the stochastic point vortices, which are solutions of the following SDEs

$$\begin{aligned} \mathrm{d}\boldsymbol{X}_s^{x,t} &= -\boldsymbol{u}(X_s^{s,t}, t-s)\,\mathrm{d}s + \sqrt{2\nu}\,\mathrm{d}\boldsymbol{W}_s, \quad s \le t, \\ X_0^{x,t} &= x. \end{aligned} \tag{6.14}$$

The deformation matrices $(U_s^{x,t})_{0\le s\le t}$ are themselves solutions to the following SDEs with random coefficients

$$\begin{aligned} \mathrm{d}U_s^{x,t} &= U_s^{x,t}\mathcal{D}_u(X_s^{x,t}, t-s)\,\mathrm{d}s, \quad s \le t, \\ U_0^{x,t} &= I, \end{aligned} \tag{6.15}$$

where $\mathcal{D}_u = 1/2(\nabla \boldsymbol{u} + \nabla \boldsymbol{u}^T)$ is the deformation tensor.

Although the probabilistic representation given above is rather complicated, and is unlikely to simplify the solution of the vorticity equation, it is nevertheless interesting. First, because it proves that there is a rigorous and exact SDE representation of the vorticity equation, in line with our intuition. Secondly, because it shows that the continuous vorticity equation has an exact representation in terms of Lagrangian point vortices. Finally, the SDE is the analytical version of the numerical vortex methods which are discussed in the following section. Thus, this SDE may help us develop alternative numerical vortex methods. It is also possible that a careful analysis of the SDE form of the vorticity equation will suggest new analytical approximations which are not obvious from the usual PDE form.

6.4 Algebraic vortices

6.4.1 *Numerical vortex methods*

We now consider the third and final class of vortices: algebraic vortices. By algebraic vortices we mean vortices used as computational elements to reduce the vorticity PDE to a system of algebraic equations. This algebraic system is then solved using the techniques of numerical linear algebra. Algebraic vortices are therefore a sort of basis for the vorticity equation. The vortices themselves generally have no internal degrees of freedom.

In a sense, the most primitive numerical vortex method is pseudo-spectral simulation. In this method the vortices are the *eddies* of the homogeneous isotropic turbulence, i.e. Fourier modes of different wavelengths. Indeed, in large eddy simulation (LES) this connection is explicit: the flow is represented by the largest (and hence most energetic) eddies.

Usually, however, the term *vortex method* is reserved for those methods that discretize the vorticity equation using computational elements in physical space, and use a Lagrangian or semi-Lagrangian version of the vorticity equations. Vortex methods date back to Prager [34] and Rosenhead [36] in the 1920s and 1930s, but were not commonly used (or justified mathematically) until the work of Chorin [6], Majda [1; 2] Leonard [26], Krasny [25] and others. Although vortex methods are usually applied to two-dimensional flows, a highly accurate three-dimensional vortex method for wake flows was developed recently by Cottet & Poncet [8]. In this section we sketch the basics of two-dimensional vortex methods, and relate these algebraic vortices to the geometric and stochastic vortices described earlier. Cottet & Koumoutsakos [7] have published a very nice overview of the subject, which the reader should consult for more details on vortex methods.

Two-dimensional vortex methods first sample the circulation of the initial vorticity field onto a finite number N of points (i.e. vortex particles). The vorticity equations are then integrated in time from this initial condition using a two-step method:

(a) **I**nviscid step. The particles are moved by Lagrangian advection due to the velocity field of all other particles (according to Kelvin's circulation theorem 6.6, the circulation of the particles is unchanged).
(b) **V**iscous step. Vorticity diffusion is modelled by exchanging circulation amongst nearby particles (particle strength exchange), or by perturbing the particle's position by white noise to model Brownian diffusion (this is much less accurate).

Since these two steps do not commute, the time accuracy of the method is limited to second-order.

To obtain a stable numerical method, one must replace the singular velocity of a true point vortex by a smoothed velocity field. Mathematically, this means replacing the singular Biot–Savart kernel

$$\boldsymbol{K}(\mathbf{x}) = \frac{1}{2\pi|\mathbf{x}|^2}(-x_2, x_1) \tag{6.16}$$

by a smoothed or *mollified* kernel, such as the Gaussian kernel,

$$\boldsymbol{K}_\epsilon(\mathbf{x}) = \boldsymbol{K} \star \zeta_\epsilon = \frac{1}{2\pi|\mathbf{x}|^2}[1 - \exp(-|\mathbf{x}|^2\epsilon^{-2})](-x_2, x_1), \tag{6.17}$$

where $\star$ denotes convolution, ζ_ϵ is called the cut-off function and ϵ is a small parameter. Note that the mollified kernel (6.17) matches the singular kernel (6.16) when $|\mathbf{x}| \gg \epsilon$. Cottet & Koumoutsakos [7] show that the accuracy of the mollified kernel approximation increases with the number of vanishing moments of the cut-off function. The N vortex particles (defined by their positions $\mathbf{x}_p(t)$ and circulations $\Gamma_p(t)$) are then advected by the velocity field

$$\boldsymbol{u}(\mathbf{x}_p, t) = \boldsymbol{u}_\infty + \sum_{p=1}^{N} \boldsymbol{K}_\epsilon \star \Gamma_p \delta(\mathbf{x} - \mathbf{x}_p) \tag{6.18}$$

where $\boldsymbol{u}_\infty$ is the mean velocity, and $\delta(\mathbf{x})$ is the delta-function (i.e. the vortices are point particles). Note that because of the mollified kernel, the particles are not self-advected. At first glance the convolution calculation in equation (6.18) should increase the complexity of the method to $O(N^2)$, but grid-free methods exist (such as the fast multipole method [12]) that reduce the cost to $O(N \log N)$, or even $O(N)$. In practice, however, it is often more efficient to interpolate the vorticity onto a regular grid and use a fast Poisson solver to find the velocity. This makes the method semi-Lagrangian. Vortex methods have proved to be accurate and efficient for a wide range of vortical flows, especially wake flows in large domains [24].

It is appropriate at this point to compare the algebraic vortices of the numerical vortex method to stochastic and geometric vortices. The probabilistic representation of the vorticity equation shares many similarities with the numerical vortex method, especially when diffusion is introduced via white noise perturbations. There are, however, two fundamental differences. First, the vortex method consists of one realization made up of many thousands of particles, whereas the probabilistic representation consists of many realizations, each containing just one particle. Secondly, the vortex method smoothes the velocity field associated with the point vortices, whereas the probabilistic representation retains the singular velocity field. The probabilistic representation also explicitly includes the deformation tensor, which is absent in the vortex methods. It would be interesting to investigate further the links between vortex methods and probabilistic representations of the vorticity equation.

There are also links between the geometric and algebraic descriptions. In fact, the inviscid vortex method is based on the geometric idealization of point vortices, and some three-dimensional vortex methods use the slender vortex filaments discussed in section 6.2 as computational elements. Vortex methods are also well-suited to exploring topological problems such as vortex reconnection [7]. On the other hand, the algebraic representation does not include any information about the coherence or incoherence of the vorticity being represented. The particles are the same whether they make up coherent vortices, or are part of the Gaussian noise. The algebraic vortices are, after all, simply computational elements.

6.4.2 *Adaptive wavelet method for the vorticity equation*

Finally, we consider an algebraic vortex method that combines some of the geometric and stochastic aspects discussed earlier. This is the WDNS method applied to the vorticity equation [41; 42]. (For WDNS applied to the Navier–Stokes equations in primitive form, please see [19].) We will see that WDNS may be classified as an Eulerian vortex method.

Grid adaptation occurs naturally in wavelet methods. Consider $\omega(\mathbf{x})$, defined on a closed two-dimensional rectangular domain Ω. We use tensor product second generation wavelets [39] constructed on a set of grids,

$$\mathcal{G}^j = \left\{ \mathbf{x}^j_{\boldsymbol{k}} \in \Omega : \boldsymbol{k} \in \mathcal{K}^j \right\}, \quad j \in \mathcal{J}, \tag{6.19}$$

where $\boldsymbol{k} = (k_1, \ldots, k_n)$ and grid points $\mathbf{x}^j_{\boldsymbol{k}} = (x^j_{1,k_1}, \ldots, x^j_{n,k_n})$ are constructed as a tensor product of uniformly spaced one-dimensional grids. The only restriction is that each individual set of one-dimensional grids is nested ($x^j_{m,k_l} = x^{j+1}_{m,2k_l}$, $m = 1, 2$), which guarantees the nestedness of the grids, *i.e.* $\mathcal{G}^j \subset \mathcal{G}^{j+1}$. The procedure for constructing two-dimensional scaling functions $\phi^j_{\boldsymbol{k}}(\mathbf{x})$ and a family of two-dimensional wavelets $\psi^{\mu,j}_{\boldsymbol{l}}(\mathbf{x})$, $\mu = 1, \ldots, 3$ on two-dimensional dyadic grid is described by Sweldens [39]. Once wavelets and scaling functions are constructed, the vorticity $\omega(\mathbf{x})$ can be decomposed as

$$\omega(\mathbf{x}) = \sum_{\boldsymbol{k} \in \mathcal{K}^0} c^0_{\boldsymbol{k}} \phi^0_{\boldsymbol{k}}(\mathbf{x}) + \sum_{j=0}^{+\infty} \sum_{\mu=1}^{3} \sum_{\boldsymbol{l} \in \mathcal{L}^{\mu,j}} d^{\mu,j}_{\boldsymbol{l}} \psi^{\mu,j}_{\boldsymbol{l}}(\mathbf{x}). \tag{6.20}$$

For functions like $\omega(\mathbf{x})$ which contain isolated small scales on a large-scale background, most wavelet coefficients are small. Thus, we retain a good approximation even after discarding a large number of wavelets with small coefficients. Intuitively, the wavelet coefficient $d^{\mu,j}_{\boldsymbol{l}}$ is small unless $\omega(\mathbf{x})$ has variation on the scale of j in the immediate vicinity of wavelet $\psi^{\mu,j}_{\boldsymbol{l}}(\mathbf{x})$. In fact, the error incurred by ignoring coefficients with magnitude lower than ϵ is $O(\epsilon)$.

A vorticity-adapted grid can therefore be constructed by filtering the vorticity in wavelet space, and then performing the inverse transform using only the significant wavelets. This automatically interpolates the vorticity onto the set of significant grid points (note that there is a unique grid point associated with each wavelet coefficient, so neglecting wavelets is equivalent to removing grid points). A localized one-dimensional function, and its corresponding multiscale adapted grid are shown in figure 6.2. Recall that in Section 6.3 we mentioned that wavelet filtering corresponds to a de-noising of the vorticity field, and that the significant wavelet modes correspond to the coherent vortices of the flow.

When solving the vorticity equations, an additional criterion for grid adaptation must be added to take into account flow dynamics. The computational grid should consist of grid points associated with wavelets whose coefficients are significant and also those which could become significant during a time step. In other words, the computational grid should also include the nearest neighbouring points in position and scale. This corresponds to a CFL criterion of one, and provides full dealiasing for quadratic nonlinearities. The grid is adapted in this way at each time step, and the equations are integrated in time using a standard technique for ordinary differential equations.

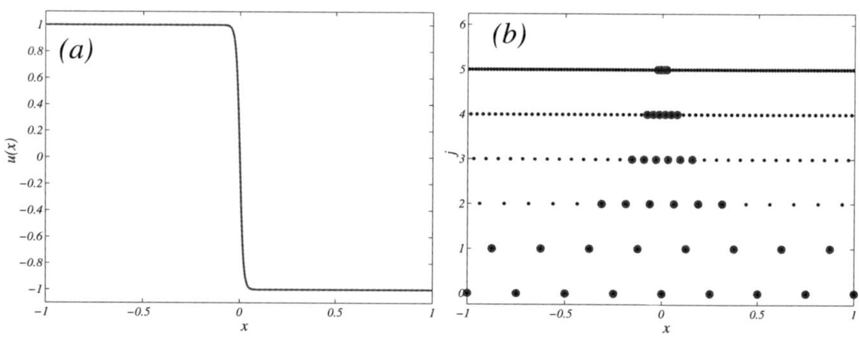

Fig. 6.2 Adaptive wavelet dyadic grid after Vasilyev & Bowman [41]. (a) Function. (b) Grid as a function of dyadic scale j and position x. Small points, uniform grid; large points, adapted grid. The actual computational grid is the union of the large points at all levels j.

When applied to the two-dimensional vorticity equations, WDNS constitutes an Eulerian vortex method. The adapted grid points may be thought of as vortex particles with circulation given by their vorticity times the volume associated to the grid point. Indeed, we even use the fast multipole method [12] to calculate the velocity field from these vortex particles. Figure 6.3 shows the vorticity and adapted grid for periodic flow past a cylinder at Re = 20 000. Because the grid adapts to local gradients of vorticity, the distribution of points in the adapted grid resembles the distribution of point vortices in a vortex method. The main difference is that the grid points are densest where the gradients of vorticity are large, rather than where vorticity itself is large. (In fact, the WDNS grid for the Navier–Stokes equations [19] more closely resembles the locations of point vortices in the vortex method, since in this case the wavelets are sensitive to velocity gradients.) Although the vortex particles are restricted to a regular Cartesian grid, the fact that this grid can refine or coarsen as required to resolve the vorticity field means it is similar in spirit to a Lagrangian vortex method. One important distinction with respect to vortex methods is that in WDNS we resolve only the coherent vortices (if the threshold ϵ is set appropriately, as explained in Section 6.3). Thus WDNS links algebraic, stochastic and geometric vortices.

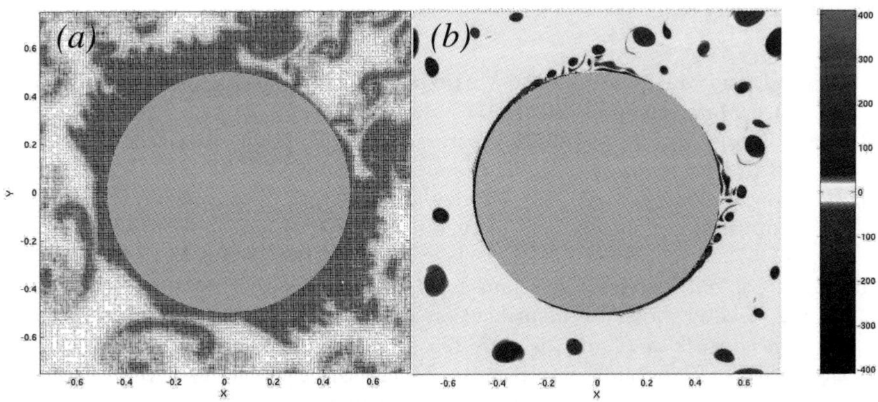

Fig. 6.3 WDNS simulation of flow past a periodic array of cylinders at Re = 20 000. (a) Adapted grid (including nearest neighbours in position and scale). (b) Vorticity.

6.5 Conclusions

This paper has looked at vortices from three different, but complementary, points of view: geometric, stochastic and algebraic. Vortices are not only beautiful esthetically and mathematically elegant, they are also at the heart of many practical engineering applications. We began by showing how vortices are central to understanding such important problems in fluid dynamics as mixing, flow-induced vibration, turbulence and noise. We then saw how the geometric, stochastic and algebraic points of view each provide insight into the structure, dynamics and function of vortices. In particular, we saw how these different points of view help us to understand the role of vortices in complicated flows, such as turbulence. Finally, the wavelet direct numerical simulation (or coherent vortex simulation) showed how each of the three mathematical approaches can be combined in a single model.

Acknowledgment

Research funding from NSERC is gratefully acknowledged. This paper was written during a visit to DAMTP, University of Cambridge, and I would like to thank them for their hospitality.

Bibliography

1. Beale, J. and Majda, A. (1982). Vortex method I: Convergence in three dimensions. *Math. Comput.* **39**, 1–27.
2. Beale, J. and Majda, A. (1982). Vortex method II: High order accuracy in two and three dimensions. *Math. Comput.* **39**, 29–52.
3. Berkooz, G., Holmes, P. and Lumley, J. (1993). The proper orthobonal decomposition in the analysis of turbulent flows. *Annu. Rev. Fluid Mech.* **25**, 539–575.
4. Bonnet, J. P., Delville, J., Glauser, M. N. *et al.* (1998). Collaborative testing of eddy structure identification methods in free turbulent shear flows. *Experiments in Fluids* **25**, 197–225.
5. Busnello, B., Flandoli, F. and Romito, M. (2003). A probabilistic representation for the vorticity of a 3D viscous fluid and for general systems of parablic equations. In *Summer program: probability and partial differential equations in modern applied mathematics.* IMA.
6. Chorin, A. (1973). Numerical study of slightly viscous flow. *J. Fluid Mech.* **57**, 785–796.
7. Cottet, G. and Koumoutsakos, P. (2000). *Vortex methods: theory and practice.* Cambridge University Press.
8. Cottet, G.-H. and Poncet, P. (2002). Particle methods for direct numerical simulations of three-dimensional wake flows. *J. Turbulence* **3**, 1–9.
9. Donoho, D. L. (1993). Wavelet shrinkage and w.v.d. – a ten-minute tour. Technical Report 416, Dept. of Statistics, Stanford University.
10. Farge, M., Schneider, K. and Kevlahan, N. K.-R. (1999). Non-gaussianity

and coherent vortex simulation for two-dimensional turbulence using an adaptive orthogonal wavelet basis. *Phys. Fluids* **11**, 2187–2201.
11. Frisch, U. (1995). *Turbulence : the legacy of A.N. Kolmogorov.* Cambridge University Press.
12. Greengard, L. and Rokhlin, V. (1987). A fast algorithm for particle simulations. *J. Comput. Phys.* **73**, 325–348.
13. Griebel, M. and Koster, K. (2000). Adaptive wavelet solvers for the unsteady incompressible Navier Stokes equations. In J. Malek, J. Necas and M. Rokyta (editors), *Lecture Notes of the Sixth International School "Mathematical Theory in Fluid Mechancis"*, Advances in Mathematical Fluid Mechanics. Springer Verlag.
14. Griebel, M. and Koster, K. (2002). Multiscale methods for the simulation of turbulent flows. In E. Hirschel (editor), *DFG/CNRS Workshop, Nice, 2001, Notes on Numerical Fluid Mechanics.* Vieweg-Verlag.
15. Hunt, J., Wray, A. and Moin, P. (1988). Eddies, stream, and convergence zones in turbulent flows. Technical Report CTR-S88, Center for Turbulence Research.
16. Jeong, J. and Hussain, F. (1995). On the identification of a vortex. *J. Fluid Mech.* **285**, 69-94.
17. Jiménez, J. and Wray, A. (1998). On the characteristics of vortex filaments in isotropic turbulence. *J. Fluid Mech.* **373**, 255–285.
18. Kambe, T., Minota, T., Murakami, T. *et al.* (1990). Oblique collision of two vortex rings and its acoustic emission. In H. Moffatt and A. Tsinober (editors), *Topological fluid mechanics: proceedings of the IUTAM symposium, Cambridge, August 13–18, 1989.* 515–524.
19. Kevlahan, N. and Vasilyev, O. (2003). An adaptive wavelet collocation method for fluid–structure interaction at high Reynolds numbers. *SIAM J. Sci. Comput.* To appear.
20. Kevlahan, N. K.-R. (1997). The vorticity jump across a shock in a non-uniform flow. *J. Fluid Mech.* **341**, 371–384.
21. Klein, R., Knio, O. and Ting, L. (1996). Representation of core dynamics in slender vortex filament simulations. *Phys. Fluids* **8**, 2415.
22. Klein, R. and Majda, A. (1991). Self-stretch of a perturbed vortex filament. *Physica D* **49**, 323.
23. Klein, R., Majda, A. J. and Damodaran, K. (1995). Simplified equations for the interactions of nearly parallel vortex filaments. *J. Fluid Mech.* **288**, 201–248.
24. Koumoutsakos, P. and Leonard, A. (1995). High-resolution simulations of the flow around an impulsively started cylinder using vortex methods. *J. Fluid Mech.* **296**, 1–38.
25. Krasny, R. (1986). A study of singularity formation in a vortex sheet by the point vortex approximation. *J. Fluid Mech.* **167**, 65–93.
26. Leonard, A. (1980). Vortex methods for flow simulation. *J. Comput. Phys.* **37**, 289–335.
27. Lesieur, M. and Métais, O. (1997). New trends in large-eddy simulations of turbulence. *Annu. Rev. Fluid Mech.* **28**, 45–82.

28. Lighthill, J. (2001). *Waves in fluids.* Cambridge University Press.
29. Lumley, J. (1967). The structure of inhomogeneous turbulent flow. In A. Yaglom and V. Tatarski (editors), *Atmospheric Turbulence and Radio Wave Propagation.* Nauka, 160–178.
30. Majda, A. and Bertozzi, A. (2002). *Vorticity and incompressible turbulence.* Cambridge University Press.
31. Nazarenko, S., Kevlahan, N. K.-R. and Dubrulle, B. (2000). Nonlinear RDT theory of near-wall turbulence. *Physica D* **139**, 158–176.
32. Noca, F. Shiels, D. and Jeon, D. (1997). Measuring instantaneous fluid dynamics forces on bodies, using only velocity fields and their derivatives. *J. Fluids Struct.* **11**, 345–350.
33. Noca, F., Shiels, D. and Jeon, D. (1999). A comparison of methods for evaluating time-dependent fluid dynamic forces on bodies, using only velocity fields and their derivatives. *J. Fluids Struct.* **13**, 551–578.
34. Prager, W. (1928). Die druckverteilung an körpern in ebener potentialströmung. *Phys. Z.* **29**, 865–869.
35. Provenzale, A. (1999). Transport by coherent barotropic vortices. *Annu. Rev. Fluid Mech.* **31**, 55–93.
36. Rosenhead, L. (1931). The formation of vortices from a surface of discontinuity. *Proc. R. Soc. London Ser. A* **134**, 170–192.
37. Saffman, P. (1992). *Vortex dynamics.* Cambridge University Press.
38. Schneider, K., Kevlahan, N. K.-R. and Farge, M. (1997). Comparison of an adaptive wavelet method and nonlinearly filtered pseudo-spectral methods for two-dimensional turbulence. *Theoret. Comput. Fluid Dynamics* **9**, 191–206.
39. Sweldens, W. (1996). The lifting scheme: A custom-design construction of biorthogonal wavelets. *Appl. Comput. Harmon. Anal.* **3**(2), 186–200.
40. Ting, L. and Klein, R. (1991). *Viscous vortical flows.* Springer.
41. Vasilyev, O. V. and Bowman, C. (2000). Second generation wavelet collocation method for the solution of partial differential equations. *J. Comput. Phys.* **165**, 660–693.
42. Vasilyev, O. V. and Kevlahan, N. K.-R. (2002). Hybrid wavelet collocation-brinkman penalization method for complex geometry flows. *Int. J. Num. Meth. Fluids* **30**, 531–538.
43. Yokokawa, M., Itakura, K., Uno, A. *et al.* (2002). 16.4-Tflops direct numerical simulation of turbulence by a fourier spectral method on the earth simulator. In *Proceedings of the IEEE/ACM SC2002 Conference.* IEEE Computer Society, 50.
44. Zhou, H. (1997). On the motion of slender vortex filaments. *Phys. Fluids* **9**, 970–981.

Chapter 7

Experimental Investigation & Numerical Simulation of Oblique Shock/Vortex Interaction

Michael Klass, Wolfgang Schröder, and Oliver Thomer
Aerodynamisches Institut
RWTH Aachen
Wüllnerstr. zw. 5 und 7, D–52062 Aachen, Germany

7.1 Dedication to Professor Lu Ting

On the occasion of Professor Lu Ting's 80th birthday, we wish to express our highest appreciation for his work and efforts on the areas of mathematics, fluid mechanics, gas dynamics and vortex dominated flows. Professor Ting is not only held in great respect by the scientific community for his contribution to science but also in great affection by his collaborators and colleagues as a man of great humanity and dignity. It is therefore a great pleasure for us to dedicate this paper to Professor Lu Ting, who was and still is much more than just a first class scientist.

7.2 Introduction

The breakdown of slender vortices, caused by positive pressure gradients, has been intensively investigated throughout the past decades. However, no general, reliable breakdown condition could be formulated neither for incompressible nor for compressible flow.

Previous investigations mainly deal with vortices in incompressible flow [Althaus and Weimer (1998)], while supersonic vortical flows are hardly analyzed. One of the first numerical attempts to establish a breakdown criterion for shock–vortex interaction can be found in [Erlebacher *et al.* (1997)]. Numerical studies on breakdown of leading edge vortices on delta wings are reported by [Kandil *et al.* (1992)].

The experimental analysis of the vortex breakdown phenomenon in supersonic flows is primarily based on qualitative flow visualizations and pressure measurements [Délery (1994); Délery *et al.* (1984); Kalkhoran and Sforza (1994); Kalkhoran *et al.* (1998); Smart and Kalkhoran (1995); Cattafesta and Settles (1992)]. Quantitative findings of the velocity field in the interaction region, e.g., in the form of Laser Doppler Velocimetry (LDV) or Particle Image Velocimetry (PIV), have rarely been performed.

Early experiments of [Zatoloka *et al.* (1978)] reveal the interaction between a free vortex filament and the shock system, which exists in an intake of a supersonic aircraft. The experiments also show that the interaction could be responsible for a reduced performance of the intake.

Recent investigations of flows in turbine engine cascades seem to indicate that shock induced vortex breakdown might be responsible for stall observed at high compressor loads. Investigations of [Schlectriem and Lötzerich (1997)] indicate that the shock–vortex interaction determines tip clearance vortex breakdown in a transonic axial compressor rotor. Although not rigorously verified, this conjecture could lead to a new understanding of the stall mechanism in turbine and compressor engines. Such ideas seem to be justified by the fact that tip vortices of canard wings also experience breakdown in supersonic flow, when they are intersected by the leading edge shock of the main wing.

A detailed overview of recent numerical and experimental investigations of supersonic vortex breakdown for oblique and normal shock–vortex interaction, respectively, was published by [Kalkhoran and Smart (2000)].

In supersonic flows the interaction between a slender vortex and an oblique (OSVI) or normal shock (NSVI) has been classified and divided into three distinct modes, namely weak, moderate and strong interaction. Weak shock–vortex interaction, primarily observed for vortices with a slight wake profile and low circulation, is characterized by a slight distortion of the vortex and a nearly steady supersonic flow field. The shock does not change its shape significantly, although an increase of the shock angle in the vicinity of the vortex axis may occur. Moderate shock–vortex interaction, however, is accompanied by an alteration of the shock and a highly distorted vortex in conjunction with a subsonic region without a stagnation point downstream of the shock. Finally, strong shock–vortex interaction, occurring for vortices with a distinct wake profile of the axial Mach number distribution and a high circulation, always leads to a vast reorganization of the vortex structure. Recent numerical investigations by [Thomer *et al.* (2000); Thomer *et al.* (2001); Krause *et al.* (2003)] show that a large subsonic region and a stagnation point plus a recirculation area can be found downstream of the shock. The shock itself shows a pronounced deformation wrapped around this recirculation area. It goes without saying that the strong shock–vortex interaction is a highly unsteady flow phenomenon.

The temporal evolution of supersonic vortex breakdown and the velocitiy and vorticity distribution in the interaction region cannot be completely explained yet. Hence, the analysis of the flow structure in the recirculation area for strong shock–vortex interaction is still connected with a considerable amount of uncertanity.

The experimental and numerical results in the present study are designed to give somewhat more insight into the structure of the flow field downstream of the shock and the dependence of the global flow field on the parameters, which can cause vortex breakdown. Furthermore, the experiments serve as a reference database for upcoming velocity measurements with non–intrusive measurement techniques like PIV and LDV.

7.3 Experimental investigation

The experimental investigation is performed in the supersonic wind tunnel of the Aerodynamisches Institut Aachen (AIA). This suction type wind tunnel operating according to the intermittent working principle is equipped with three nozzle sets for supersonic flow ($Ma_\infty = 1.5$, 2.1 and 2.9) and one nozzle set for subsonic flow in the Mach number range $Ma_\infty = 0.3$ to 0.9. The test section has a cross sectional area of 15×15 cm^2 and a length of 18 cm. Measurement times range from $t_{meas} = 20$ s to 40 s depending on the Mach number. The specifications of the wind tunnel, a schematic of which is shown in Fig. 7.1, are summarized in Table 7.1.

Table 7.1 Technical data of the wind–tunnel facility.

Mach numbers	Ma_∞	[-]	$0.3 - 0.9, 1.5, 2.1, 2.9$
Reynolds numbers per meter[1]	$\frac{Re}{m}$	[-]	$6.7 - 15.2, 15.7, 12.8, 8.6 \times 10^6$
Test section	A_{meas}	[cm^2]	15×15
measurement time	t_{meas}	[s]	max. 30 (at $Ma_\infty = 2.1$)
stagnation pressure/temperature			ambient conditions T_0, p_0

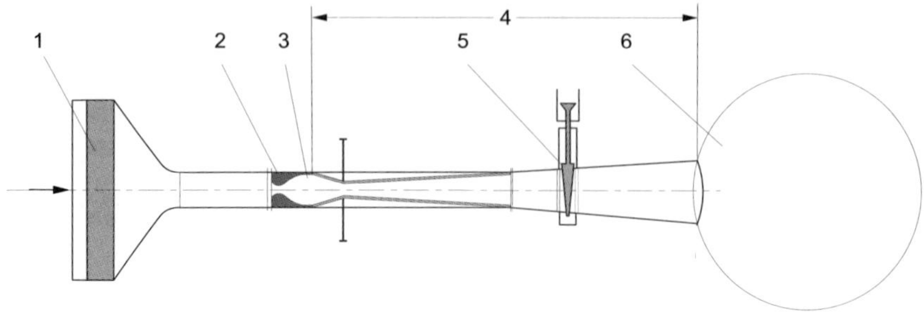

Fig. 7.1 Supersonic wind tunnel.

7.3.1 *Experimental setup*

The experimental arrangement is depicted in Fig. 7.2. Two rectangular wings with half–angles $\beta_{VG} = 8°$ and $\beta_{VG} = 12°$, respectively, are used to generate a slender streamwise tip vortex. Both vortex generators have a diamond–shaped cross section, a span of 75 mm, a chord length of 25 mm and can be positioned at angles of attack ranging from $\alpha_{SG} = 0°$ to 20°. They are continuously adjustable within this range and thereby provide the possibility to modifiy the circulation — and thus the strength of the vortex — over a wide range.

The generation of an oblique shock is realized by using several two-dimensional wedge sections with wedge angles of $\beta_{SG} = 10°$, 20° and 25°. Each shock generator has a span of 150 mm, which equals the width of the test section. The strength of the oblique shock can be varied by changing the angle attack of the ramp. It can be altered from $\alpha_{SG} = -10°$ to 20°, such that total flow deflection angles between $\beta_{fd} = 0°$ and 45° can be realized.

Additionally, the vertical distance y_{SG} between the tip of the rectangular wing (located on the wind tunnel centerline) and the wedge as well as the axial distance z_{SG} between the wing (measured from the trailing edge) and the shock generator can be varied over a wide range. On the one hand, this ensures that the shock generator can be positioned in an area downstream

[1]for $T_0 = 15\ °C$ und $p_0 = 1015\ hPa$

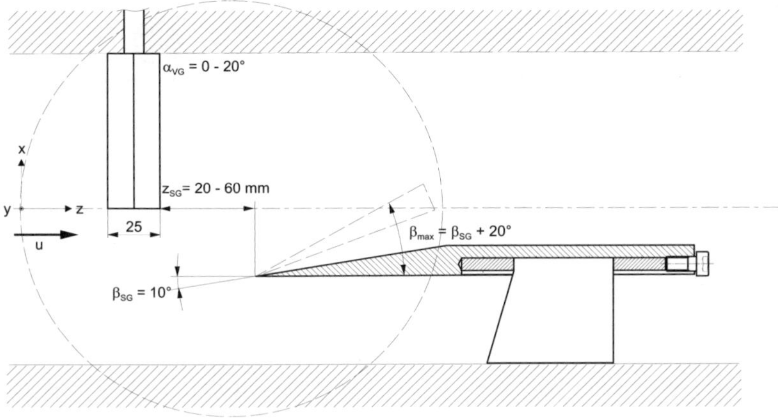

Fig. 7.2 Experimental arrangement.

of the vortex generator, where disturbances in the shock shape originating from the rectangular wing can be neglected. On the other hand, the shock generator can be intentionally positioned such that the interaction region gets close to the wedge. Thus, the influence of the distance between the interaction region and the wedge can be studied.

The capability of vertical and horizontal adjustment of the wedge also ensures that velocity measurements of the undisturbed vortex and the interaction region are performed in the same position. This allows the comparison between the structure of the undistorted vortex and its shape downstream of the shock in the case of shock–vortex interaction.

7.3.2 *Measurement techniques*

Two different flow visualization techniques are applied to investigate the temporal evolution of the shock structure and the vortex breakdown region downstream of the shock.

To obtain information about the global structure of the flow field and the unsteady flow phenomena the flow is visualized using color schlieren along with a digital video camera. A standard z–type Toepler color schlieren system as shown in Fig. 7.3 is used.

The light source of this system is a high–pressure discharge lamp. A tricolored circular filter mask equipped with filters especially adapted to the wavelengths of the discharge lamp and an iris diaphragm serving as schlieren stop are used to visualize the density gradients of the flow field

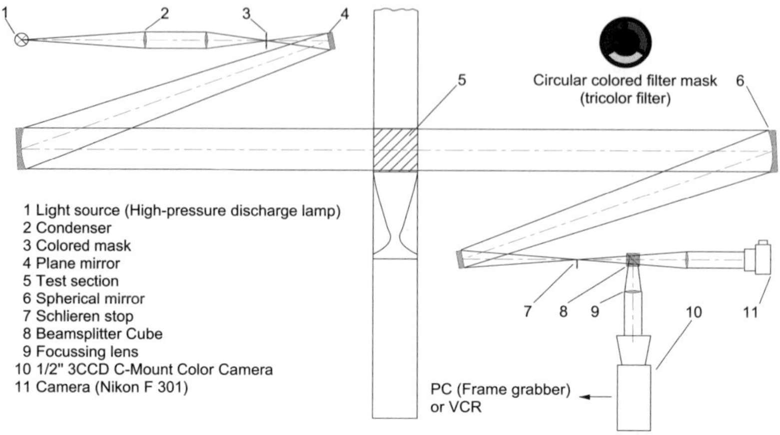

Fig. 7.3 Optical arrangement for color schlieren technique.

and their direction. The images obtained with this system are recorded directly to harddisk using a frame grabber card (Fast AV Master) and a digital 3–CCD colour camera (Hitachi HV–C20A, 1/2" CCD chip) with a resolution (effective number of pixels) of 752×582 pixels, an electronic shutter of 1/10,000th second and a maximum frame rate of 25 Hz.

For a more detailed analysis of the shock–vortex interaction the differential interferometry is applied. The differential interferometry, Fig. 7.4, uses a similar setup like the color schlieren technique. However, instead of the tricolored mask and the schlieren stop the interferometric system utilizes two Wollaston–prisms in conjunction with a polarization filter each.

7.3.3 *Experimental program*

The experiments are conducted at nominal test section Mach numbers of $Ma_\infty = 2.0$ and $Ma_\infty = 2.9$. Since the stagnation pressure and temperature are determined by the ambient conditions (in this case $p_0 = 1.015 \cdot 10^5\ Pa$ and $T_0 = 288\ K$), the Reynolds numbers per meter are approx. $Re/m = 13.3 \cdot 10^6$ per meter and $Re/m = 8.6 \cdot 10^6$ per meter, respectively.

As summarized in Table 7.2 four different parameters have been varied to investigate weak, moderate and strong shock–vortex interaction modes. The axial velocity profile of the slender vortex can be varied by exchanging the vortex generator, while the vortex strength is changed via the angle of attack. The flow deflection angle, i.e., the shock strength, is modified by

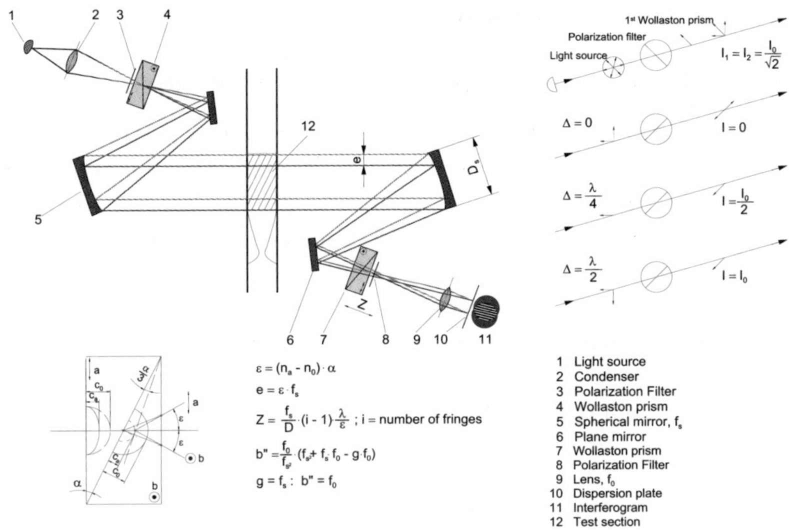

Fig. 7.4 Basic setup for differential interferometry.

changing the angle of attack of the shock generator.

Table 7.2 Parameters of oblique shock–vortex interaction.

Parameter		Value
Ma_∞	[–]	1.5, 2.0, 2.9
α_{VG}	[°]	0, 4, 6, 10, 12, 20, 25
β_{fd}	[°]	0, 10, 15, 20, 25, 30, 35
β_{VG}	[°]	8, 12

Not all possible combinations of the values listed below are investigated. Flows with a detached shock in front of the shock generating wedge are not considered. Furthermore, since the transition from weak to moderate interaction (and from moderate to strong interaction, respectively) is of major interest to develop a breakdown criterion, a measurement series is finished once the point of transition between the modes is found.

7.4 Numerical investigation

7.4.1 *Computational setup and boundary conditions*

A wedge-like grid was generated. Although OSVI can be simulated using a rectangular grid in combination with the Rankine-Hugoniot equations, a wedge aligned grid resembles the experimental setup much closer. Due to the Prandtl-Meyer expansion caused by the convex shape at the aft end of the wedge stronger vortices (larger wake or higher circulation) are necessary to enforce breakdown. The breakdown also depends on the wedge geometry, i. e., the interchange between shock and expansion fan forces. Near the vortex axis the grid is clustered such that the vortex core ($-1 < r < 1$) is resolved with approximately 21 grid points in the radial direction. The inflow conditions are assumed to be described by a slender streamwise vortex centered at x=y=0 together with a radial distribution of the Gaussian axial velocity. Controlled by the sign of the parameter $\delta \in [0..1]$ the axial velocity component in the inflow plane is modeled as a 'wake–like' profile given by

$$v_z(r) = v_{z_\infty} \cdot (1 - \delta e^{-\mu_w \cdot r^2}). \tag{7.1}$$

The quantity $r_i = \pm\sqrt{1/(2\mu_w)}$ describes the inflection point position for the Gaussian axial velocity profile. The circumferential velocity distribution is modeled as a slender iso–energy Burgers vortex with

$$v_\varphi(r) = \Gamma_0 \cdot (1 - e^{\frac{-r^2}{4r_0^2}})/(2\pi r). \tag{7.2}$$

The ratio of the azimuthal and the axial velocity components is defined as the helix angle

$$\tau = \frac{v_\varphi}{v_z} \tag{7.3}$$

at the vortex core $r = 1$. For a given azimuthal velocity distribution the quantity τ increases if either the circulation Γ_0 and the wake parameter δ are increased or the Mach number Ma_∞ and the width of the profile μ_w are decreased. The inflow conditions are also used as initial conditions for the entire flow field.

Since the flow is supersonic near the exit, a simple Taylor approximation is sufficient at the outflow boundary. Due to the inviscid theory all characteristics leave the computational domain. The approximation for the lateral boundary conditions is obtained by assuming the shape of the boundaries to be identical with streamline surfaces such that the kinematic flow condition can be implemented. The normal derivatives of the solution vector with respect to the slope of the local streamline vanish, if the flow is assumed to be inviscid. In the computational method, the differentials in time and space are discretized to second order accuracy. The temporal integration is performed using an optimized 5–step Runge–Kutta scheme. A node–centered scheme is used to approximate the fluxes via an explicit finite–volume scheme. The convective fluxes are discretized using the $AUSM+$ approach by Liou.

7.5 Results

7.5.1 *Experimental results*

7.5.1.1 *Weak/moderate interaction*

Figure 7.5 shows color schlieren images for four different combinations of vortex strength and shock strength at a nominal test section Mach number of $Ma_\infty = 2.9$.

Two different vortex strengths are generated by altering the angle of attack of the vortex generator between $\alpha_{VG} = 10°$ and $\alpha_{VG} = 15°$. Both positions can be considered to produce a strong vortex. This agrees with previous experimental investigations of [Smart and Kalkhoran (1995)] , where a vortex generated at $\alpha = 10.4°$ was classified as a strong vortex.

When the vortex at $\alpha_{VG} = 10°$ interacts with the oblique shock caused by a flow deflection angle of $\beta_{fd} = 10°$, (Fig. 7.5a) no significant change of the shock structure can be observed. Since the vortex is barely visible downstream of the shock, the structure of the vortex inside and downstream of the interaction region is not revealed by the color schlieren images. Nevertheless, due to the little impact on the shock shapeit can be assumed that the basic structure of the vortex remains unchanged as it passes through the oblique shock. According to the aforementioned classification of shock–vortex interaction, this combination represents the weak interaction mode.

Increasing either the vortex strength (Fig. 7.5b) or the shock strength (Fig. 7.5c) while keeping the other parameter unchanged leads to a deformed shock. While the shock angle remains constant in the outer regions near the sidewalls of the wind tunnel, it increases at the position where the vortex intersects the shock. Since the color schlieren technique shows an integrated two–dimensional image of a three–dimensional shock surface, the undisturbed parts of the shock and the distorted inner region occur in the same projected plane.

Increasing the shock strength and vortex strength simultaneously, Fig. 7.5d, yields a shock system that shows the strongest deviation of the shock pattern of the interaction region from the undisturbed flow field.

These results agree qualitatively with the numerical investigations from [Nedungadi and Lewis (1995)]. They found weak and moderate interaction modes for a freestream Mach number of $Ma_\infty = 3.0$ in combination with a circulation value of $\Gamma_0 = 0.4m^2/s$ and two different centerline to freestream velocity ratios, which represent two axial Mach number profiles of the vortex.

Figure 7.6 displays the experimental visualization of the strong OSVI at a freestream Mach number of $Ma_\infty = 2.0$ for a flow deflection angle of $\beta_{fd} = 11.5°$. The weak vortex is generated by the double–wedge wing at an angle of attack of $\alpha_{VG} = 6°$. To visualize the flow field we apply differential interferometry adjusted to infinite fringe spacing. This technique reacts more sensitively to densitiy gradients than the color schlieren method. Furthermore, differential interferometry has a higher light efficiency and thus permits lower shutter times of the camera. The images presented here were recorded at 25 fps with a shutter time of $t = 100\mu s$.

Likewise the color schlieren images, the differential interferometry provides a two–dimensional image of a three–dimensional flow structure. Although the shock front is highly distorted in the central area above the wedge, it remains almost at its original angle (which is prescribed by the Mach number and the deflection angle) in the outer regions of the shock surface. Only in the symmetry plane of the wind tunnel is the shock almost vertical.

The pictures extracted from a video illustrate the temporal evolution of the strong shock–vortex interaction and the highly unsteady development of the shock front. As in the numerical results the shock is distorted into an s-shaped shock front. Since the vortex centerline is no straight line but a slightly curved path, the center of the shock bubble is located above the wind tunnel centerline and therefore above the tip of the rectangular wing. The time series evidences the strong time–dependent behavior of the flow. This temporal development becomes obvious when the axial extension of the shock bubble is observed. The size of this bubble varies between 3 mm (Fig. 7.6c) and almost 15 mm (Fig. 7.6d). Moreover, not only its size but also its position changes during the measurement. A detailed investigation shows a variation of its location of ± 5mm during one experiment.

7.5.2 *Numerical results*

Figure 7.7 shows the flow field in the vertical x–z center plane of the computational domain for a strong interaction for $Ma_\infty = 1.48$, $\Gamma = 2.69$, $\delta = 0.245$, $\mu = 0.4$ and $\beta = 11.5°$ wedge angle. The MACH number distribution shows that the shock angle is particularly inclined in the vicinity of the vortex core. The originally straight oblique shock is distorted into an 's'–shaped shock front near the vortex core. Downstream of the inclined shock a region of subsonic flow occurs. With progressing time a small cone–like shape emerges, which is of the order of the vortex diameter. Compared to the normal shock case [Althaus and Weimer (1998)] this cap size is smaller. Since breakdown of the vortex is necessarily connected with the formation of a stagnation point on or near the axis, the increased shock angle is a must for the locally subsonic flow downstream of the shock. As a second condition the pressure increases across the deformed part of the shock has to be sufficient to initiate breakdown. The streamlines inside the burst part of the vortex indicate recirculation and a stagnation point.

The parameter study for OSVI shows that the shock deformation in the vicinity of the vortex core depends primarily on the magnitude of the axial velocity deficit in conjunction with the free–stream shock strength $Ma_\infty \sin \sigma$ and the wedge angle. Secondly, the vortex circulation seems to determine the pressure on the vortex axis. For a slender isentropic vortex the radial momentum equation and the energy equation are decoupled. In this case the influence of circulation is only linear on the axial momentum while the wake is of quadratic order. Using this concept it is possible to catalogue the already explained OSVI modes: $\beta_A < \beta_{crit}(Ma_A) \wedge \Upsilon < 1$ or $\Upsilon > 1$ leads to a weak shock vortex interaction mode. The quantity Υ is the relation between forces upstream and downstream of the shock on the vortex axis. Both shock and vortex are only slightly influenced. The flow field remains supersonic everywhere. The relation $\beta_A > \beta_{crit}(Ma_A) \wedge \Upsilon > 1$ stands for a moderate interaction mode. The shock resembles a strong solution in the vortex core vicinity. A small subsonic pocket is generated downstream of the shock. Elsewhere the flow remains supersonic. The vortex is influenced without being destroyed. No stagnation point is observed. In the computational investigation of [Nedungadi and Lewis (1995)] the streamwise vortex tube was divided by the shock. A strong interaction is denoted by $\beta_A > \beta_{crit}(Ma_A) \wedge \Upsilon < 1$. The shock has a similar form to that observed for moderate interaction. Additionally, a larger subsonic region is formed and a stagnation point can be realized.

7.6 Conclusions

Experimental and numerical investigations of oblique shock–vortex interaction were conducted. The numerical solutions comprised results for the strong oblique shock–vortex interaction and the experiments also covered the weak and moderate interaction modes.

On the one hand, the results confirm previous investigations, showing a significant time–dependence of the strong OSVI flow field. On the other hand, the computations and the experimental flow visualizations evidence the occurrence of a locally normal shock in the vicinity of the vortex core, which causes vortex breakdown.

Furthermore, the numerical investigation shows a small region of reversed flow with a free stagnation point downstream from the locally s–shaped shock when breakdown occurs. A combined bubble– and spiral–like flow structure can be observed in the numerical data downstream of the discontinuity where the flow is slightly oscillating. The investigation shows that breakdown is enhanced by a reduction of the axial velocity or an increase of the azimuthal velocity component. Moreover, breakdown is more likely to occur at strong shocks and high circulation.

Although flow visualization techniques like the differential interferometry and the color schlieren method do not reveal the completely three–dimensional structure of the vortex downstream of the shock, both techniques are suitable means to analyze the shock deformation and to determine the interaction mode. For further investigations of the interaction region and especially the vortex pattern in the vicinity of the shock, quantitative measurement techniques must be applied to validate the numerical values and to have a sound data base to develop a general reliable breakdown criterion for shock–vortex interaction.

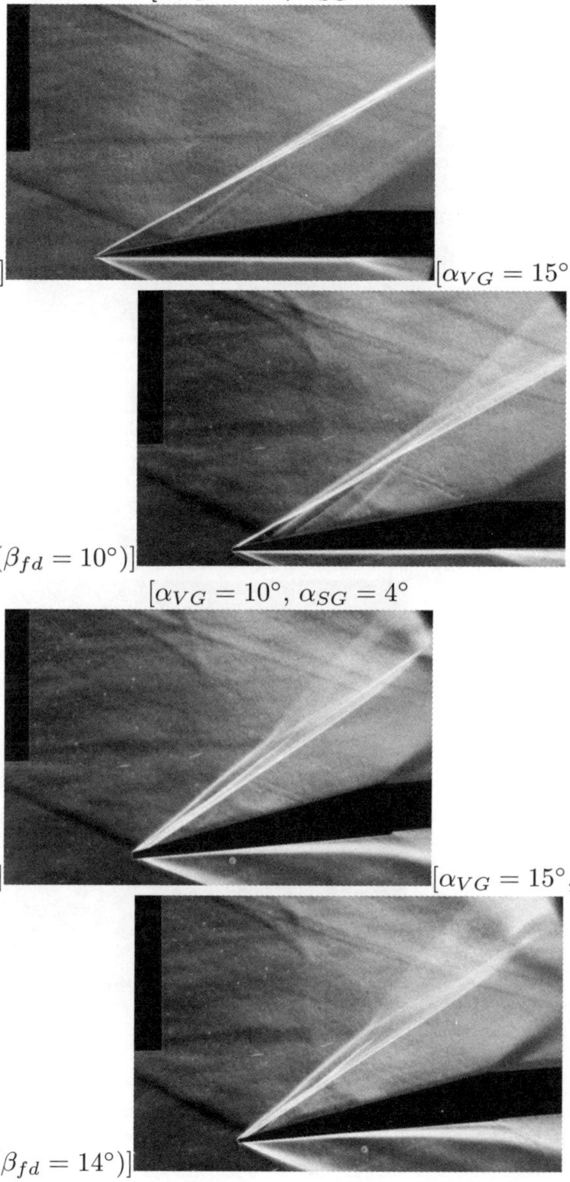

Fig. 7.5 Color schlieren images of oblique shock–vortex interaction at $Ma = 2.9$, variation of vortex strength α_{VG} and shock strength β_{fd}.

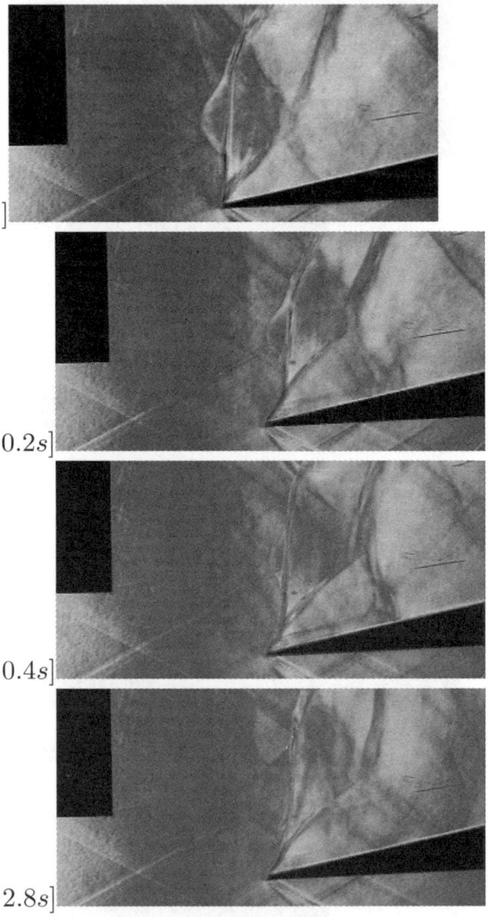

[$t = t_1$]

[$t_2 = t_1 + 0.2s$]

[$t_3 = t_1 + 0.4s$]

[$t_4 = t_1 + 2.8s$]

Fig. 7.6 Time series of oblique shock–vortex interaction at $Ma_\infty = 2.0$, $\beta_{fd} = 11.5°$, $\alpha_{VG} = 6°$. Visualization technique: differential interferometry.

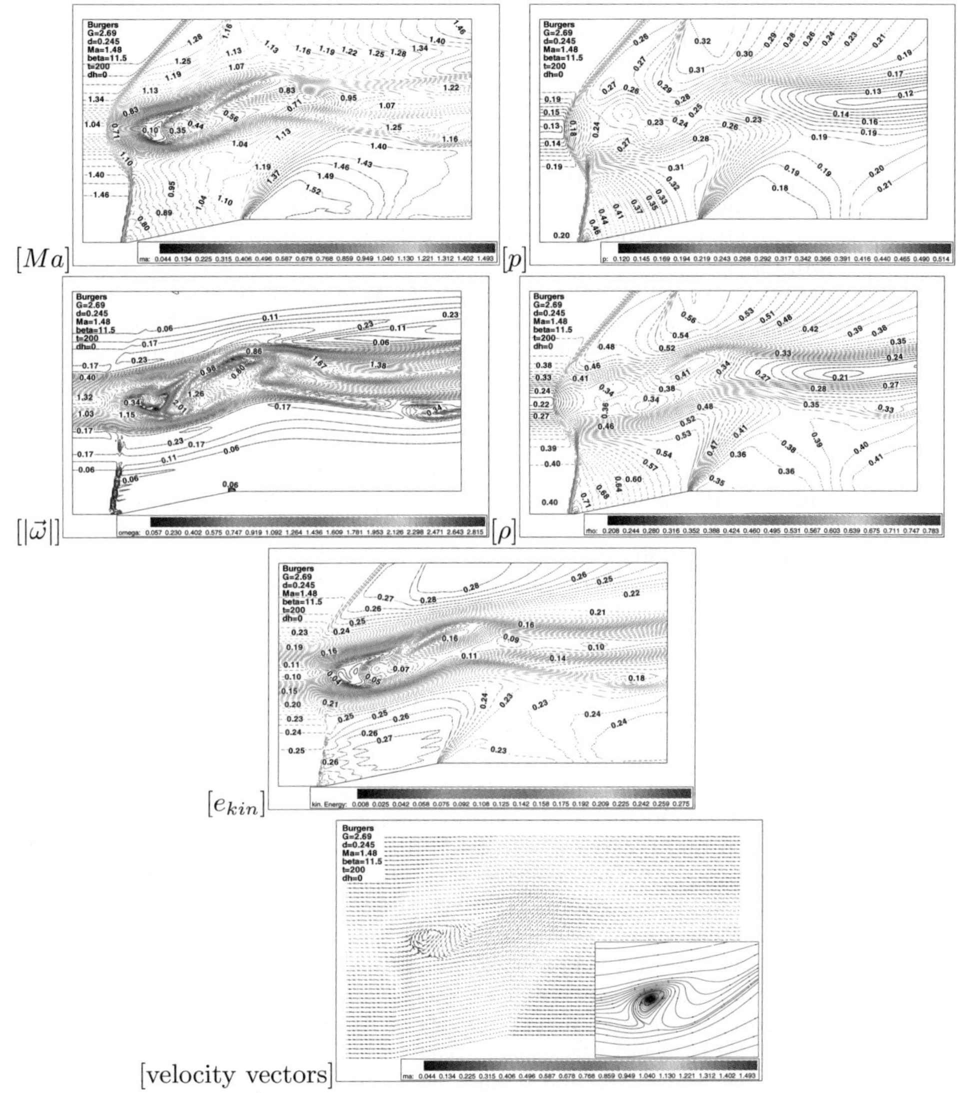

Fig. 7.7 $Ma-$, $|\vec{\omega}|-$, $e_{kin}-$, $p-$, and ρ–iso lines and velocity vectors in the y–$z$middle plane of the computational domain for OSVI ($\Gamma = 2.69$, $\delta = 0.245$, $\mu = 0.4$, $Ma = 1.48$, $\beta = 11.5°$, t=200). Flow direction is from left to right.

Bibliography

Althaus, W. and Weimer, M. (1998). Review of the Aachen Work on Vortex Breakdown, in *IUTAM Symposium on Dynamics of Slender Vortices*, (editors) E. Krause and K. Gersten, Kluwer Academic Publishers, pp. 331–344.

Cattafesta, L. N. and Settles, G. S. (1992). Experiments on Shock/Vortex Interaction, *AIAA Paper* **92-0315**.

Délery, J., Horowitz, E., Leuchter, O., and Solignac, J.-L. (1984). Fundamental studies on vortex flows, *La Recherche Aérospatiale* **2**, pp. 81–104.

Délery, J. M. (1994). Aspects of vortex breakdown, *Progress in Aerospace Sciences* **30**, pp. 1–59.

Erlebacher, G., Hussaini, M. Y., and Shu, C.-W. (1997). Interaction of a Shock with a Longitudinal Vortex, *Journal of Fluid Mechanics* **337**, pp. 129–153.

Kalkhoran, I. M. and Sforza, P. M. (1994). Airfoil Pressure Measurements During Oblique Shock-Wave/Vortex Interaction in a Mach 3 Stream, *AIAA Journal* **32** (4), pp. 783–788.

Kalkhoran, I. M. and Smart, M. K. (2000). Aspects of shock wave-induced vortex breakdown, *Progress in Aerospace Sciences* **36**, pp. 63–95.

Kalkhoran, I. M., Smart, M. K., and Wang, F. Y. (1998). Supersonic vortex breakdown during vortex/cylinder interaction, *Journal of Fluid Mechanics* **369**, pp. 351–380.

Kandil, O. A., Kandil, H. A., and Liu, C. H. (1992). Shock/Vortex Interaction and Vortex Breakdown Modes, in *IUTAM Symposium on Fluid Dynamics of High Angle of Attack*, (editors) R. Kawamura and Y. Aihara, Springer-Verlag, pp. 192–212.

Krause, E., Thomer, O., and Schröder, W. (2003). Strong Shock-Vortex Interaction – a Numerical Study, *Computational Fluid Dynamics Journal* **12(2)**(33), pp. 266–278.

Nedungadi, A. and Lewis, M. J. (1995). A Computational Study of Three-Dimensional Oblique Shock/Vortex Interactions, *AIAA Paper* **95-2770**.

Schlechtriem, S. and Lötzerich, M. (1997). Breakdown of Tip Leakage Vortices in Compressors at Flow Conditions Close to Stall, *ASME Paper* **97-GT-041**.

Smart, M. K. and Kalkhoran, I. M. (1995). Effect of Shock Strength on Oblique Shock-Wave/Vortex Interaction, *AIAA Journal* **33**(11), pp. 2137–2143.

Thomer, O., Krause, E., Schröder, W., and Meinke, M. (2000). Computational Study of Normal and Oblique Shock–Vortex Interactions, in *European Congress on Computational Methods in Applied Sciences and Engineering (ECCOMAS 2000)*, Barcelona, Spain.

Thomer, O., Schröder, W., and Krause, E. (2001). Normal and Oblique Shock–Vortex Interaction, in *International Conference on Recent Developments in Applied Mathematics and Mechanics (RDAMM–2001)*, Akademgorodok, Novosibirsk, Russia.

Zatoloka, V. V., Ivanyushkin, A. K., and Nikolayev, A. V. (1978).Interference of Vortices with Shocks in Airscoops. Dissipation of Vortexes, *Fluid mechanics: Soviet research* **7**(4), pp. 153–158.

Chapter 8

Breakdown of Slender Vortices: The State of the Art

Egon Krause
Aerodynamisches Institut
RWTH Aachen
Aachen, Germany

Dedicated to Professor Lu Ting on the occasion of his eightieth birthday.

Abstract

The flow near the axis of slender vortices is investigated with particular emphasis on the deceleration of the axial flow, leading to a stagnation point and bursting or breakdown of the vortex. The Euler equations are reduced for steady, inviscid, incompressible, axially symmetric flow in a stream tube of small radius. The angular velocity is found to be directly proportional to the axial velocity component, implying that a deceleration of the axial flow can cause the angular velocity to vanish and breakdown to occur. The extension of the analysis to viscous incompressible flow identifies a thin viscous layer immediately upstream of the stagnation point, its thickness being inversely proportional to the Reynolds number. An approximate solution for the viscous flow shows that the angular velocity does not vanish at the stagnation point of the axial flow. The results are compared with experimental and numerical data of the recent literature.

Keywords: Slender vortices; breakdown; formation of stagnation point.

8.1 Introduction

This article addresses a problem that has intensely attracted the interest of research in vortex dynamics. For a long time it was attempted to find a formulation of the conditions to be satisfied for vortex breakdown to occur. Until now a condition generally valid could not be derived, and experts estimate that another two decades of research will be necessary for a better understanding of the process. Despite this critical situation, the present paper attempts to interpret available experimental and numerical data and draw some more general conclusions from the results. Needless to say, that a final answer cannot be given; only the simplest cases will be studied. Nevertheless, it is hoped, that the results will enhance our understanding of the breakdown process, as discussed in the recent literature, which extends over a huge number of contributions, hinting at the complexity of the problem.

Some of the references pertaining to the problem to be considered may be found in [Krause, E., Gersten, K. (Eds.) (1998)]. It is well known that there are many different aspects of analyzing vortex flows, which are, for example, reflected in the topics of the main chapters reference [Krause, E., Gersten, K. (Eds.) (1998)]: The introductory chapters are devoted to methods used for analyzing vortical flows, the asymptotic methods, experimental and numerical methods. The following two chapters discuss vortices in shear layers and their mutual interaction. The next chapter is devoted to the topic of vortex breakdown, and the last two to the problem of vortex sound and of aircraft and helicopter vortices, a great variety of problems, all centered on the flow in vortices.

In the chapter on vortex breakdown the bursting process of slender vortices is discussed in several contributions: P. Billant *et al.* in [Billant, P., *et al.* (1998)] investigated instabilities and vortex breakdown in swirling jets and wakes. They also derived a breakdown criterion for incompressible flow, which will be referred to in the present study. T. Sarpkaya and F. Novak studied experimentally turbulent vortex breakdown in tubes at high Reynolds numbers [Sarpkaya, T., Novak, F. (1998)], and C. H. Brücker also experimentally investigated the breakdown process in spinning tube flow [Brücker, C. H. (1998)]. R. E. Spall and T. B. Gatski numerically simulated the breakdown process of the flow in a slightly diverging tube with a numerical solution of the Reynolds-averaged Navier-Stokes equations [Spall, R. E., Gatski, T. B. (1998)], while V. Shtern and F. Hussain considered vortex breakdown as a catastrophe in [Shtern, V., Hussain, F.

(1998)]. W. Althaus and M. Weimer in [Althaus, W., Weimer, M. (1998)] reviewed the experimental and numerical work carried out on the bursting problem at the RWTH Aachen, where O. Thomer, somewhat later, studied the interaction of oblique and normal shocks with longitudinal vortices. He derived a simplified breakdown criterion for inviscid supersonic flow [Thomer, O. (2003)].

One of the more recent experimental investigations of the bursting process was carried out by Th. Helming in 1988 at the RWTH Aachen [Helming, Th. (1988)], who used a tip vortex of a model of a half wing and enforced breakdown with the aid of a diffuser, co-axially aligned with the vortex. It was also shown in the experiment, that the burst part of the vortex could almost completely regain its initial distribution of its azimuthal velocity component. These data will be used to check a relation derived first in [Krause, E. (2000)] and later in [Krause, E. (2002)], and extended in the first part of this paper. According to this relation the angular velocity Ω near the axis of a vortex in inviscid, incompressible, axially symmetric flow is directly proportional to the axial velocity component u along the axis. In the following the considerations are extended to a Rankine vortex, and it is shown, that the breakdown criterion given in [Billant, P., *et al.* (1998)] also follows from the momentum equations reduced for $r \rightarrow 0$, by requiring Ω to vanish. Acceleration and deceleration of the axial flow far away from the axis of the vortex can also be included.

In private communications [Okulov, V. L. (2002)] V. L. Okkulov of the Institute of Thermophysics of the Russian Academy of Sciences in Akademgorodok agreed to initiate an experimental and numerical study to further substantiate the validity of the relations derived. The flow investigated was generated in a cylindrical container with a rotatable top for two Reynolds numbers. In the experiment the distribution of the axial velocity component was measured with the particle-image velocimetry in the meridional plane. In [Okulov, V. L., *et al.* (2004)] V. L. Okulov *et al.* also compared the measurements with flow computations, obtained by the authors with a numerical solution of the Navier-Stokes equations given in [Serre, E., Pulicani, J. P. (2001)] for the boundary conditions of the experiment. The computations yielded the radial distributions of the azimuthal velocity component. All data were published in [Okulov, V. L., *et al.* (2004)].

In the second part of this paper it is shown, that the relation $\Omega = Cu$, mentioned earlier completely agrees with the experimental and computed data for that part of the flow, in which the viscous forces do not play a role.

However, in the vicinity of the stagnation point of the axial flow Ω begins to deviate from the curve $Cu(x,0)$, an effect which must be attributed to the fact, that the Stokes stresses come into play, when the axial velocity component approaches zero. An approximate solution of the reduced azimuthal momentum equation with a power-series expansion will be shown to yield a non-vanishing angular velocity near the axis. According to the data of [Okulov, V. L., *et al.* (2004)] the axial velocity component remains close to zero downstream from the stagnation point, while the angular velocity differs from zero. A third solution, an approximate asymptotic solution of the differential equation for Ω is offered for this part of the flow, which qualitatively confirms the behavior of the flow in this region. The difference between the two flows investigated in [Helming, Th. (1988)] and in [Okulov, V. L., *et al.* (2004)] must be attributed to the difference in the boundary conditions. Further studies are necessary to clarify this point.

No attempts were made to explain the transition from bubble- to spiral-type breakdown. This does not seem to be advisable as long as the behavior of axially symmetric flows is not completely clarified.

8.2 Inviscid incompressible flow

In this first chapter the flow in an axially symmetric slender vortex is assumed to be steady, inviscid and incompressible so that the interplay between the azimuthal and the axial flow can be studied by employing the Euler equations. If ρ denotes the density, p the pressure, u, v, w the axial, radial, and the azimuthal velocity components, respectively, and x and r the axial and radial coordinates, the continuity equation and the momentum equations can be written as

$$\begin{aligned} u_x + (rv)_r/r &= 0 \\ uu_x + vu_r &= -p_x/\rho \\ uv_x + vv_r - w^2/r &= -p_r/\rho \\ uw_x + vw_r + vw/r &= 0 \quad . \end{aligned} \tag{8.1}$$

In Eqs. (8.1) the subscripts indicate partial differentiation with respect to x and r. The boundary conditions for the solution of Eqs. (8.1) are specified in the inflow cross-section, assumed to be given by the known functions $f_1(r) - f_4(r)$ at $x = x_i$, i.e.

$$x = x_i, \quad 0 \leq r \qquad u(r) = f_1(r),\ v(r) = f_2(r), \\ w(r) = f_3(r),\ p(r) = f_4(r), \tag{8.2}$$

along the center line by the symmetry condition

$$r = 0, x_i \leq x: \qquad u_r = v = w = p_r = 0, \tag{8.3}$$

and for large radial distances

$$r \to \infty, x_i \leq x: \quad u = u_\infty(x),\ v = w = 0, \text{ and} \\ p = f[u_\infty(x), p_{0\infty}]. \tag{8.4}$$

The boundary condition for $x \to \infty$ is usually not known, and approximations have to be introduced.

With the aid of the continuity equation and the azimuthal momentum equation it can be shown, that near the axis, $r \to 0$, the angular velocity Ω is linearly proportional to the axial velocity component u. To this end the azimuthal velocity component w is expressed in a power series in terms of the radial coordinate r, with the coefficients being a function of the axial coordinate x. If only the leading term is retained, $w(x, r \to 0)$ is approximated by the product of the angular velocity $\Omega(x)$ and the radius r, the rigid-body rotation

$$w(x, r \to 0) = \Omega(x)r + \quad \cdots \quad . \tag{8.5}$$

By taking the limit for $r \to 0$ and eliminating the term $2v_r$ in the continuity equation and in the azimuthal momentum equation of Eqs. (8.1) the following relation results for $r \to 0$:

$$\Omega_x = \Omega u_x / u, \tag{8.6}$$

which after integration in the x-direction yields

$$\Omega(x, r \to 0) = \Omega_i(x_i, r \to 0)u(x, r \to 0)/u_i(x_i, r \to 0). \tag{8.7}$$

As before the index i in Eq. (8.7) denotes the inflow cross section.

Equations (8.6) and (8.7) were first derived in [Krause, E. (2000)] and [Krause, E. (2002)]. These relations show that if breakdown or bursting of the vortex is defined by requiring that $\Omega(x, r \to 0)$ to vanish, then the axial velocity component $u(x, r \to 0)$ must also vanish, and it can be concluded, that breakdown is associated with the formation of a stagnation point on the axis of the vortex. This result has long been questioned. It is further supported by considering the stream tube for $r \to 0$. The slope of the projection of the streamlines on the meridional plane is given by the differential equation

$$dr/dx = v/u. \tag{8.8}$$

If Eq. (8.5) is inserted into the azimuthal momentum equation in Eqs. (8.1) and solved for v/u, Eq. (8.8) can be integrated to yield

$$\begin{aligned} r(x) &= r_i(x_i)[\Omega_i(x_i, r \to 0)/\Omega(x, r \to 0)]^{1/2} \\ &= r_i(x_i)[u_i(x_i, r \to 0)/u(x, r \to 0)]^{1/2}. \end{aligned} \tag{8.9}$$

The last equation shows that if the angular velocity Ω or the axial velocity component u is decreased, the radius of the stream tube increases, a flow behavior observed upstream of a stagnation point.

An experiment carried out earlier in [Helming, Th. (1988)] confirms the validity of Eq. (8.9): In the experiment a slender vortex was generated with the tip of a half model of a wing in a water tunnel. The axial flow in the vortex was decelerated by positioning a transparent diffuser with its axis coaxially aligned with the axis of the vortex. The flow was visualized with the hydrogen-bubble technique. The bubbles were injected in the inflow cross-section in a meridional plane, and their paths were photographed from a right angle to the plane. The flow patterns are depicted in the photograph in the upper part of Fig. 8.1.

The flow is from left to right, and the core of the vortex is identified by the white horizontal line, generated by agglomerated hydrogen bubbles. The increasing cross-sectional area of the diffuser - the walls of which can be recognized in the left part of the picture - generates a positive axial pressure gradient, decelerating the flow. The stream tube near the axis opens up, visualized by the spreading-out hydrogen bubbles, and bursting occurs, after a stagnation point is formed on the axis. The traces of the hydrogen bubbles show that the swirl of the flow near the axis is reduced as predicted by Eq. (8.9).

A transparent confuser, a tube with a decreasing cross-section was mounted downstream from the diffuser, again coaxially aligned with the axis of the vortex, as can be seen on the right of the photograph in Fig. 8.1. The confuser accelerates the axial flow downstream from the burst part of the vortex, as predicted by Eq. (8.9), so that it can regain swirl.

Measurements of the axial and azimuthal velocity components with Laser-Doppler-Anemometry, reported in [Helming, Th. (1988)] confirm the flow visualization studies just described. The results are depicted in the six diagrams shown in Fig. 8.1. The first three diagrams show the radial profiles of the axial velocity component u (in the notation of [Helming, Th. (1988)] designated as V_x) for the entrance (cross-section A), the exit of the diffuser (cross-section B), and the exit of the confuser, (cross-section C). The positive axial pressure gradient in the diffuser decelerates the axial flow and a wake-like profile is formed, with a small back-flow near the axis at the exit of the diffuser (diagram in the middle, cross-section B). After the acceleration in the confuser the axial velocity component has almost recovered its original profile at the exit of the confuser (diagram on the right, cross-section C).

The radial profiles of the azimuthal velocity component w measured in [Helming, Th. (1988)] (in the notation of [Helming, Th. (1988)] designated by V_t) are shown in the lower part of Fig. 8.1. It is seen, that the assumption of rigid-body rotation near the axis is confirmed by the diagrams on the left and on the right. The diagram in the middle shows, that the angular velocity Ω, given by the slope of w with respect to r, vanishes, as predicted by Eq. (8.7), when u vanishes. The diagram on the right also shows that the swirl is not completely recovered: The angular velocity in the cross-section C is not as large as in the cross-section A.

So far it was shown that the angular velocity near the axis is directly proportional to the axial velocity component u. But it is clear from the formulation of the problem, that near the axis $u(x, r \to 0)$ is also influenced by the lateral boundary conditions given by Eq. (8.4). In order to show this dependence, a Rankine vortex is assumed with a radial distribution of the azimuthal velocity component given by rigid-body rotation, Eq. (8.5), and a potential vortex for the flow outside of the core, given by

$$w(x, r) = \Gamma/(2\pi r). \tag{8.10}$$

The quantity Γ is the circulation of the vortex. It can be expressed by the angular velocity Ω, if Eq. (8.5) is matched with Eq. (8.10) at the core radius $r = R = D/2$, being defined by the location of the maximum of w

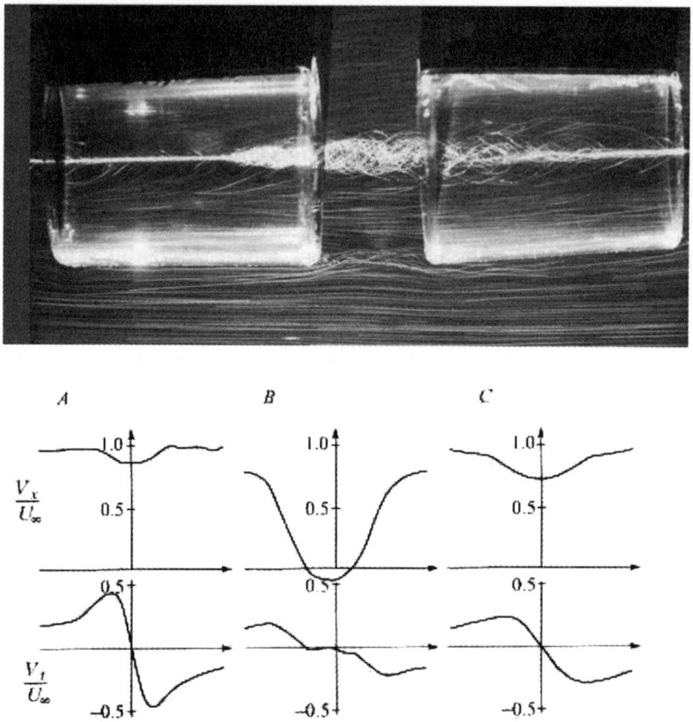

Fig. 8.1 Experimental verification, described in [Helming, Th. (1988)], of Eq. (8.7), $\Omega(x) = (\Omega_i/u_i)u(x)$: Deceleration and subsequent acceleration of the axial and azimuthal velocity components, $u(x)$ and $w(x)$ along the axis, with $p_x > 0$ and $p_x < 0$, respectively, enforced by a divergent and convergent flow cross-section. Visualization and velocity measurements are described in [Helming, Th. (1988)].

$$w_{max} = \Omega R = 2\Gamma/(\pi D^2). \tag{8.11}$$

Since only slender vortices are considered, i.e. $r/u \ll 1$, the radial momentum equation in Eqs. (8.1) can be simplified to $p_r = w^2/r$ and integrated for the azimuthal velocity profile of the Rankine vortex, given by Eqs. (8.5) and (8.10). Then the pressure on the axis can be expressed by the pressure for $r \to \infty$ and the maximum value of the azimuthal velocity component w_{max}:

$$p(x, r \to 0) = p(x, r \to \infty) - \rho\Omega^2 R^2$$
$$= p(x, r \to \infty) - \rho w_{max}^2, \qquad (8.12)$$

a relation already used by Prandtl in [Prandtl, L., Tietjens, O. G. (1957)].

Equation (8.12) can now be used to study the influence of the distribution $p(x, r \to \infty)$ on the angular velocity near the axis. To show this dependence, Eq. (8.6) is written in terms of the axial pressure gradient p_x by reducing the axial momentum equation for $r \to 0$ and inserting it into Eq. (8.6):

$$\Omega_x = -\Omega p_x/(\rho u^2). \qquad (8.13)$$

Equation (8.13) can be integrated, if u^2 is eliminated with the aid of Eq. (8.7). There results

$$\Omega^2 = \Omega_i^2 + 2[(p_i - p)\Omega_i^2]/(\rho u_i^2). \qquad (8.14)$$

The pressure difference $p_i - p$ on the axis of the vortex can now be expressed by relations corresponding to Eq. (8.12). With a slight change in notation they are

$$p_i = p_{i\infty} - \rho\Omega_i^2 R^2 = p_{i\infty} - \rho w_{imax}^2 \quad \text{and}$$
$$p = p_\infty - \rho\Omega^2 R^2 = p_\infty - \rho w_{max}^2. \qquad (8.15)$$

In Eq. (8.15) $p_{i\infty}$ and p_∞ are the static pressures far away from the axis of the vortex, $p_{i\infty}$ at the initial x-station, and p_∞ at the x-station considered further downstream; the maximum values of the azimuthal velocity are denoted as w_{imax} and w_{max}. Since the axial velocity component far away from the axis varies with x, $p_{i\infty}$ and p_∞ can be written in terms of the stagnation pressure $p_{0\infty}$ and the axial velocity components $u_{i\infty}$ and u_{infty}, according to Bernoullis equation:

$$p_{i\infty} = p_{0\infty} - \rho u_{i\infty}^2/2 \quad \text{and} \quad p_\infty = p_{0\infty} - \rho u_\infty^2/2. \qquad (8.16)$$

With the aid of Eq. (8.15) and Eq. (8.16), Eq. (8.14) can be cast into the following form

$$\Omega/\Omega_i = [1 + (u_\infty^2 - u_{i\infty}^2)/(u_i^2 - 2R^2\Omega_i^2)]^{1/2}. \tag{8.17}$$

If the angular velocity Ω is to vanish, i. e. if breakdown is to occur, the square-bracketed term on the right-hand side of Eq. (8.17) has to vanish. This condition is satisfied for

$$R\Omega_i/u_i = \{(1/2)[1 + (u_\infty^2 - u_{i\infty}^2)/u_i^2]\}^{1/2}. \tag{8.18}$$

The critical value of the ratio $R\Omega_i/u_i$, for which breakdown for variable axial flow occurs, is given in the following table.

Table 8.1 Critical values of $(R\Omega_i/u_i)$ for which the angular velocity Ω near the axis of the vortex vanishes and breakdown occurs as a function of the ratio $(u_\infty^2 - u_{i\infty}^2)/u_i^2$.

$(u_\infty^2 - u_{i\infty}^2)/u_i^2$	-1.0	-0.75	-0.5	-0.25	0.0	0.5	1.0
$(R\Omega_i/u_i)_{crit.}$	0.0	0.354	0.5	0.61	0.71	0.87	1.0

Equation (8.18) without the term $(u_\infty^2 - u_{i\infty}^2)/u_i^2$ was first stated in the form $R\Omega_i/u_i = 1/2^{1/2}$ as breakdown criterion given in [Billant, P., *et al.* (1998)]. The criterion was derived by postulating that the pressure difference along the axis be equal to the radial pressure difference. The above considerations show that the criterion can directly be obtained from simplified momentum considerations by requiring that the angular velocity has to vanish, including variable axial flow far away from the axis of the vortex.

It was also stated in [Billant, P., *et al.* (1998)], that for uniform axial flow for $r \to \infty$, Eq. (8.18), in the form $R\Omega_i/u_i = 1/2^{1/2}$, agrees well with available experimental data. Figure 8.2 taken from [Billant, P., *et al.* (1998)], depicted below, shows the reciprocal of the critical value $S = (R\Omega_i/u_i)^{-1/2}$ as a function of the Reynolds number Re of the experiments described in [Billant, P., *et al.* (1998)].

The abbreviation S is the reciprocal of the velocity ratio w_{max}/u_i in the nomenclature of [Billant, P., *et al.* (1998)]. The diagram indicates, that the results are almost independent of the Reynolds number Re.

Since the results discussed so far were obtained for inviscid flow, it is interesting to study the influence of the Stokes stresses on the variation of the angular velocity near the axis of the vortex. This is the subject of the following chapter.

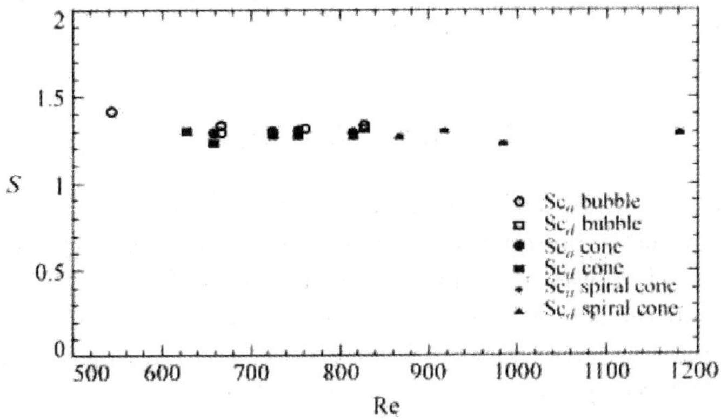

Fig. 8.2 Comparison of the predicted value of S for vortex breakdown given in [Billant, P., (1998)] with experimental data of the same reference.

8.3 Viscous incompressible flow

In order to study the influence of the Stokes stresses on the azimuthal and axial flow near the axis, the azimuthal momentum equation of the Navier-Stokes equations is reduced for $r \to 0$. If the temperature of the flow is assumed to be constant, the momentum equation for axially symmetric flow

$$\rho(uw_x + vw_r + vw/r) = \mu r^{-2}\{r^2[r(w/r)_r]\}_r + \mu w_{xx} \tag{8.19}$$

with $w = \Omega r$ can be cast into the form

$$u\Omega_x + 2v_r\Omega = v\Omega_{xx}. \tag{8.20}$$

Using the continuity equation in Eqs. (8.1), the last equation for $r \to 0$ takes on the form

$$u\Omega_x - u_x\Omega = v\Omega_{xx}. \tag{8.21}$$

In a first approach to a solution Eq. (8.7) is inserted in Eq. (8.21). It is noted, that Eq. (8.7) is also a solution of Eq. (8.21) if the second derivative of the axial velocity component u with respect to x vanishes.

If this special case is excluded, it is seen that when Eq. (8.21) is multiplied by R and written in the form

$$(R\Omega/u)_x = (\upsilon R/u^2)\Omega_{xx}, \tag{8.22}$$

that the left-hand side of Eq. (8.22) represents the slope of the ratio $(R\Omega/u)$, which was found to be equal to $2^{-1/2}$, if Ω is to vanish for constant uniform axial flow u_∞ far away from the axis of the vortex. Equation (8.22) shows, however, that the derivative of $(R\Omega/u)$ with respect to x does not vanish as long as the second derivative of Ω with respect to x is different from zero. The viscous forces per unit volume therefore have to be taken into account when the axial velocity u tends to zero and Ω remains different from zero. An estimate for the thickness δ of the layer, in which the viscous forces cannot be neglected is obtained by writing Eq. (8.22) in dimensionless form so that all terms are of order unity. Then the thickness δ, nondimesionalized with the core radius R is

$$\delta/R = O(1/Re), \qquad \text{with} \qquad Re = \Omega_r R^2/\upsilon. \tag{8.23}$$

The Reynolds number Re is defined by a reference angular velocity Ω_r, the core radius R and the viscosity ν. In laboratory experiments Re is of the order of a few thousand, and hence the thickness of the viscous layer just upstream of the stagnation point is very small in comparison to the core radius.

In order to proceed with the solution of Eq. (8.22) an assumption must be introduced for the axial velocity component u. If the outer edge of the viscous layer is assumed to be at $x = x_\delta$, where the axial velocity component $u = u_\delta$ and its derivative $u_x = u_{x\delta}$ are assumed to be known, then because of the small thickness of the viscous layer the axial velocity can be assumed to vary linearly with x.

$$u = u_\delta + u_{x\delta}(x - x_\delta). \tag{8.24}$$

The x-coordinate of the stagnation point is then given by

$$x_S = x_\delta - u_\delta/u_{x\delta}. \tag{8.25}$$

With Eqs. (8.24) and (8.25) the change of the angular velocity Ω in the viscous layer can be determined. If Ω is expanded in a Taylor series in terms of $x - x_\delta$, i.e.

$$\Omega = a_0 + a_1(x - x_\delta) + a_2(x - x_\delta)^2 + a_3(x - x_\delta)^3 + \cdots, \tag{8.26}$$

and is together with Eq. (8.24) inserted into Eq. (8.21), the following expression is obtained:

$$\begin{aligned}\Omega = \Omega_\delta + \Omega_{x\delta}(x - x_\delta) + (u_\delta\Omega_{x\delta} - u_{x\delta}\Omega_\delta)(x - x_\delta)2/(2\upsilon)\\ + u_\delta(u_\delta\Omega_{x\delta} - u_{x\delta}\Omega_\delta)(x - x_\delta)^3/(6\upsilon^2) + \cdots.\end{aligned} \tag{8.27}$$

The value of the angular velocity Ω at the stagnation point x_S is determined by setting $x = x_S$ in the above series expression and inserting Eq. (8.25) into Eq. (8.27)

$$\Omega_S = (u_\delta\Omega_{x\delta} - u_{x\delta}\Omega_\delta)(1/u_{x\delta})[1 - u_\delta^2/(2\upsilon u_{x\delta}) + u_\delta^4/(6\upsilon^2 u_{x\delta}^2) + \cdots]. \tag{8.28}$$

Since $u_{x\delta} < 0$, the sum of the square-bracketed terms on the right-hand sight of Eq. (8.28), representing the leading terms of the truncated series is larger than zero. It is noted that with $u_{x\delta} < 0$ the expression $-u_\delta^2/(\nu u_{x\delta})$ is equal to a Reynolds number based on the velocity u_δ at the edge of the viscous layer and its thickness

$$-u_\delta^2/(\upsilon u_{x\delta}) = u_\delta(x_S - x_\delta)/\upsilon = Re_S, \tag{8.29}$$

which again can be rewritten in terms of the definition given in Eq. (8.23). In Eq. (8.28) the term $u_\delta\Omega_{x\delta} - u_{x\delta}\Omega_\delta$ is equal to $-\nu\Omega_{xx\delta}$. Then, in general, when Ω is decreasing, $\Omega_{xx\delta}$ is larger than zero, and Ω_S is also larger than zero at the stagnation point. Thus in contrast to inviscid flow the angular velocity Ω does not vanish, when the axial velocity does.

In order to check this result V. L. Okulov of the Institute of Thermophysics of the Russian Academy of Sciences in Novosibirsk agreed to study the incompressible viscous flow in a cylindrical container with a rotating top in a specially designed experiment and in a comparison computation with a numerical solution of the Navier-Stokes equations in the stream function-vorticity formulation. In the study particular emphasis was placed on obtaining results for the flow near the axis of the container. A brief description of the investigation and some of the results were provided by V. L. Okulov in a private communication [Okulov, V. L. (2002)]; a detailed description can be found in [Okulov, V. L., *et al.* (2004)].

The container consisted of a circular cylinder with inner diameter of 288 mm, the top and bottom of which could be rotated, the height could be varied between 0 and 600 mm. The results presented here were obtained for the case that only the top was rotated, and the cylinder height was fixed at an aspect ratio $H/R = 2$. Figure 8.3 shows the experimental set-up and the PIV recording system described in [Okulov, V. L., *et al.* (2004)].

The fluid used in the experiments was a mixture of glycerin and water with different combinations of the constituents for the various tests. As the viscosity of the fluid was very sensitive to temperature variations the viscosity-temperature dependence was carefully controlled during the experiments so that the overall error in the Reynolds number $Re = \Omega R^2/nu$ did not exceed ± 10 percent. In the experiments only the top was rotated, and the frequency of rotation could be steadied within an error of less than 0.1 percent, according to [Okulov, V. L., *et al.* (2004)].

The cylindrical wall of the container was fabricated of transparent plexiglas with optical quality. The container was immersed in a water bed to minimize optical aberrations during the measurements. For the flow visualization Rhodamin B particles with a mean diameter of approximately 30 microns were seeded into the test fluid in the container. The particles have an almost neutral buoyancy and their size is sufficiently small to match the photographic requirements. The illumination was provided with a Nd: YAG laser. The flow patterns were recorded with a CCD video camera placed at a right angle to the measuring plane. The videotaped recordings of the flow motion were analyzed frame by frame. Standard Dantec equipment was used for the flow diagnosis with particle-image velocimetry.

The method of E. Serre and J. P. Pulican [Serre, E., Pulicani, J. P. (2001)] was used for the numerical solution of the Navier-Stokes equations. It was deviced to reach steady-state flow conditions for large times, and the computations were carried out for axially symmetric boundary conditions obtained from the PIV-measurements. In Fig. 8.4 the numerical simulations of the axial flow are compared with the flow visualizations obtained from the PIV-measurements. Shown are stream-tube profiles, with the experimental visualizations on the left. The upper pictures correspond to a Reynolds number $Re = 1560$, and the lower to $Re = 2043$. The axial flow is directed upward.

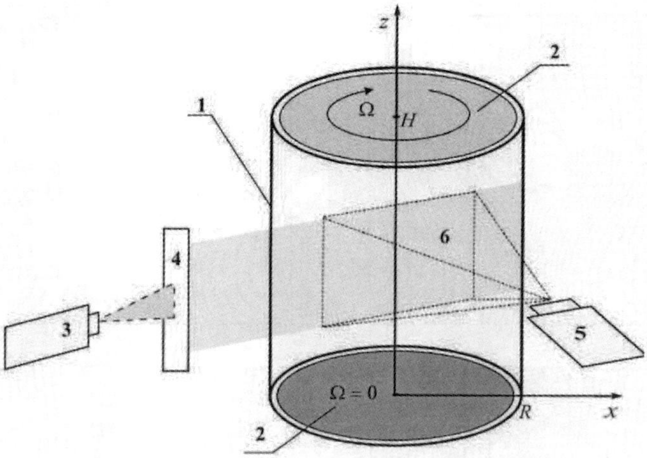

Fig. 8.3 Experimental set-up used by V. L. Okulov *et al.* in [Okulov, V. L. (2004)]: 1 cylindrical container; 2 rotatable top and bottom disks; 3 Nd: YAG laser; 4 optical prisms; 5 CCD video camera; 6 part of the meridional plane, in which measurements were obtained, bounded by the window frame coordinates $-0.25 \cdot H \leq x \leq 0.25 \cdot H$ and $0.1 \cdot H \leq z \leq 0.6 \cdot H$.

In the flow with the Reynolds number $Re = 1560$ on the top a stagnation point is formed in the middle of the window and a closed recirculation region, shaped like a bubble, is generated just downstream from the stagnation point. The agreement between experiment and numerical solution with regard to location of the stagnation point, axial and radial extent of the bubble and separated region on the container wall is entirely satisfactory. Moreover, the two data sets confirm that the flow in the experiment was as close as possible to axial symmetry, and also that steady flow conditions prevailed in the experiment. Obviously small disturbances, noted during the experiment, were not amplified to cause instabilities.

Similar remarks can be made for the flow with Reynolds number $Re = 2043$, shown on the bottom of Fig. 8.4. The stagnation point is moved upstream, now being closer to the bottom of the container. Also the recirculation bubble is larger in diameter and in axial extent than in the flow with $Re = 1560$. A second smaller recirculation region is noticed in the experimental and also in the numerical data downstream from the bubble.

In Fig. 8.5 the radial profiles of the computed azimuthal (left) and axial

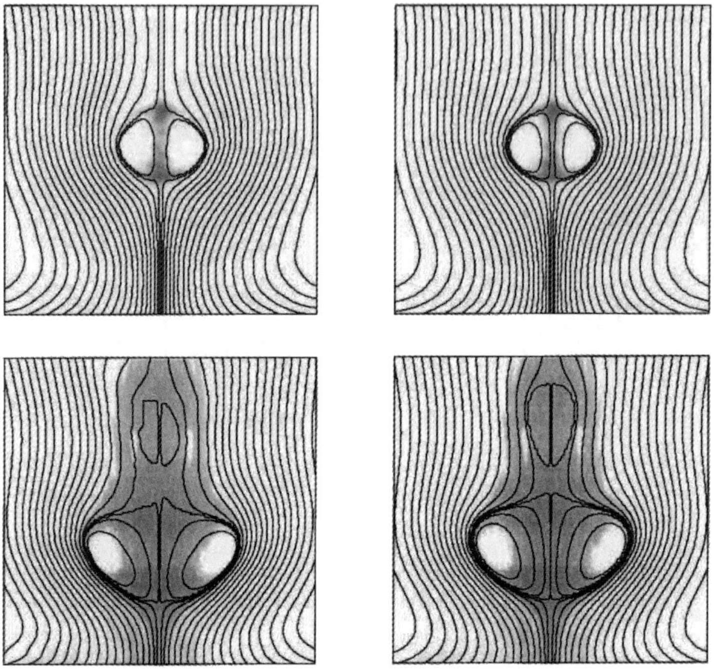

Fig. 8.4 Flow in a cylindrical container with rotating top, directed upward. Comparison of flow visualizations obtained by PIV-measurements of the axial flow (on the left) with numerical simulations (right) by [Okulov, V. L. (2004)]. The computations for steady flow conditions confirm steady axially symmetric flow in the experiment. The Reynolds numbers are $Re = 1560$ (top) and $Re = 2043$ (bottom).

velocity components (right) are shown for ten axial locations in the test window. Based on the agreement in the previous comparison with the experimental data, it can be concluded that the actual flow in the experiment exhibits rigid-body rotation near the axis as does the numerically simulated flow, and the assumption of rigid-body rotation near the axis introduced earlier is confirmed.

Figure 8.5 also shows that the slope of the radial profiles of the azimuthal velocity component and hence the angular velocity near the axis changes in the axial direction, as the axial velocity component does, implied in Eq. (8.7) ; but Ω, given by the slope of the profiles in the radial direction does not approach zero when the axial velocity component vanishes. All profiles show a value larger than zero. The influence of the Stokes stresses

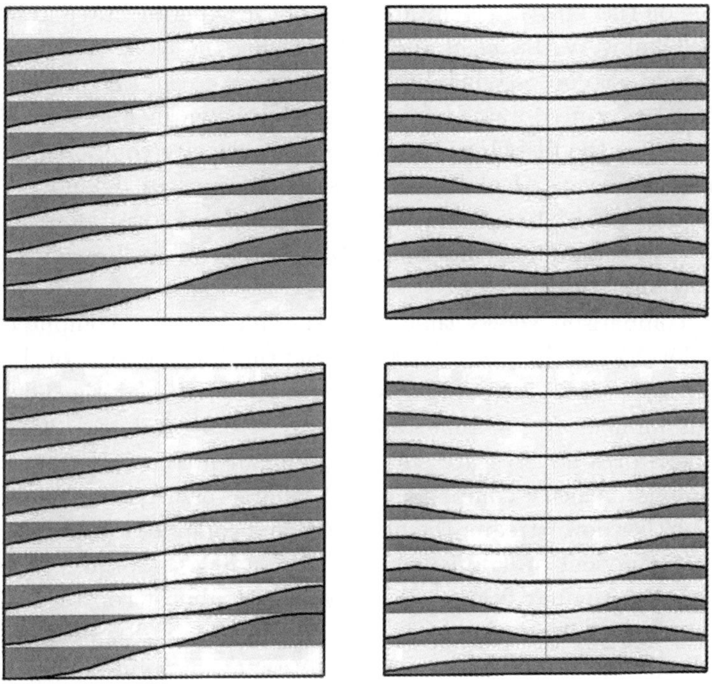

Fig. 8.5 Flow in cylindrical container with rotating top [Okulov, V. L. (2004). Radial profiles of the azimuthal (left) and the axial velocity component (right) computed for different axial locations. The azimuthal velocity component attains its smallest value when the axial velocity component approaches zero or becomes negative, for Reynolds numbers Re = 1560 (top) and Re = 2043 (bottom).

as indicated by the result obtained with Eq. (8.28) seems to be confirmed by the experimental and numerical data. Note that aside from the first two profiles near the bottom of the window the axial velocity component near the axis is close to zero in all other profiles. Again, the profiles shown in the upper part of Fig. 8.5 were obtained for the Reynolds number $Re = 1560$ and those in the lower for $Re = 2043$.

Since Eq. (8.7) describes the dependence of the angular velocity Ω on the axial velocity component u along the axis of the vortex for inviscid flow, it would be most desirable to extract a corresponding data set from the numerical and experimental data given in [Okulov, V. L., *et al.* (2004)]. The results obtained in Eq. (8.28) for viscous flow stated that the angular velocity would not approach zero when a stagnation point of the axial flow is formed on the axis, and a suitable plot of the experimental and numerical data of [Okulov, V. L., *et al.* (2004)] showing the axial dependence of Ω and u would also be very helpful in this case.

Such plots were obtained from the data shown in Fig. 8.5. The next Fig. 8.6 shows the distributions of the axial velocity component u and of the angular velocity Ω along the axis as obtained in [Okulov, V. L., *et al.* (2004)]. The data are plotted in a normalized form $u(x)/u_i(x_i)$ and $\Omega(x)/\Omega_i(x_i)$, again for the flows investigated for the two Reynolds numbers $Re = 1560$ and $Re = 2043$.

The comparison shows that the experimental and computed results of [Okulov, V. L., *et al.* (2004)] confirm the validity of Eq. (8.7), $\Omega(x)/\Omega_i(x_i) = u(x)/u_i(x_i)$, for values of $x \leq 0.5$ for the flow with a Reynolds number $Re = 1560$, and for $x \leq 0.3$ for $Re = 2043$. Obviously the Stokes stresses are sufficiently small in the regions mentioned, and the axial velocity component $u(x)$ is large enough, so that Eq. (8.7) holds. Further downstream, the measured and computed values of the axial velocity component of [Okulov, V. L., *et al.* (2004)] are close to zero, but the computed angular velocity is shifted by an almost constant value and does not vanish. This result was obtained in Eq. (8.28), stating that the angular velocity Ω does not vanish at the stagnaion point of the axial flow.

Downstream from $x = 0.5$ for $Re = 1560$ and $x = 0.3$ for $Re = 2043$, where the axial velocity component is close to zero, the curves $\Omega(x)/\Omega_i(x_i)$ differ for increasing values of the axial coordinate x by an almost constant value from the curves $u(x)/u_i(x_i)$, up to a dimensionless coordinate $x \approx 1.2$. The angular velocity Ω does not approach zero, in contrast to the results obtained in [Helming, Th. (1988)]. Obviously both cases $\Omega = 0$ and $\Omega \neq 0$ occur. The different flow behavior observed in [Helming, Th. (1988)] and in [Okulov, V. L., *et al.* (2004)] must be attributed to the difference in the boundary conditions of the two flows.

A simple approximation for the region in which $u \approx 0$ can be constructed with the ansatz

$$u(x) = R_0\Omega(x) - u_0, \tag{8.30}$$

which, with R_0 and u_0 assumed to be constants, reduces Eq. (8.21) to

$$u_0\Omega_x = -\upsilon\Omega_{xx}. \tag{8.31}$$

Integration of Eq. (8.31) yields

$$\Omega(x) = \Omega_i(x_i) + [\Omega_{xi}(x_i)\upsilon/u_0][1 - e^{-u_0(x-x_i)/\upsilon}] \quad . \tag{8.32}$$

For large values of x the angular velocity attains the value

$$\Omega(x \to \infty) = \Omega_i(x_i) + \Omega_{xi}(x_i)\upsilon/u_0. \tag{8.33}$$

This result is qualitatively in agreement with the numerical and experimental data shown in Fig. 8.6 of [Okulov, V. L., *et al.* (2004)]. While the axial flow may generate a stagnation point on the axis of the vortex, the angular velocity in the vicinity of the axis does not vanish.

8.4 Concluding remarks

In the present study the flow near the axis of slender vortices was investigated with simplified relations of the governing equations. By reducing the Euler equations for steady incompressible inviscid flow for the vicinity of the axis of the vortex, it could be shown that the angular velocity is directly proportional to the axial velocity component. This result was compared with the experiment carried out by Th. Helming in 1988. The measurements clearly confirmed the relation derived. The analysis also substantiated the breakdown criterion derived by P. Bilant *et al.*, and extended it to variable axial flow far away from the axis of the vortex.

The influence of the Stokes stresses was investigated for the immediate vicinity of the stagnation point. A thin viscous layer was detected immediately upstream of the stagnation point. With the assumption of a linear variation of the axial velocity component in the viscous layer, the reduced azimuthal momentum equation for viscous flow could be solved with a power-series expansion. The solution showed that the angular velocity does not vanish at the stagnation point of the axial flow. This result deviates from the findings of Th. Helming and P. Bilant *et al.*, but is in agreement with the experimental and numerical data of V. L. Okulov *et al.*

In the experiment of V. L. Okulov *et al.* the axial velocity component was close to zero downstream from the stagnation point of the axial flow, while the angular velocity component remained nearly constant at a value different from zero. An asymptotic solution of the reduced azimuthal momentum equation could also be found for this region, again yielding qualitative agreement with the experimental and numerical data mentioned.

Which of the two flow situations occurs, either the angular velocity near the axis vanishes when the axial velocity component vanishes, or the angular velocity remains different from zero while the axial velocity component vanishes, seems to depend on the boundary conditions of the flow investigated. Additional experiments and flow computations are necessary for a better understanding of the problem.

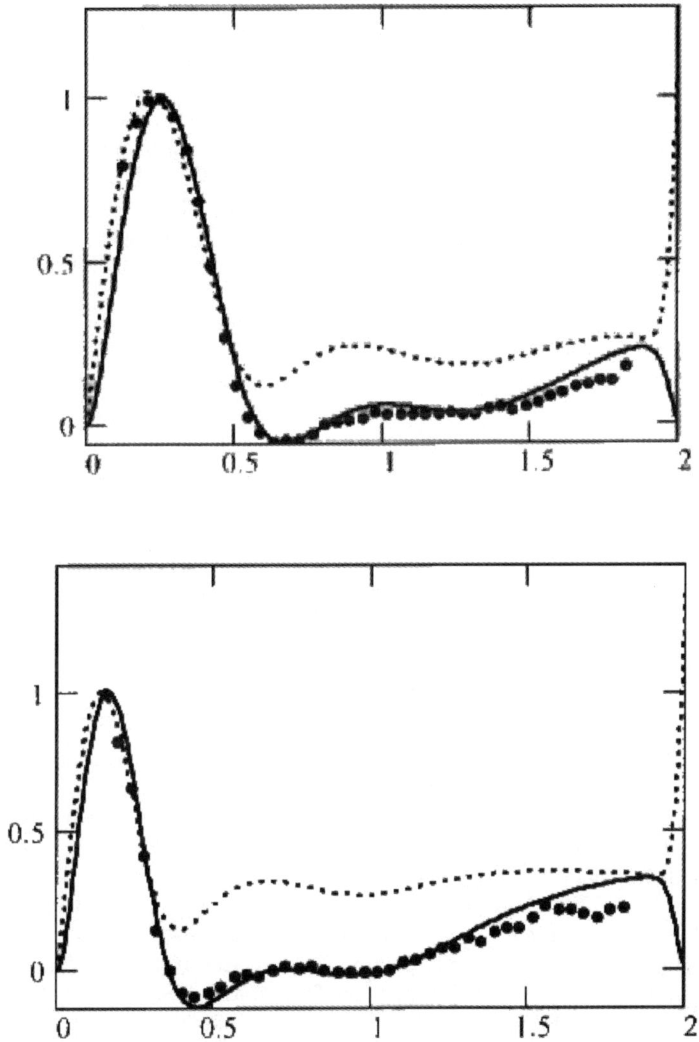

Fig. 8.6 Comparisons of measured and computed distributions of the normalized axial velocity component $u(x)/u_i(x_i)$ and of the normalized angular velocity $\Omega(x)/\Omega_i(x_i)$ along the axis of the container for the Reynolds numbers of the experiment [Okulov, V. L. (2004)], $Re = 1560$ (top) and $Re = 2043$ (bottom): The solid lines ____ represent the numerical simulations of $u(x)/u_i(x_i)$, and the solid circles ... represent the values of the PIV-measurements for the axial velocity. The dotted lines $- - -$ represent the computed distributions of the angular velocity according to [Okulov, V. L. (2004)]

Bibliography

Althaus, W., Weimer, M. (1998). Review of the Aachen Work on Vortex Breakdown. In: Krause, E., Gersten, K., eds. IUTAM Sympos. on Dynamics of Slender Vortices. Dordrecht: Kluwer, pp. 331-344.

Billant, P., Chomaz, J. M., Delbende, I.,Huerre,P., Loiseleux, T., Olendraru, C., Rossi, M., Sellier, A. (1998). Instabilities and Vortex Breakdown in Swirling Jets and Wakes. In: Krause, E., Gersten, K., eds. IUTAM Sympos. on Dynamics of Slender Vortices. Dordrecht: Kluwer, pp. 267-286.

Brücker, C. H. (1998). Breakdown of Spinning Tube Flows. In: Krause, E., Gersten, K., eds. IUTAM Sympos. on Dynamics of Slender Vortices. Dordrecht: Kluwer, pp. 321-330.

Helming, Th. (1988). Untersuchungen zum Aufplatzen und zur Instabilität von Randwirbeln, Diss. RWTH Aachen.

Krause, E., Gersten, K. (Eds.) eds. (1998). Dynamics of Slender Vortices, Proc. IUTAM Sympos, Aachen, Germany, 31 Aug.-3 Sept. 1997, Kluwer.

Krause, E. (2000). Shock Induced Vortex Breakdown. Proc. Int. Conf. on Methods of Astrophysical Research, 9-16 July, 2000, Novosibirsk, Publ. House of Russian Acad. of Sciences-Siberia, Pt. II, pp. 109-114.

Krause, E. (2002). Axial Flow in Slender Vortices, J. Engineering Thermophysics, Vol. 11. No. 3, pp. 229-242; review in: Bathe K. J., ed. (2003). Computational Fluid and Solid Mechanics. Oxford: Elsevier, pp. 984-987.

Okulov, V. L. (2002). Private communication.

Okulov, V. L., Sorensen, J. N., Varlamova, E., A. (2004). Emergence of Asymmetry and Unsteadiness in Laboratory Simulation of the Hydrodynamic Structure of a Tornado, Izv., Atmospheric and Oceanic Phys., Vol. 40, No. 2, 169-182.

Prandtl, L., Tietjens, O. (1957). Fundamentals of Hydro- and Aeromechanics. Dover.

Sarpkaya, T., Novak, F. (1998). Experiments in Tubes at high Reynolds numbers. In: Krause, E., Gersten, K., eds. IUTAM Sympos. on Dynamics of Slender Vortices. Dordrecht: Kluwer, pp. 287-296.

Serre, E., Pulicani, J. P. (2001). A three-dimensional pseudospectral method for rotating flows in a cylinder. Comput. Fluids 30(4), 491-519.

Shtern, V., Hussain, F. (1998). Vortex Breakdown as a Catastrophe. In: Krause, E., Gersten, K., eds. IUTAM Sympos. on Dynamics of Slender Vortices. Dordrecht: Kluwer, pp.297-306.

Spall, R. E., Gatski, T. B. (1998). Turbulent Vortex Breakdown: A Numerical Study. In: Krause, E., Gersten, K., eds. IUTAM Sympos. on Dynamics of Slender Vortices. Dordrecht: Kluwer, pp. 307-320.

Thomer, O. (2003). Numerische Untersuchung der Wechselwirkung von Längswirbeln mit senkrechten und schrägen Verdichtungsstößen, Diss. RWTH Aachen.

Chapter 9

A Numerical Analysis of Vortex Dislocation in Wake-type Flow with Different Spanwise Nonuniformity

Guocan Ling [1]
The State Key Laboratory of Nonlinear Mechanics, Institute of Mechanics, Chinese Academy of Sciences, Beijing, 100080, China

Dedicated to Professor Lu Ting on his eightieth birthday

Abstract

Two distinct types of vortex dislocation generated in wake-type flows with different spanwise nonuniformities are numerically studied by DNS. A local spanwise disturbance to the velocity of the coming flow leads to generation of consecutive twisted chain-like vortex dislocations in the middle downstream, which is mainly caused by phase difference between vortex shedding cells, while a stepped variation of velocity in the coming flow yields a periodic vortex splitting-reconnection to form a spot-like dislocation, which is mainly caused by frequency difference between the shedding cells. Dynamics and main features of these dislocations, especially the vortex linkages over the dislocations are clearly described by tracing the substantial modification of vorticity lines and other key flow quantities. Local irregularities in the variations of velocity, vorticity and frequency and transition behavior of the wake flows with vortex dislocations are analyzed.

Keywords: two distinct types of vortex dislocation; wake-type flow; transition behavior; Direct Numerical Simulation

[1] linggc@imech.ac.cn

9.1 Introduction

It has been confirmed that for low Reynolds number the vortex dislocation is a fundamental characteristic of three-dimensionalities of bluff body wakes. Depending on different three-dimensional flow conditions, either in cylinder geometry or in the coming flow, the vortex dislocation may present various fascinating modes or patterns, for instance, vortex "splitting and reconnection", vortex "kinks" or "holes", or a kind of large-scale spot-like vortex dislocations and so forth. Those phenomena can be found in cylinder wakes with diameter variation or nonuniform coming flows, or in the wake of stepped cylinder, as well as in the wake of the cylinder with end effects. Detailed presentations of them can be found in the experimental studies and reviews of Gerrard (1966), Eisenlohr & Eckelmann (1989), Bearman (1992), Lewis & Gharib (1992) , Papangelou (1992), Williamson (1992, 1996), among others. Again, the vortex dislocations may be characterized by "a local break of continuity appearing on the "spinal column" of a main vortex row", which occurs randomly in the natural transition of cylinder wakes (Braza *et al.*, 2001). In those previous works, flow visualization and measurement have provided much information on the generation and evolution of vortex dislocation. Some preliminary analyses on the mechanism based on the inviscid vortex dynamics have been given too. Recently there are few numerical studies on the forced vortex dislocations in wake-type flow (Ling & Xiong, 2001, Ling & Zhao, 2003), and the natural vortex dislocations in the wake transition (Braza *et al.*, 2001). The generation and the influence of the dislocation on the flow transition were reported. The occurrence of vortex dislocations is closely related to the break-up to turbulence and is considered a new mechanism of the wake transition. Moreover, it is associated with the variation of forces acting on the cylinders. Besides, the phenomenon of vortex dislocation can also be observed in three-dimensional evolution of a mixing layer (Dallard & Broward, 1993) and some of other nonlinear waves. Therefore, study of the vortex dislocation has received a great deal of attention.

However, due to its very complex three-dimensional nature the detailed information on the formation and the features of vortex linkages in dislocations are still not yet known thoroughly. The mechanistic explanation given for the vortex dislocation in previous studies was basically based on an inviscid vortex dynamics, which is not capable of offering a complete understanding for the real viscous flows. The physical understanding and the description are far from complete. Moreover, concerning the influence of the nonuniformity in the coming flow on the vortex dislocation we now only have a little knowledge. Thus, it is believed that to make further numerical study to address those issues is significant.

The purpose of the present work is to study the character of vortex dislocations generated in wake-type flows with different spanwise nonuniformity in coming flow by direct numerical simulation approach. Based on the DNS results the dynamical scenario of the formation of vortex dislocation and the basic features of the dislocations are described by analyzing the substantial modification of voriticity line tracks as well as key quantities of the flow field. Vortex linking, vortex splitting and reconnection in the dislocation produced in real viscous flows are illustrated in detail. The influence of the nonuniformity imposed in the coming flow on the dislocation is reported. Transition behavior and the irregularity of the flow are also studied.

9.2 Numerical simulation and method

In the present calculations, two cases of the nonuniform coming flow are considered, which is taken as

$$U(y, z) = 1.0 - a(z)(2.0 - \cosh(b(z)y^2))\exp(-c^2(z)y^2). \qquad (9.1)$$

It is a kind of time-averaged streamwise velocity profile in a cylinder wake at the position where the flow is most unstable. The parameters b and c are determined by referring to DNS results of cylinder wake flow given by Karniadakis and Triantafyllou (1992), and the experimental measurements of Nishioka and Sato (1974). Here $a(z)$ is introduced to express the nonuniformity in momentum defect in the coming flow along the span. For case one, a local spanwise nonuniformity is imposed, and $a(z) = 1.1 + 0.4\exp(-z^2)$, $b = 1.1$, $c = 1.2$. The local disturbance to coming flow is introduced near the center of the span and exponentially decayed with the increase of the distance. For case two, a stepped variation of the coming flows is considered, and $a(z) = 1.1$, $b = 0.9$, $c = 0.9818$ for $|z| < 2.5$ and $a(z) = 1.1$, $b = 1.1$, $c = 1.2$ for $2.5 \leq |z| \leq 15$, respectively. It is believed that these wake-type flows are unstable. Development of the instability will lead to different vortex street flows, and evolves vortex dislocations, which may have some similar behavior, to some extent, to those in real cylinder wakes with diameter variation.

In the present DNS, compact finite difference-Fourier spectral hybrid method is used for solving three-dimensional Navier-Stokes equations. A detailed presentation of these equations and the numerical method can be found in Xiong and Ling (1996), Ling & Xiong (2001). The procedure of the method is summarized as follows. In the spanwise direction of the flow, the periodic boundary conditions are assumed. And then we expand all the flow variables into a truncated Fourier series as

$$\varphi(x,y,z,t) = \sum_{m=-N/2}^{N/2-1} \varphi_m(x,y,t)e^{-im\beta z}, \tag{9.2}$$

where x, y and z are the streamwise, vertical, and spanwise directions, respectively. N is the cutoff, and β is the spanwise wavenumber. The three-dimensional incompressible Navier-Stokes equations are written, in primitive variable formation, as

$$\frac{\partial \vec{V}}{\partial t} + (\vec{V}\cdot\nabla)\vec{V} = -\nabla p + \frac{1}{Re}\nabla^2\vec{V}, \tag{9.3}$$

where $\vec{V} = \{u, v, w\}$ are the velocity components in x, y, z directions, respectively, and p is the pressure. The characteristic length D and the velocity scales U_0 are introduced, where U_0 is specified to be the streamwise wake-type inflow velocity $U(y,z)$ at $y = \pm\infty$, and D is chosen such that in non-dimensional variables, $U(D/2,\infty) = \frac{1}{2}\left[U(0,\infty) + U(\infty,\infty)\right]$. The Reynolds number is defined as $Re = U_0 D/\nu$, where ν is the kinematic viscosity. Substituting (9.2) into (9.3), we find a system of equations for the m-th harmonic in a two-dimensional (x, y) domain as

$$\frac{\partial \vec{V}_m}{\partial t} + F_m\left[\left(\vec{V}\cdot\nabla\right)\vec{V}\right] = -\nabla_m p_m + \frac{1}{Re}\nabla_m^2\vec{V}_m, \tag{9.4}$$

where $\nabla_m = \{\partial/\partial x, \partial/\partial y, -im\beta\}$, $\nabla_m^2 = \partial^2/\partial x^2 + \partial^2/\partial y^2 - m^2\beta^2$, and $F_m\left[\left(\vec{V}\cdot\nabla\right)\vec{V}\right]$ is the Fourier transformation of the nonlinear terms. For the time discretization of the equation a third-order mixed explicit-implicit scheme is used. The solution procedure at every time step is split into the following three substeps:

$$\begin{aligned}
\frac{\vec{u}'_m - \sum_{q=0}^{J_i-1}\alpha_q \vec{u}_m^{n-q}}{\Delta t} &= -\sum_{q=0}^{J_e-1}\beta_q F_m\left[\left(\vec{u}^{n-q}\cdot\nabla\right)\vec{u}^{n-q}\right],\\
\frac{\vec{u}''_m - \vec{u}'_m}{\Delta t} &= -\nabla_m p_m^{n+1},\\
\frac{\gamma_0\vec{u}_m^{n+1} - \vec{u}''_m}{\Delta t} &= \frac{1}{Re}\nabla_m^2\vec{u}_m^{n+1},
\end{aligned} \tag{9.5}$$

where $\vec{u}'_m$, $\vec{u}''_m$ are intermediate velocities; J_i, J_e are parameters for the order of the scheme, and α_q, β_q, γ_0 are appropriately chosen weights. For the third order case, the values of these coefficients, following the study of Karniadakis et al. (1994), are cited as follows:

$$J_e = 3, J_i = 3, \alpha_0 = 3, \alpha_1 = -\tfrac{3}{2}, \alpha_2 = \tfrac{1}{3}, \beta_0 = 3, \beta_1 = -3, \beta_2 = 1, \gamma_0 = \tfrac{11}{6}.$$

To evaluate the nonlinear terms in the split equations, the pseudo-spectral method is adopted, and the fifth order upwind compact scheme is used to approximate the terms. The derivatives of the Fourier coefficients u_m are calculated by the sixth-order center compact schemes. For solving Helmholtz equations for pressure and velocity, a nine-point compact scheme of fourth-order central compact scheme is derived and used for nonhomogeneous term calculations, which is

$$\frac{10(\varphi_1+\varphi_3)-2(\varphi_2+\varphi_4)+(\varphi_5+\varphi_6+\varphi_7+\varphi_8)-20\varphi_0}{(\Delta x)^2}$$

$$+\frac{10(\varphi_2+\varphi_4)-2(\varphi_1+\varphi_3)+(\varphi_5+\varphi_6+\varphi_7+\varphi_8)-20\varphi_0}{(\Delta y)^2}$$

$$-(8\varphi_0+\varphi_1+\varphi_2+\varphi_3+\varphi_4)b = (8f_0+f_1+f_2+f_3+f_4).$$

For the pressure boundary conditions, with formulas (9.4) and (9.5), the semi-discretized representation can be written as

$$\begin{aligned}\frac{\partial p_m^{n+1}}{\partial n} &= \vec{n}\cdot\left\{-\frac{\partial \vec{u}_m}{\partial t}+\frac{1}{Re}\nabla^2\vec{u}_m^{n+1}-\sum_{q=0}^{J_e-1}\beta_q F_m\left[(\vec{u}^{n-q}\cdot\nabla)\vec{u}^{n-q}\right]\right\}\\ &= \vec{n}\cdot\left\{-\frac{\partial \vec{u}_m}{\partial t}-\frac{1}{Re}F_m\left[\nabla\times\vec{\Omega}^n\right]-\sum_{q=0}^{J_e-1}\beta_q F_m\left[(\vec{u}^{n-q}\cdot\nabla)\vec{u}^{n-q}\right]\right\}\end{aligned} \tag{9.6}$$

where $\nabla^2\vec{u} = \nabla D - \nabla\times\vec{\Omega}$, $\vec{\Omega}\equiv\nabla\times\vec{u}$, $D = \nabla\cdot\vec{u}$. By forcing $D^{n+1}\equiv 0$ in each time step and replacing $\nabla\times\vec{\Omega}^{n+1}$ by $\nabla\times\vec{\Omega}^n$, the divergence-free constraint can be well satisfied on the boundaries.

For a finite computing domain, the non-reflecting boundary condition for outflow in physical space is suggested:

$$\frac{\partial\vec{u}}{\partial t}+u\frac{\partial\vec{u}}{\partial x} = \frac{1}{Re}\left(\frac{\partial^2\vec{u}}{\partial y^2}+\frac{\partial^2\vec{u}}{\partial y^2}\right). \tag{9.7}$$

The corresponding spectral form used in the computation is

$$\frac{\partial \vec{u}_m}{\partial t} + F_m\left[u\frac{\partial \vec{u}}{\partial x}\right] = \frac{1}{Re}\left(\frac{\partial^2 \vec{u}_m}{\partial y^2} - m^2\beta^2\vec{u}_m\right). \tag{9.8}$$

Equation (9.8) must be solved simultaneously with the same third-order mixed explicit-implicit schemes adopted for Navier-Stokes equations in the inner domain.

In the present computation, Reynolds number is taken as 200. The computational domains are 100, 30, 30 in streamwise, vertical and spanwise directions, respectively. The cutoff of the truncated Fourier series is $N = 64$ and the corresponding grid points in $x - y$ plane is 202×62. The numerical code used in the present work has been verified first. It shows that evolution of the wake-type flow without the local spanwise non-uniformity results in a normal Kármán vortex street with a Strouhal number of 0.189, which is well compared with both numerical simulation of wake-type flow evolution ($St = 0.195$, see Karniadakis and Triantafyllou, 1992) and DNS results of flow around cylinder ($St = 0.179$, see Triantafyllou and Karniadakis, 1990).

9.3 Numerical results

The present DNS results have shown that spatial temporal evolution of the coming flows leads to two different types of vortex street flow with two distinct types of vortex dislocation generated in the middle downstream of the flows, respectively. Variation of the vorticity line tracks, isosurface of vorticity, vorticity component contours and fluctuating velocity distributions are used to describe the whole picture of the dislocation and its development in space.

9.3.1 *Local spanwise nonuniformity*

For this case, a series of symmetric twisted chain-like vortex dis locations is generated in the middle downstream as shown by the isosurface of vorticity in Fig. 9.1. The vortex dislocation occurs consecutively based on the background shedding flow, and is characterized by an undulated spanwise vortex connected by streamwise and vertical vortex branches. Behaviors of representative vorticity lines emitting from different positions in main vortex show that, as the flow travels downstream, these lines undergo a substantial modification from spanwise direction to streamwise and vertical ones. In particular, in the upstream region, the spanwise vortex lines start to undulate in the spanwise direction, then undergo a stronger distortion near the central area (Fig. 9.2). In the middle downstream, a set of vortex lines pass across the span with a big distortion near the central area, and the others bend into upstream, then after turning 180 degree go into the adjacent main vortex. It is interesting to note that some vortex lines starting from the different vertical positions of the same vortex will go back to downstream and join the next spanwise vortex (Fig. 9.3). That is the phenomenon of vortex splitting and reconnection occurred in the vortex dislocation. Calculation shows that the real behavior of the phenomenon is some vorticity lines, located at the outer region of a vortex roll (vorticity blob), changing their directions, under the interaction with adjacent vortices, from original spanwise direction to streamwise, vertical direction and further joining to their neighboring main vortices with opposite sign. In present calculation, the initial phase and strength of shedding vortices are evaluated. Due to the local spanwise nonuniformity of the incoming flow, there are pronounced differences in phase and strength between shedding cells. The results demonstrated that these differences make spanwise vorticity lines curve to produce vorticity component in other two directions. Moreover the vortex splitting and reconnection are only produced at some positions where the rate of phase difference reaches a maximum value; meantime three-dimensional vortex dislocation structures are formed. It is in agreement with the experimental study of Eisenlohr & Eckelman (1989).

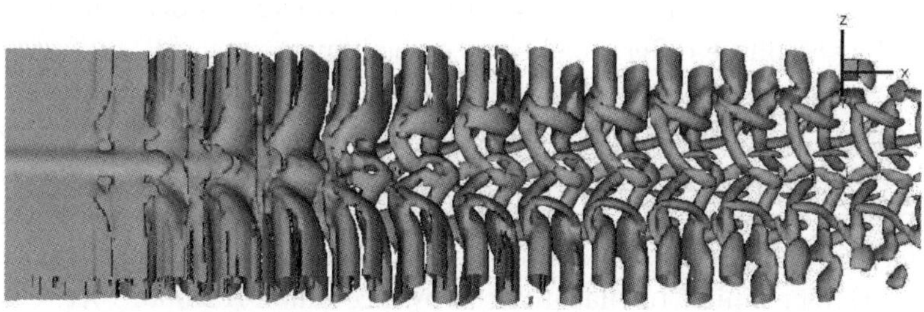

Fig. 9.1 Symmetric twisted chain-like vortex dislocation pattern, visualized by Iso-surface of vorticity $|\omega| = 0.12$, at $t = 260$.

In summary, for the present case, the basic features of vortex linking in vortex dislocation, as shown in Figs. 9.2, 9.3 and 9.4, can be briefly described in the following outline: (1) large undulated spanwise vortex roll;

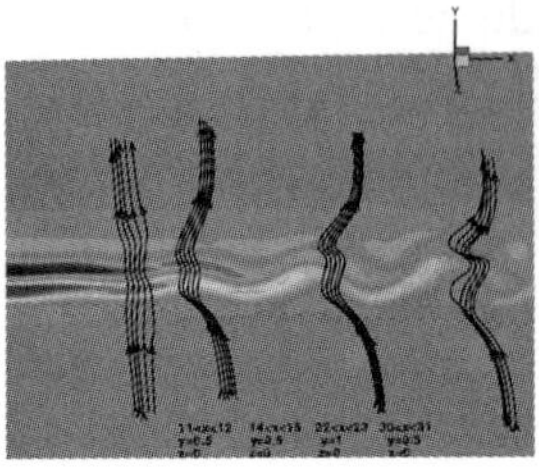

Fig. 9.2 Spatial variation of vorticity line tracks

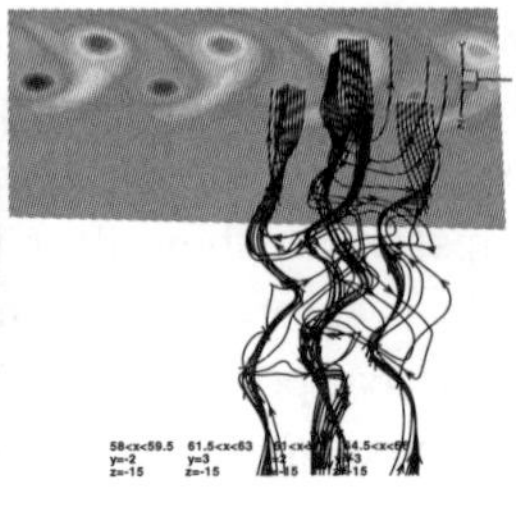

Fig. 9.3 The vortex linkages in vortex dislocation. Vortex splitting and reconnection with adjacent primary spanwise vortices are shown by variation of vorticity line tracks.

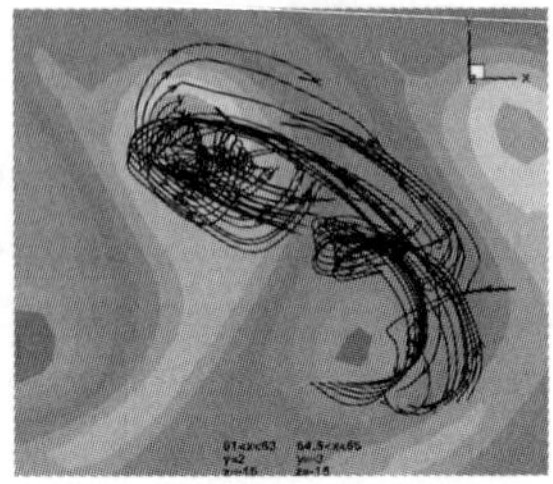

Fig. 9.4 End view of vortex linkages in vortex dislocation. Vorticity lines wind round the neighboring vortices.

(2) some split spanwise vortex lines diverting their direction and connecting with adjacent vortices by a cross-vortex street mode; (3) the distorted vortex lines wind round their neighboring spanwise vortices. These features coincide with those shown in Fig. 9.1. The vortex dislocation is mainly caused by the phase difference between shedding vortices.

The influence of vortex dislocation on the flow transition is examined by irregular variation of some key quantities of flow field. Time series of fluctuation velocity components, at the positions near the center part of the span present a pronounced irregularity and accompanied by a low frequency modulation, while near both sides of the span the variation of velocity is quasi-periodic, similar to that induced by a common Kármán vortex street.

In spectra of velocity components several peak values are found at a basic frequency, its harmonics and at a low frequency. The basic one corresponds the vortex shedding of the base flow. Near the vortex dislocation the spectra show the variation of frequency is irregular. In addition to those, the vorticity line tracks in vortex dislocation, as shown in Fig. 9.4, also present a chaotic behavior. It is clear that the appearance of the vortex dislocation and its breakdown make flow irregular, which is closely related to the break up to the local turbulence and the spatial chaos of the flow.

9.3.2 *Stepped spanwise nonuniformity*

For this case, the vortex dislocation occurs periodically near the steps of the incoming flow velocity in the span and caused mainly by the frequency difference between shedding cells over the steps.The main features are visualized by isosurfaces of vorticity and vorticity component in stream direction, as shown in Figs. 9.5 and 9.6. The vortex dislocation is characterized by undulated spanwise vortex roll and vortex splitting from it. The split spanwise vorticity lines turn their direction towards upstream becoming streamwise vortex branch and then turn further joining a neighboring main vortex. Several vortex patterns of this kind constitute a periodic spot-like vortex dislocation and travel downstream periodically.

(a)$t = 316$ (b)$t = 334$

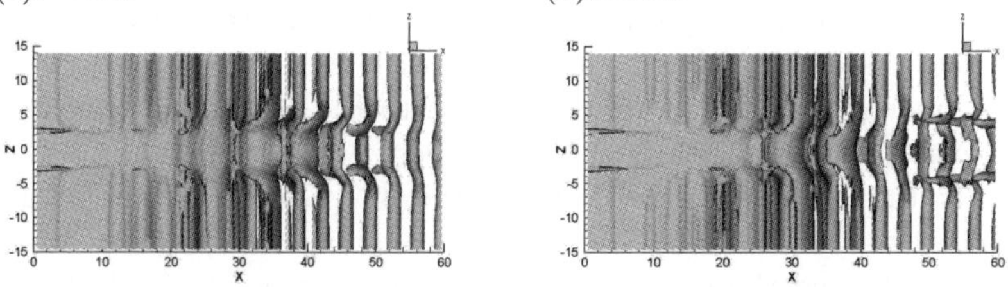

Fig. 9.5 Visualization of spot-like dislocation pattern, which consists of several vortex splitting and reconnections shown by the isosurface of vorticity $|\omega| = 0.15$.

(a)$t = 316$ (b)$t = 334$

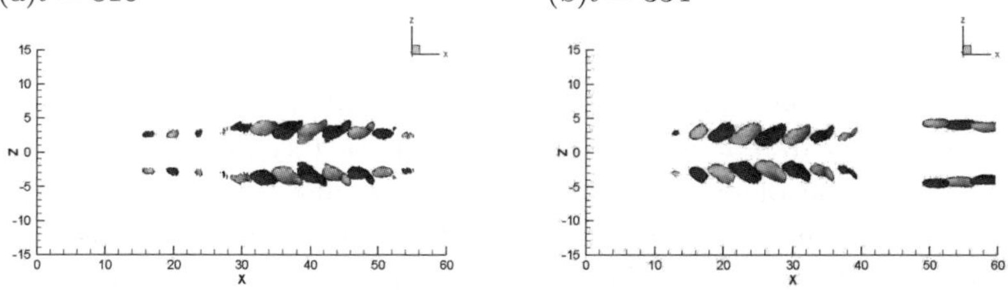

Fig. 9.6 Iso-surface of streamwise vorticity in vortex dislocation $\omega_x = 0.11$ (gray) and $\omega_x = -0.11$ (black).

Time series of fluctuation velocity components and their spectra are examined. A sample of these behaviors is shown in Figs. 9.7, 9.8 and 9.9, respectively. In the center region of the span, (i.e. for $|z| \leq 3$), a pronounced velocity modulation occurs periodically over the time series, which corresponds to the appearance of the vortex dislocation, while away from the center region, the velocity presents a regular variation like that produced by a common Kármán vortex street. In the spectra of vertical velocity component (Fig. 9.8), there are two pronounced peaks at $f_1 = 0.117$ (for $z = 0$ to 4) and $f_2 = 0.141$ (for $z \geq 4$). They are incommensurable, and correspond to the vortex shedding in the center region and outer region, respectively. A peak at the difference of these two frequencies is found in low frequency area. It is the 'beat' between the two sets of vortex shedding and is just the period of the spot-like vortex dislocation. It is the result of nonlinear interaction between these two frequencies. However in spectra for spanwise velocity component as shown in Fig. 9.9 multi-peak values at different frequencies are presented. The evolution with time of the velocity is irregular. The characteristic of time-frequency of velocity components at different positions in flow field is further studied by using wavelet analysis. A sample of these results is given by Fig. 9.10. It shows that at a point in the flow field the dominant frequency varies with time and shifts between two main frequencies. This variation is due to the generation of vortex dislocation and its passage through the position.

Based on the DNS results the Proper Orthogonal Decomposition (POD) analysis of wake-type flow with vortex dislocation has been performed too. The nonlinearity and the transition behavior of the flow are further analyzed. For details, reader is referred to the work of Zhao & Ling (2003).

9.4 Concluding remarks

The present study has shown that two kinds of vortex dislocation, the symmetric twisted chain-like vortex dislocation and the spot-like dislocation, are generated in wake-type flows. The characteristic and dynamics of the vortex dislocations are dependent on the coming flow nonuniformity in spanwise direction. The two kinds of dislocation are mainly caused by the differences in phase and in frequency, respectively. The formation and the structure of the vortex dislocations, as well as the real linkages over the dislocations, have been clearly described by substantial modification of vorticity field, especially, by the vortex line behavior, and the other flow field quantities. Large distortion and splitting of the original spanwise vortex roll, and the reconnection in viscous vortex dislocation are shown in detail. Frequency variation in wake flow is examined. The local irregularities in variation of velocity, vorticity and frequency, characterizing the transition behavior of the two wake flows with vortex dislocations, are reported.

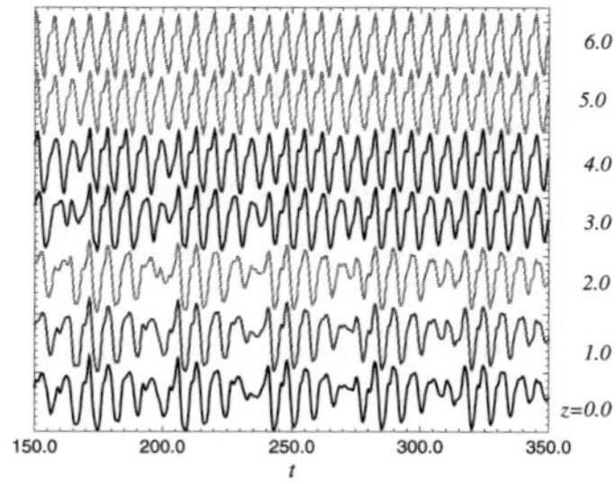

Fig. 9.7 Time series of vertical velocity component along the span at $(x, y) = (30, 0.5)$.

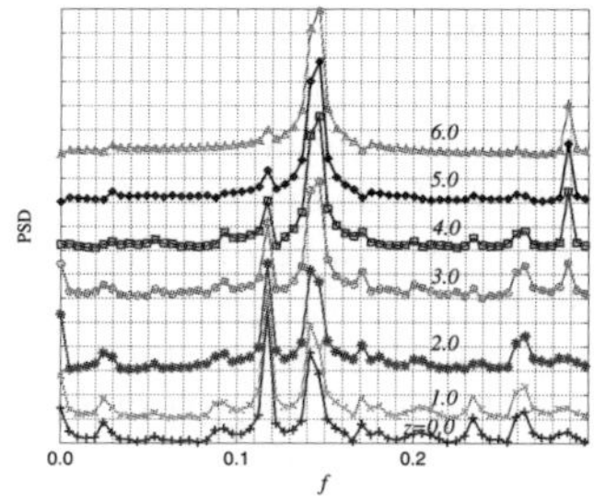

Fig. 9.8 Spectra of vertical velocity component along the span at $(x, y) = (30, 0.5)$.

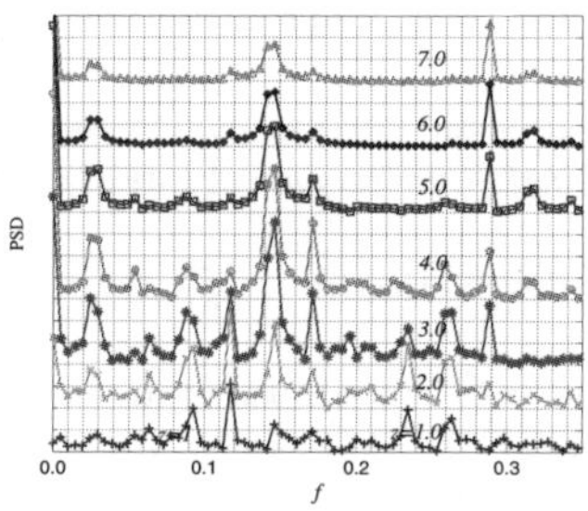

Fig. 9.9 Spectra of spanwise velocity component along the span at $(x, y) = (30, 0.5)$.

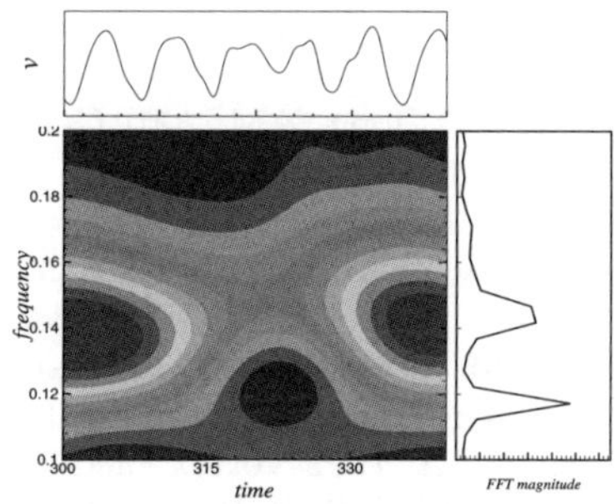

Fig. 9.10 Wavelet analysis of vertical velocity component at $(x, y, z) = (45, 0.5, 0)$.

Acknowledgment

This work is supported by the National Natural Science Foundation of China under Grant No.10272104, and partially by CAS Innovation projects "study on Complex Flows". The calculation in present work is finished by Dr. Hongliang Zhao. His contribution is also acknowledged.

Bibliography

Bearman, P.W. 1992 Challenging problems in bluff body wakes. *Bluff-Body Wakes, Dynamics and Instabilities, IUTAM Symposium*. Göttingen: Springer-Verlag.

Braza, M., Faghani, D. & Persillson, H. 2001 Successive stages and the role of natural vortex dislocation in three-dimensional wake transition. *J. Fluid Mech.* **439**:1-41.

Dallard, T. & Browand, F.K. 1993 The growth of large scales at defect sites in the plane mixing layer. *J. Fluid Mech.* **247**:339-368.

Eisenlohr, H. & Eckelemann, H. 1989 Vortex splitting and its consequences in vortex street wake of cylinders at low Reynolds number. *Phys. Fluids A* **1**:189-192.

Gerrard J.H. 1966 The three-dimensional structure of the wake of a circular cylinder. *J. Fluid Mech.* **25**:143-164.

Karniadakis, G.E. & Triantafyllou, G.S. 1992 Three-dimensional dynamics and transition to turbulence in the wake of bluff objects. *J. Fluid Mech.* **238**:1-30.

Kaniadakis, G.E., Israeli, S.A. & Orzag, S.A. 1994 High-order splitting methods for the incompressible Navier-Stokes equations. J. Comput. Phys. **97**:414.

Lewis, C.G. & Gharib, M. 1992 An exploration of wake three-dimensionalities caused by a local discontinuity in cylinder diameter. *Phys. Fluids A* **4**:104-117.

Ling, G.C. & Xiong, Z.M. 2001 Generation of large-scale vortex dislocations in a three-dimensional wake-type flow. *Science in China (series A)* **4**(12):1585-1595.

Ling, G.C. & Zhao H.L. 2003 Structure of vortex dislocation in a wake-type flow. CHIN. PHYS. LETT. **20**(3):383-385.

Nishioka, M. & Sato, H. 1974 Measurements of velocity distributions in the wake of a circular cylinder at low Reynolds numbers. *J. Fluid Mech.* **65**:97-112.

Papangelou, A. 1992 Vortex shedding from slender cones at low Reynolds numbers. *Bluff-Body Wakes, Dynamics and Instabilities, IUTAM Symposium.* Göttingen: Springer-Verlag.

Triantafyllou, G.S., Karniadakis, G.E. 1990 Computational reducibility of unsteady viscous flow. *Phys. Fluids A* **2**:653-658.

Willamson, C.H.K. 1992 The natural and forced formation of spot-like vortex dislocations in the transition of a wake. *J. Fluid Mech.* **243**:393-441.

Willamson, C.H.K. 1996 Vortex dynamics in the cylinder wake. *Annu. Rev. Mech.* **28**:477-539.

Xiong, Z.M. & Ling, G.C. 1996 Compact finite difference-Fourier spectral method for three-dimensional incompressible Navier-Stokes equations. *ACTA. Mechanica Sinica* **12**(4):296-306.

Zhao, H.L. & Ling, G.C. 2003 Low-dimensional dynamical systems for a wake-type shear flow with vortex dislocations. *Fluid Dynamics Research* **33**:299-312.

Chapter 10

Vortex Dipole Coordinates on the Sphere

Paul K. Newton and Houman Shokraneh
Department of Aerospace & Mechanical Engineering and Mathematics
University of Southern California, Los Angeles CA 90089-1191, USA

To Lu Ting on the occasion of his 80th birthday, with admiration and well wishes

Abstract

A dipole coordinate system is introduced for a system of $2N$ point vortices on the sphere that are grouped into N pairs of dipoles, i.e. the vortex strengths associated with each pair are related via $\Gamma_{i+N} = -\Gamma_i$, for $i = 1, ..., N$. The coordinates are associated with the centers-of-vorticity and centroids of each of the dipoles and are particularly useful for studying dipole interactions. Three canonical dipole scattering interactions are then studied for the case of two dipoles (i.e. $N = 2$). We call them *e*xchange scattering, *non-exchange scattering*, and *l*oop-scattering.

10.1 Introduction

Vortex dipoles play an important role in the transport and mixing of particles (both passive and active) in oceanographic and atmospheric flows, yet rarely do they appear in complete isolation. Typically, they are forced to interact with a background vorticity field due to the rotating atmosphere, jetstreams, or ocean currents (see, for example, [10]) and often they interact with each other or with higher order structures such as tripoles or quadrupoles [8]. Despite this fact, a study of their interactions on the

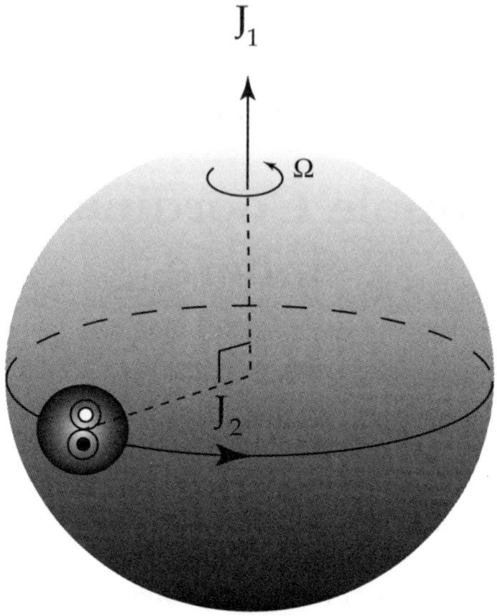

Fig. 10.1 An isolated dipole with the *d*ipole coordinates $(\mathbf{J}_1, \mathbf{J}_2)$. In these coordinates, $\mathbf{J}_1$ = constant, $\|\mathbf{J}_2\|$ = constant, while $\mathbf{J}_2$ rotates around $\mathbf{J}_1$ with frequency $\Omega = \|\mathbf{J}_1\|$, tracing out a great circle path.

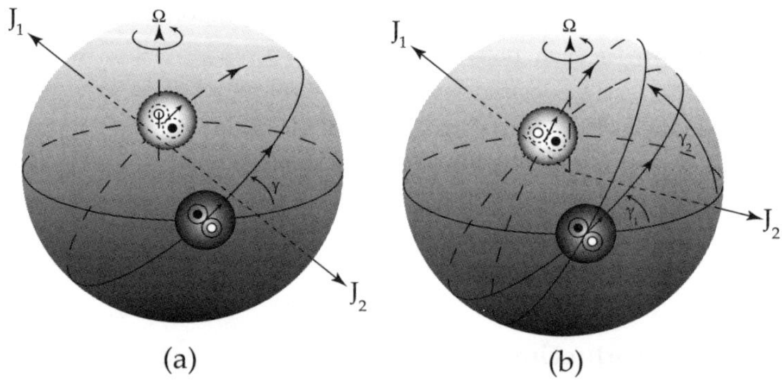

Fig. 10.2 Twin dipole scattering schematic. (a) Head-on collision: $\mathbf{J} = 0$; (b) Non-aligned collision: $\mathbf{J} \neq 0$.

sphere has not been carried out. In the unbounded plane, a pioneering study of the scattering process is described in [2]. Unlike the unbounded plane, since the sphere is compact, the dipoles cannot escape to infinity and the dynamics typically involves multiple 'scattering events'. When there is only one dipole on the sphere, it moves along a closed geodesic [5; 6], allowing the assignment of a 'natural' frequency, Ω_i, to each dipole.

In this paper we introduce a special coordinate system called the *dipole coordinates*, that are particularly well suited to study the general dipole interaction problem in which $2N$ point vortices are grouped into N pairs with strengths $\Gamma_{i+N} = -\Gamma_i$, $(i = 1, ..., N)$. We study in detail a two-dipole system, which is a special case of a 4-vortex problem [6] in which the vortex strengths appear in equal and opposite pairs: $\Gamma_1 = -\Gamma_2$; $\Gamma_3 = -\Gamma_4$. The point vortex equations in Cartesian coordinates are given by

$$\dot{\mathbf{x}}_\alpha = \frac{1}{4\pi}\sum_{\beta=1}^{4}{}' \Gamma_\beta \frac{\mathbf{x}_\beta \times \mathbf{x}_\alpha}{(1 - \mathbf{x}_\alpha \cdot \mathbf{x}_\beta)} + \Omega \hat{\mathbf{e}}_z \times \mathbf{x}_\alpha \qquad (\alpha = 1, 2, 3, 4) \quad (10.1)$$

$$\mathbf{x}_\alpha \in \mathbb{R}^3, \quad \|\mathbf{x}_\alpha\| = 1.$$

The prime on the summation indicates that the singular term $\beta = \alpha$ is omitted. Initially, the vortices are located at positions $\mathbf{x}_\alpha(0) \in \mathbb{R}^3$, $(\alpha = 1, 2, 3, 4)$ and we group them into two dipoles initially located on the equator at the front and back of the sphere, as shown in figure 10.2. The denominator in (10.1) is the chord distance between vortex Γ_α and Γ_β since $\|\mathbf{x}_\alpha - \mathbf{x}_\beta\|^2 = 2(1 - \mathbf{x}_\alpha \cdot \mathbf{x}_\beta)$ and Ω is the solid-body rotational frequency of the sphere.

It follows from equations (10.1) that the *center of vorticity* vector, $\mathbf{J}$, defined as

$$\mathbf{J} = \sum_{\alpha=1}^{4} \Gamma_\alpha \mathbf{x}_\alpha = \left(\sum_{\alpha=1}^{4} \Gamma_\alpha x_\alpha, \sum_{\alpha=1}^{4} \Gamma_\alpha y_\alpha, \sum_{\alpha=1}^{4} \Gamma_\alpha z_\alpha \right) = (J_x, J_y, J_z) \quad (10.2)$$

is constant on the non-rotating sphere $(\Omega = 0)$ and breaks into two conserved quantities on the rotating sphere $(\Omega \neq 0)$: $J_x^2 + J_y^2 = C_1$, $J_z = C_2$, reflecting the fact that $\mathbf{J}$ rotates about the z-axis with frequency Ω. This quantity is monitored in order to measure the extent of the numerical error accumulated as the scattering processes unfold. We use a variable time-step 7th/8th order Runge-Kutta solver that works quite effectively. In all cases, the deviation of $\mathbf{J}$ from its conserved value is negligible over the time scales considered.

Although not conserved, the individual centers-of-vorticity of each of the two dipoles, defined as

$$\mathbf{J_1} = \Gamma_1 \mathbf{x}_1 + \Gamma_2 \mathbf{x}_2 = \left(\sum_{\alpha=1}^{2} \Gamma_\alpha x_\alpha, \sum_{\alpha=1}^{2} \Gamma_\alpha y_\alpha, \sum_{\alpha=1}^{2} \Gamma_\alpha z_\alpha \right) = (J_x^{(1)}, J_y^{(1)}, J_z^{(1)}) \tag{10.3}$$

and

$$\mathbf{J_2} = \Gamma_3 \mathbf{x}_3 + \Gamma_4 \mathbf{x}_4 = \left(\sum_{\alpha=3}^{4} \Gamma_\alpha x_\alpha, \sum_{\alpha=3}^{4} \Gamma_\alpha y_\alpha, \sum_{\alpha=3}^{4} \Gamma_\alpha z_\alpha \right) = (J_x^{(2)}, J_y^{(2)}, J_z^{(2)}) \tag{10.4}$$

determine the orientation of the dipoles (as shown in figure 10.2) through the angles γ_1 and γ_2 which measure the initial tilt of the plane in which the respective dipoles move, normal to $\mathbf{J_1}$ and $\mathbf{J_2}$. For the two-dipole interaction problem, these are two of the four coordinates that constitute the *dipole coordinate system* defined in the next section.

The scattering of two identical dipoles, called 'twin dipoles' (we take the strengths to be $\Gamma_1 = \Gamma_3 = 1; \Gamma_2 = \Gamma_4 = -1$) is shown schematically in figure 10.2. When the dipoles are perfectly aligned so that $\mathbf{J} = 0$, the problem is integrable and the general case is shown in figure 10.2(a). When the twin dipoles are not perfectly aligned, so that $\mathbf{J} \neq 0$, as shown in figure 10.2(b), generically the problem is non-integrable as discussed in [9]. We will describe three fundamental scattering events both for the non-rotating sphere and for the rotating sphere. In the latter case, the ratio of the natural dipole frequency, Ω_i, (i = 1,2) to the solid-body frequency, Ω, plays an important role, as does the orientation of the $\mathbf{J}$ vector with the axis of rotation. The trajectories shown in this study are all integrable, thus they are either periodic, or quasi-periodic. See [9] for a more thorough discussion.

10.2 The dipole coordinate system

Consider a system of $2N$ point vortices on the non-rotating sphere ($\Omega = 0$)

$$\dot{\mathbf{x}}_\alpha = \frac{1}{4\pi} \sum_{\beta=1}^{2N}{}' \Gamma_\beta \frac{\mathbf{x}_\beta \times \mathbf{x}_\alpha}{(1 - \mathbf{x}_\alpha \cdot \mathbf{x}_\beta)} \qquad (\alpha = 1, ..., 2N) \tag{10.5}$$

paired up into a collection of N dipoles where we group the pairs so that $(\mathbf{x}_i, \mathbf{x}_{i+N})$ form a dipole, $(i = 1, ..., N)$. Hence, we assume that $\Gamma_{i+N} =$

$-\Gamma_i$. From (10.5), it is straightforward to show that

$$\mathbf{J} = \sum_{\beta=1}^{2N} \Gamma_\beta \mathbf{x}_\beta = \text{constant}. \tag{10.6}$$

We now define N centers-of-vorticity coordinates for each of the N dipoles, i.e.

$$\mathbf{J}_1 \equiv \Gamma_1 \mathbf{x}_1 + \Gamma_{N+1}\mathbf{x}_{N+1} = \Gamma_1(\mathbf{x}_1 - \mathbf{x}_{N+1}), \tag{10.7}$$

$$\vdots$$

$$\mathbf{J}_N \equiv \Gamma_N \mathbf{x}_N + \Gamma_{N+N}\mathbf{x}_{N+N} = \Gamma_N(\mathbf{x}_N - \mathbf{x}_{2N}).$$

Their sum is the total center-of-vorticity of the system, i.e. $\mathbf{J} \equiv \sum_{\alpha=1}^{N} \mathbf{J}_\alpha$ is a constant. We also define N *centroid* coordinates associated with each of the N dipoles

$$\mathbf{J}_{N+1} \equiv \Gamma_1 \mathbf{x}_1 - \Gamma_{N+1}\mathbf{x}_{N+1} = \Gamma_1(\mathbf{x}_1 + \mathbf{x}_{N+1}), \tag{10.8}$$

$$\vdots$$

$$\mathbf{J}_{2N} \equiv \Gamma_N \mathbf{x}_N - \Gamma_{N+N}\mathbf{x}_{N+N} = \Gamma_N(\mathbf{x}_N + \mathbf{x}_{2N}).$$

The coordinate transformation

$$(\mathbf{x}_1, ..., \mathbf{x}_N, \mathbf{x}_{N+1}, ..., \mathbf{x}_{2N}) \mapsto (\mathbf{J}_1, ..., \mathbf{J}_N; \mathbf{J}_{N+1}, ..., \mathbf{J}_{2N})$$

from the original Cartesian coordinates to the dipole coordinates is now complete. It is defined from

$$\mathbf{x}_i = \frac{1}{2\Gamma_i}(\mathbf{J}_i + \mathbf{J}_{i+N}), (i = 1, ..., N) \tag{10.9}$$

$$\mathbf{x}_{i+N} = -\frac{1}{2\Gamma_i}(\mathbf{J}_i - \mathbf{J}_{i+N}). \tag{10.10}$$

In matrix form, the transformation is

$$\hat{\mathbb{J}} = \mathbb{M}\hat{\mathbf{x}} \tag{10.11}$$

where

$$\hat{\mathbb{J}} \equiv \begin{pmatrix} \mathbf{J}_1 \\ \vdots \\ \mathbf{J}_N \\ \mathbf{J}_{N+1} \\ \vdots \\ \mathbf{J}_{2N} \end{pmatrix}; \hat{\mathbf{x}} \equiv \begin{pmatrix} \mathbf{x}_1 \\ \vdots \\ \mathbf{x}_N \\ \mathbf{x}_{N+1} \\ \vdots \\ \mathbf{x}_{2N} \end{pmatrix} \tag{10.12}$$

and $\mathbb{M}$ is a $2N \times 2N$ matrix with block structure:

$$\mathbb{M} \equiv \begin{pmatrix} \mathbf{M}_\Gamma & -\mathbf{M}_\Gamma \\ \mathbf{M}_\Gamma & \mathbf{M}_\Gamma \end{pmatrix}. \tag{10.13}$$

$\mathbf{M}_\Gamma \in \mathbb{R}^N \times \mathbb{R}^N$ is the diagonal matrix given by

$$\mathbf{M}_\Gamma = \begin{pmatrix} \Gamma_1 & \ldots & 0 \\ \vdots & \vdots & \vdots \\ 0 & \ldots & \Gamma_N \end{pmatrix}. \tag{10.14}$$

We would like to transform (10.5) to the dipole coordinates (10.7), (10.8). This transformation applied to an isolated dipole helps clarify why the new coordinate system is so useful.

10.3 One and two dipoles

Example 1: N = 1 (One dipole) The equations for the two vortices making up a single dipole are

$$\dot{\mathbf{x}}_1 = \frac{\Gamma_2}{4\pi} \frac{\mathbf{x}_2 \times \mathbf{x}_1}{(1 - \mathbf{x}_1 \cdot \mathbf{x}_2)}, \qquad \dot{\mathbf{x}}_2 = \frac{\Gamma_1}{4\pi} \frac{\mathbf{x}_1 \times \mathbf{x}_2}{(1 - \mathbf{x}_1 \cdot \mathbf{x}_2)} \tag{10.15}$$

with $\Gamma_2 = -\Gamma_1$. The dipole coordinates are

$$\mathbf{J}_1 = \Gamma_1(\mathbf{x}_1 - \mathbf{x}_2); \qquad \mathbf{J}_2 = \Gamma_1(\mathbf{x}_1 + \mathbf{x}_2), \tag{10.16}$$

with inverses given by

$$\mathbf{x}_1 = \frac{1}{2\Gamma_1}(\mathbf{J}_1 + \mathbf{J}_2); \qquad \mathbf{x}_2 = -\frac{1}{2\Gamma_1}(\mathbf{J}_1 - \mathbf{J}_2). \tag{10.17}$$

Note that $\mathbf{J}_1 \perp \mathbf{J}_2$ since

$$\begin{aligned}\mathbf{J}_1 \cdot \mathbf{J}_2 &= \Gamma_1^2(\mathbf{x}_1 - \mathbf{x}_2) \cdot (\mathbf{x}_1 + \mathbf{x}_2) \\ &= \Gamma_1^2(\mathbf{x}_1 \cdot \mathbf{x}_1 - \mathbf{x}_2 \cdot \mathbf{x}_1 + \mathbf{x}_1 \cdot \mathbf{x}_2 - \mathbf{x}_2 \cdot \mathbf{x}_2) \\ &= \Gamma_1^2(\|\mathbf{x}_1\|^2 - \|\mathbf{x}_2\|^2) = 0. \end{aligned} \tag{10.18}$$

Also,

$$\mathbf{x}_1 \times \mathbf{x}_2 = \frac{1}{2\Gamma_1^2}\mathbf{J}_1 \times \mathbf{J}_2 \tag{10.19}$$

$$\mathbf{x}_1 \cdot \mathbf{x}_2 = -\frac{1}{4\Gamma_1^2}(\|\mathbf{J}_1\|^2 - \|\mathbf{J}_2\|^2). \tag{10.20}$$

Using (10.17) in (10.15) yields

$$\dot{\mathbf{J}}_1 + \dot{\mathbf{J}}_2 = \frac{\Gamma_1^2}{\pi}\frac{\mathbf{J}_1 \times \mathbf{J}_2}{(4\Gamma_1^2 + \|\mathbf{J}_1\|^2 - \|\mathbf{J}_2\|^2)} \tag{10.21}$$

$$\dot{\mathbf{J}}_1 - \dot{\mathbf{J}}_2 = -\frac{\Gamma_1^2}{\pi}\frac{\mathbf{J}_1 \times \mathbf{J}_2}{(4\Gamma_1^2 + \|\mathbf{J}_1\|^2 - \|\mathbf{J}_2\|^2)}. \tag{10.22}$$

When added and subtracted, the system decouples in the dipole coordinates:

$$\dot{\mathbf{J}}_1 = 0, \tag{10.23}$$

$$\dot{\mathbf{J}}_2 = \frac{1}{4\pi}\frac{\mathbf{J}_1 \times \mathbf{J}_2}{\left(1 + \frac{\|\mathbf{J}_1\|^2 - \|\mathbf{J}_2\|^2}{4\Gamma_1^2}\right)}. \tag{10.24}$$

This immediately yields the result that $\mathbf{J}_1 =$ const. and taking the dot product of equation (10.24) with $\mathbf{J}_2$ yields $\|\mathbf{J}_2\| =$ const. It is then easy to show that $\mathbf{J}_2$ rotates around $\mathbf{J}_1$ with frequency $\Omega \equiv \|\mathbf{J}_1\|$, as shown in figure 10.1. For the N-dipole system, if each dipole were uncoupled to all the others, each of the $\mathbf{J}_{i+N}$ vectors would rotate around its corresponding $\mathbf{J}_i$ vector with frequency $\Omega_i \equiv \|\mathbf{J}_i\|$, $i = 1, ..., N$ tracing geodesic paths on the sphere. The extent to which the dipoles are coupled, and hence the deviation of $\mathbf{J}_{i+N}$ from a geodesic path, can be seen most clearly for a system of two dipoles whose equations in dipole coordinates we show next.
Example 2: N = 2 (Two dipoles) In dipole coordinates, the two-dipole

system is given by:

$$\dot{\mathbf{J}}_1 = \frac{\Gamma_1\Gamma_2}{4\pi}\left[\frac{(\mathbf{J}_1\times\mathbf{J}_2+\mathbf{J}_1\times\mathbf{J}_4+\mathbf{J}_3\times\mathbf{J}_2+\mathbf{J}_3\times\mathbf{J}_4)}{4\Gamma_1\Gamma_2-(\mathbf{J}_1\cdot\mathbf{J}_2+\mathbf{J}_1\cdot\mathbf{J}_4+\mathbf{J}_3\cdot\mathbf{J}_2+\mathbf{J}_3\cdot\mathbf{J}_4)} + \frac{(\mathbf{J}_1\times\mathbf{J}_2+\mathbf{J}_1\times\mathbf{J}_4-\mathbf{J}_3\times\mathbf{J}_2-\mathbf{J}_3\times\mathbf{J}_4)}{4\Gamma_1\Gamma_2+(\mathbf{J}_1\cdot\mathbf{J}_2+\mathbf{J}_1\cdot\mathbf{J}_4-\mathbf{J}_3\cdot\mathbf{J}_2-\mathbf{J}_3\cdot\mathbf{J}_4)} + \frac{(\mathbf{J}_1\times\mathbf{J}_2-\mathbf{J}_1\times\mathbf{J}_4+\mathbf{J}_3\times\mathbf{J}_2-\mathbf{J}_3\times\mathbf{J}_4)}{4\Gamma_1\Gamma_2+(\mathbf{J}_1\cdot\mathbf{J}_2-\mathbf{J}_1\cdot\mathbf{J}_4+\mathbf{J}_3\cdot\mathbf{J}_2-\mathbf{J}_3\cdot\mathbf{J}_4)} + \frac{(\mathbf{J}_1\times\mathbf{J}_2-\mathbf{J}_1\times\mathbf{J}_4-\mathbf{J}_3\times\mathbf{J}_2+\mathbf{J}_3\times\mathbf{J}_4)}{4\Gamma_1\Gamma_2-(\mathbf{J}_1\cdot\mathbf{J}_2-\mathbf{J}_1\cdot\mathbf{J}_4-\mathbf{J}_3\cdot\mathbf{J}_2+\mathbf{J}_3\cdot\mathbf{J}_4)}\right] \tag{10.25}$$

$$\dot{\mathbf{J}}_2 = \frac{\Gamma_1\Gamma_2}{4\pi}\left[\frac{(\mathbf{J}_1\times\mathbf{J}_2+\mathbf{J}_1\times\mathbf{J}_4+\mathbf{J}_3\times\mathbf{J}_2+\mathbf{J}_3\times\mathbf{J}_4)}{4\Gamma_1\Gamma_2-(\mathbf{J}_1\cdot\mathbf{J}_2+\mathbf{J}_1\cdot\mathbf{J}_4+\mathbf{J}_3\cdot\mathbf{J}_2+\mathbf{J}_3\cdot\mathbf{J}_4)} + \frac{(\mathbf{J}_1\times\mathbf{J}_2+\mathbf{J}_1\times\mathbf{J}_4-\mathbf{J}_3\times\mathbf{J}_2-\mathbf{J}_3\times\mathbf{J}_4)}{4\Gamma_1\Gamma_2+(\mathbf{J}_1\cdot\mathbf{J}_2+\mathbf{J}_1\cdot\mathbf{J}_4-\mathbf{J}_3\cdot\mathbf{J}_2-\mathbf{J}_3\cdot\mathbf{J}_4)} + \frac{(\mathbf{J}_1\times\mathbf{J}_2-\mathbf{J}_1\times\mathbf{J}_4+\mathbf{J}_3\times\mathbf{J}_2-\mathbf{J}_3\times\mathbf{J}_4)}{4\Gamma_1\Gamma_2+(\mathbf{J}_1\cdot\mathbf{J}_2-\mathbf{J}_1\cdot\mathbf{J}_4+\mathbf{J}_3\cdot\mathbf{J}_2-\mathbf{J}_3\cdot\mathbf{J}_4)} + \frac{(\mathbf{J}_1\times\mathbf{J}_2-\mathbf{J}_1\times\mathbf{J}_4-\mathbf{J}_3\times\mathbf{J}_2+\mathbf{J}_3\times\mathbf{J}_4)}{4\Gamma_1\Gamma_2-(\mathbf{J}_1\cdot\mathbf{J}_2-\mathbf{J}_1\cdot\mathbf{J}_4-\mathbf{J}_3\cdot\mathbf{J}_2+\mathbf{J}_3\cdot\mathbf{J}_4)}\right] \tag{10.26}$$

$$\dot{\mathbf{J}}_3 = \frac{1}{4\pi}\frac{\mathbf{J}_1\times\mathbf{J}_3}{\left(1+\frac{\|\mathbf{J}_1\|^2-\|\mathbf{J}_3\|^2}{4\Gamma_1^2}\right)} - \frac{\Gamma_1\Gamma_2}{4\pi}\left[\frac{(\mathbf{J}_1\times\mathbf{J}_2+\mathbf{J}_1\times\mathbf{J}_4+\mathbf{J}_3\times\mathbf{J}_2+\mathbf{J}_3\times\mathbf{J}_4)}{4\Gamma_1\Gamma_2-(\mathbf{J}_1\cdot\mathbf{J}_2+\mathbf{J}_1\cdot\mathbf{J}_4+\mathbf{J}_3\cdot\mathbf{J}_2+\mathbf{J}_3\cdot\mathbf{J}_4)} + \frac{(\mathbf{J}_1\times\mathbf{J}_2-\mathbf{J}_1\times\mathbf{J}_4+\mathbf{J}_3\times\mathbf{J}_2-\mathbf{J}_3\times\mathbf{J}_4)}{4\Gamma_1\Gamma_2+(\mathbf{J}_1\cdot\mathbf{J}_2-\mathbf{J}_1\cdot\mathbf{J}_4+\mathbf{J}_3\cdot\mathbf{J}_2-\mathbf{J}_3\cdot\mathbf{J}_4)} - \frac{(\mathbf{J}_1\times\mathbf{J}_2+\mathbf{J}_1\times\mathbf{J}_4-\mathbf{J}_3\times\mathbf{J}_2-\mathbf{J}_3\times\mathbf{J}_4)}{4\Gamma_1\Gamma_2+(\mathbf{J}_1\cdot\mathbf{J}_2+\mathbf{J}_1\cdot\mathbf{J}_4-\mathbf{J}_3\cdot\mathbf{J}_2-\mathbf{J}_3\cdot\mathbf{J}_4)} - \frac{(\mathbf{J}_1\times\mathbf{J}_2-\mathbf{J}_1\times\mathbf{J}_4-\mathbf{J}_3\times\mathbf{J}_2+\mathbf{J}_3\times\mathbf{J}_4)}{4\Gamma_1\Gamma_2-(\mathbf{J}_1\cdot\mathbf{J}_2-\mathbf{J}_1\cdot\mathbf{J}_4-\mathbf{J}_3\cdot\mathbf{J}_2+\mathbf{J}_3\cdot\mathbf{J}_4)}\right] \tag{10.27}$$

$$\dot{\mathbf{J}}_4 = \frac{1}{4\pi}\frac{\mathbf{J}_2 \times \mathbf{J}_4}{\left(1 + \frac{\|\mathbf{J}_2\|^2 - \|\mathbf{J}_4\|^2}{4\Gamma_2^2}\right)} \tag{10.28}$$
$$+ \frac{\Gamma_1\Gamma_2}{4\pi}\left[\frac{(\mathbf{J}_1 \times \mathbf{J}_2 + \mathbf{J}_1 \times \mathbf{J}_4 + \mathbf{J}_3 \times \mathbf{J}_2 + \mathbf{J}_3 \times \mathbf{J}_4)}{4\Gamma_1\Gamma_2 - (\mathbf{J}_1 \cdot \mathbf{J}_2 + \mathbf{J}_1 \cdot \mathbf{J}_4) + \mathbf{J}_3 \cdot \mathbf{J}_2 + \mathbf{J}_3 \cdot \mathbf{J}_4)}\right.$$
$$+ \frac{(\mathbf{J}_1 \times \mathbf{J}_2 + \mathbf{J}_1 \times \mathbf{J}_4 - \mathbf{J}_3 \times \mathbf{J}_2 - \mathbf{J}_3 \times \mathbf{J}_4)}{4\Gamma_1\Gamma_2 + (\mathbf{J}_1 \cdot \mathbf{J}_2 + \mathbf{J}_1 \cdot \mathbf{J}_4 - \mathbf{J}_3 \cdot \mathbf{J}_2 - \mathbf{J}_3 \cdot \mathbf{J}_4)}$$
$$- \frac{(\mathbf{J}_1 \times \mathbf{J}_2 - \mathbf{J}_1 \times \mathbf{J}_4 + \mathbf{J}_3 \times \mathbf{J}_2 - \mathbf{J}_3 \times \mathbf{J}_4)}{4\Gamma_1\Gamma_2 + (\mathbf{J}_1 \cdot \mathbf{J}_2 - \mathbf{J}_1 \cdot \mathbf{J}_4 + \mathbf{J}_3 \cdot \mathbf{J}_2 - \mathbf{J}_3 \cdot \mathbf{J}_4)}$$
$$\left.- \frac{(\mathbf{J}_1 \times \mathbf{J}_2 - \mathbf{J}_1 \times \mathbf{J}_4 - \mathbf{J}_3 \times \mathbf{J}_2 + \mathbf{J}_3 \times \mathbf{J}_4)}{4\Gamma_1\Gamma_2 - (\mathbf{J}_1 \cdot \mathbf{J}_2 - \mathbf{J}_1 \cdot \mathbf{J}_4 - \mathbf{J}_3 \cdot \mathbf{J}_2 + \mathbf{J}_3 \cdot \mathbf{J}_4)}\right].$$

Note that the right-hand sides of equations (10.25) and (10.26) contain only interaction terms between the two dipoles, while the first terms in equations (10.27) and (10.28) are the same as for the uncoupled system which would cause each to move along a great circle in the absence of other terms. The additional coupling terms in equations (10.27) and (10.28) cause the centroid paths of $\mathbf{J}_3$ and $\mathbf{J}_4$ to deviate from geodesic motion. The effect of these coupling terms is our main focus.

10.4 The fundamental interactions

Exchange scattering mode

The simplest example of an exchange scattering process is shown in figure 10.3. This is a symmetric case of exchange scattering of twin dipoles which initially have orientation $\gamma_1 = \gamma_2 = \pi/2$. In this case, $\mathbf{J} = 0$. Two exchange events take place (i.e. the dipoles exchange partners) at antipodal points during one periodic cycle. Figure 10.3(a) shows the exchange scattering on the non-rotating sphere, while figure 10.3(b) shows the same event on the rotating sphere, where the dipole frequency, $\Omega_1 = \Omega_2$, to the solid-body frequency is a 1:1 ratio.

Non-exchange scattering mode

An example of non-exchange scattering is shown in figure 10.4 for twin dipoles. Figure 10.4(a) shows the basic interaction where the dipole orientations are $\gamma_1 = 90\,\text{deg}$, $\gamma_2 = 80\,\text{deg}$, hence $\mathbf{J} \neq 0$, and the sphere is not rotating. Although there is clear interaction between the two dipoles as indicated by the deviation of the dipole paths from great circle paths, no

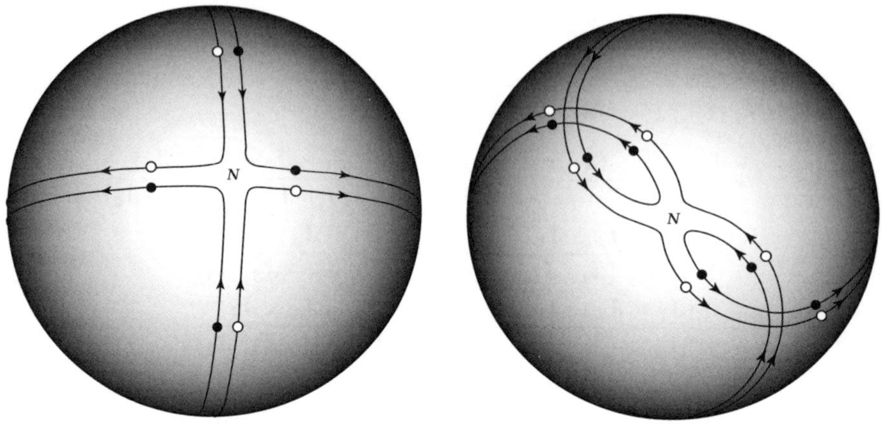

(a) (b)

Fig. 10.3 Symmetric exchange scattering ($\mathbf{J} = 0$) of twin dipoles on the sphere. Two exchange events take place at antipodal points during one periodic cycle. (a) Non-rotating sphere; (b) Rotating sphere 1:1 frequency ratio.

partner exchange takes place throughout the period and the dipoles avoid a direct collision. Figure 10.4(b) shows the long-time trajectory on the rotating sphere, while figure 10.4(c) shows the evolution of the tips of the $\mathbf{J}_1$ and $\mathbf{J}_2$ vectors during this event. Whether the orbit shown in figure 10.4(b) ultimately closes, or densely covers a portion of the sphere, crucially depends on the orientation angles of the two dipoles, and both periodic orbits and quasi-periodic orbits co-exist. Figure 10.4(d) shows a non-exchange scattering event on the rotating sphere where the frequency ratio $\Omega_1 = \Omega_2 : \Omega$ is $1 : 1$. The orientation angles for this case are $\gamma_1 = 90\,\text{deg}$, $\gamma_2 = 0\,\text{deg}$.

Loop-scattering mode

Shown in figure 10.5 is an example of a loop-scattering interaction for two unequal dipoles in which $\Omega_1{:}\Omega_2$ is 3:1. Figure 10.5(a) shows a case with dipole strengths $\Gamma_1 = -\Gamma_3 = 1.0$, $\Gamma_2 = -\Gamma_4 = 3.0$ with orientations $\gamma_1 = \gamma_2 = \pi/2$ on the non-rotating sphere. The dipoles perform a sequence of loops as they travel around the sphere, the number of loops depends on the frequency ratio of the unequal dipoles. On each loop, one of the dipoles loops inside the other, which accomodates the passage by splitting around the inner loop. Within the loop, the dipoles exchange partners, forming two new dipoles comprised of vortices of opposite sign but unequal magnitude. As a result, they move along curved trajectories within the loop. Figure 10.5(b) shows the loop-exchange scattering process on the rotating sphere.

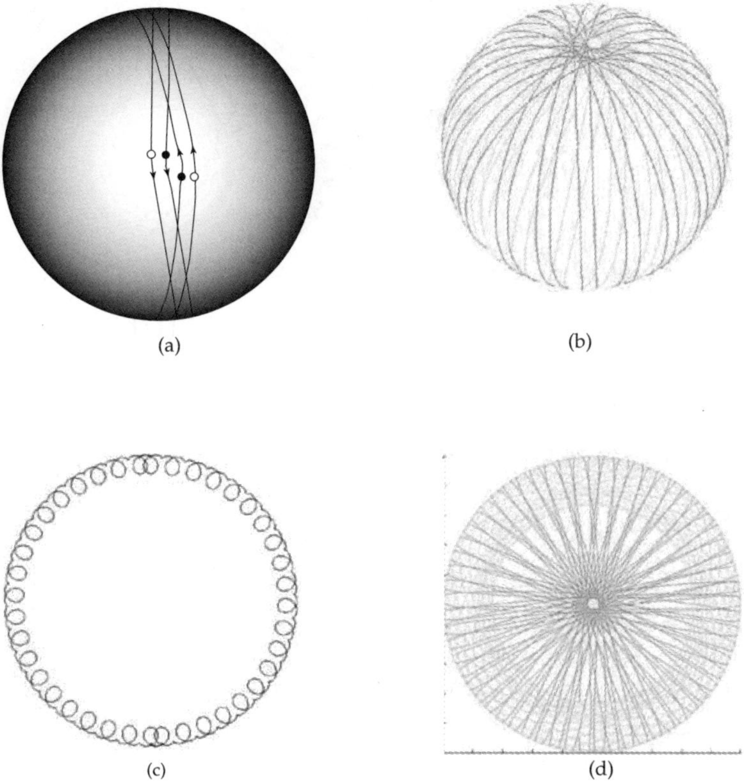

Fig. 10.4 Non-exchange scattering of twin dipoles. Each dipole retains its partner throughout the cycle. (a) Non-rotating sphere (northern hemisphere): $\gamma_1 = 90\,\text{deg}$, $\gamma_2 = 80\,\text{deg}$; (b) Long-time trajectory on the rotating sphere; (c) Tips of the $\mathbf{J}_1$ and $\mathbf{J}_2$ vectors; (d) Rotating sphere (northern hemisphere) with 1:1 frequency ratio, $\gamma_1 = 90\,\text{deg}$, $\gamma_2 = 0\,\text{deg}$.

Shown is a case with unequal dipoles with frequency ratio $\Omega_1 : \Omega_2 = 2 : 1$ and $\Omega_1 : \Omega = 1 : 1$. While the dipole trajectories are relatively complex, the tips of the $\mathbf{J}$ (circle), $\mathbf{J}_1$, and $\mathbf{J}_2$ vectors move on closed periodic orbits as shown inside the sphere revealing the integrable nature of the interaction.

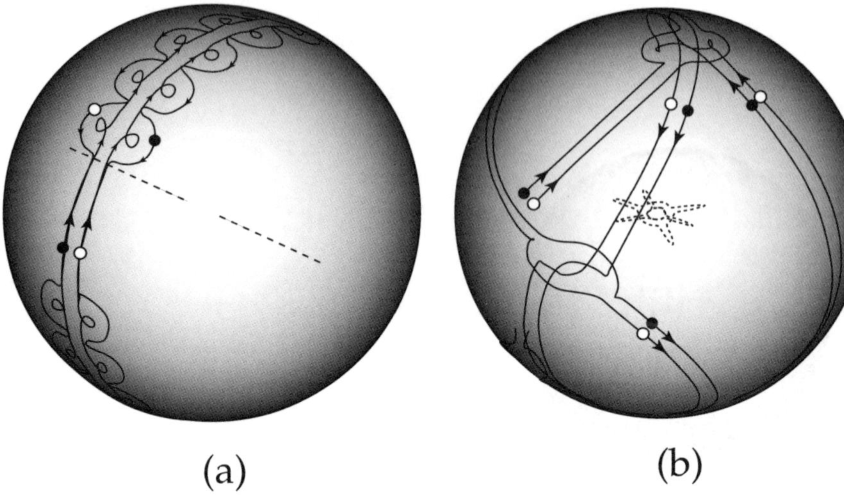

Fig. 10.5 Loop-exchange scattering events for unequal dipoles. Vortex trajectories are shown as solid curves, tips of $\mathbf{J}_1$, $\mathbf{J}_2$ and $\mathbf{J}$ vectors are shown as dashed curves. (a) Non-rotating sphere; (b) Rotating sphere.

10.5 Concluding remarks

Although the scattering trajectories shown in this paper are integrable, we believe they represent the key nonlinear modes that commonly appear under a wide range of conditions for both integrable as well as chaotic processes. Understanding these fundamental interactions is part of an ongoing effort to model turbulent vorticity fields using multipolar ballistic elements in both a dynamical setting [7] and a kinetic theory setting [8] in which interactions are used to produce rational closure schemes based on combinatorial liklihood. The multi-dipole scattering process is quite rich dynamically, with more complex interactions occurring, particularly in the chaotic regime. A more thorough study of these interactions will be reported elsewhere where the dipole coordinate system will be exploited further.

Bibliography

[1] Aref H., P.K. Newton, M.A. Stremler, T. Tokieda, D.L. Vainchtein. Vortex Crystals. *Advances in Applied Mechanics* (2003); **V**ol. 39; 1–79.
[2] Eckhardt B., H. Aref Integrable and chaotic motion of four vortices. II. Col-

lision dynamics of vortex pairs. *Phil. Trans. R. Soc. Lond. A* (1988); **3**26; 655–696.

[3] Kidambi R., P.K. Newton. Motion of three point vortices on a sphere. *Physica D* (1998); **1**16; 143–175.

[4] Kidambi R., P.K. Newton. Streamline topologies for integrable vortex motion on a sphere. *Physica D* (2000); **1**40; 95–125.

[5] Kimura Y. Vortex motion on surfaces with constant curvature. *Proc. Roy. Soc. Lond.* **A** 455, (1999); 245–259.

[6] Newton P.K. **T**he N-Vortex Problem: Analytical Techniques. Applied Math. Sci. **145**, Springer-Verlag, New York, (2001).

[7] Newton P.K. The dipole dynamical system. *Discrete and Cont. Dyn. Sys. Series B* Proceedings of the Fifth International Conference on Dynamical Systems and Differential Equations, June (2004); 1–8.

[8] Newton P.K., I. Mezic. Non-equilibrium statistical mechanics for a vortex gas. *Journal of Turbulence* (2002); 3.

[9] Newton P.K., Shokraneh H. The N-vortex problem on a rotating sphere: I. Multi-frequency states. University of Southern California preprint, (2004).

[10] Ting L., C. Tung. Motion and decay of a vortex in a nonuniform stream. *Phys. Fluids* (1965); **8**; 1309-1351.

Chapter 11

Magneto-Fluid-Dynamic Flow Control

J.S. Shang
Mechanical and Materials Engineering Department
Wright State University, Dayton, OH 45435

11.1 Dedication

I dedicate my work to my teacher, Prof. Lu Ting, on the occasion of his eightieth birthday. He has instilled the discipline to my lifetime learning experience; for that I am truly grateful.

11.2 Abstract

The present effort demonstrates an innovative hypersonic flow control mechanism using magneto-fluid-dynamic interaction. An electromagnetic perturbation is introduced to increase the growth rate of boundary-layer displacement thickness near a sharp leading edge. The ensuing viscous-inviscid interaction induces a high-pressure plateau for flow control. The electromagnetic force of low-temperature plasma is modeled by the drift-diffusion theory. This approach introduces added physical dimensions in the Joule heating and Lorentz force for flow field modification. Numerical results in the absence of an externally applied magnetic field are fully substantiated by experimental observations. However, the effectiveness of this mode of magneto-fluid-dynamic interaction still requires further validation.

11.3 Introduction

Recent magneto-fluid-dynamics (MFD) research has made impressive progress in basic understanding for interdisciplinary science [1,2]. Using electromagnetic force to enhance the aerodynamic performance was advocated by Sears et al. [3] in the late 1950's and just a few years ago by the Ajax Program of Russia [4]. Revolutionary and attractive ideas have been put forward but are limited by our ability to analyze these extremely complex physics. The assessment of these revolutionary concepts at best is inconclusive. However, the rejuvenate interest in MFD offers some realizable opportunities for flow control. In these endeavors, the flow medium must be electrically conducting. Since weakly ionized gas is often encountered in hypersonic flow, applications of MFD for flow control are naturally concentrated in this flow regime.

The energy required to generate volumetric plasma that has a sufficient charged particle number density for strong MFD interaction is enormous [5-8]. The ionization potential always underestimates the energy requirement because the nonequilibrium energy cascades to vibration excitation, recombination and attachment processes [7,8]. In the past, the strong MFD interaction is further enhanced by an externally applied magnetic field. The relative magnitude of the electromagnetic force and the fluid inertia of the interacting phenomenon is described by an interaction parameter, the Stuart number, $S = \sigma B^2 L/\rho u$ [5,6]. Since the magnetic field intensity is maximal at the magnetic poles and the inertia is minimal in the inner region of a shear layer, the most intensive MFD interaction always occurs near a solid surface.

The description of weakly ionized gas is the most challenging issue for both experimental and computational research efforts [9-11]. In most laboratory experiments, the plasma is generated by electron collision with embedded electrodes on control surfaces. The generated plasma consists of electrons in a highly excited state, but the heavy ions retain the thermodynamic condition of its surroundings [7]. Unfortunately, all the well-known methods in high-temperature chemical kinetics that have developed for hypersonic flow are inapplicable for describing ionization by the electron collision process. Recently a drift-diffusion plasma model by Surzhikov and Shang has successfully simulated a glow discharge in a magnetic field [12]. The numerical results substantiate the classic observations that an order of magnitude greater electrical conductivity exists above the plasma sheath region [7]. This condition of a combined high electrical conductivity and strong applied magnetic field greatly ensures a strong electromagnetic-fluid-dynamic interaction [10,13].

Classic hypersonic flow theory describes the inviscid-viscous interaction over a sharp leading edge by Hayes and Probstein as the pressure interaction [14]. The magnitude of the induced pressure can be calculated by a single interaction parameter $\overline{\chi} = M^3(C/Re)^{1/2}$. The cubic power dependence of the free stream Mach number strongly amplifies the pressure interaction in hypersonic flows. Several experimental and computational investigations have been exploiting this possibility by introducing the electromagnetic effect for flow control [9-11]. In short, this control mechanism can be instantaneously actuated in microseconds to deflect the flow over a fixed control surface, emulating a deflected flap.

To pursue a promising innovation, the present effort attempts to describe a basic concept that may renew a frontier for fluid dynamics research. In this process, some accomplishments for developing an effective plasma actuator for hypersonic flow control and possible research opportunities are delineated.

11.4 Governing equations

The governing equations of Magnetohydrodynamics (MHD) constitute of the time-dependent compressible Navier-Stokes equations and the Maxwell equations in the time domain. The plasma is a gas medium in which the long range Coulomb force dominates the global behavior of the charged particles. A critical length scale of plasma is the Debye shielding length. Within this distance, the surrounding charged particles effectively shield any two particles from interacting with each other. As a consequence, the plasma is globally neutral and the conductive electrical current, D(t), is negligible up to the microwave frequency [5]. Invoking these traditional MHD approximations, the governing equations become:

$$\frac{\partial \rho}{\partial t} + \nabla \cdot (\rho u) = 0 \tag{11.1}$$

$$\frac{\partial B}{\partial t} + \nabla \cdot (uB - Bu) = -\nabla \times [(\nabla \times B/\mu_e)/\sigma] \tag{11.2}$$

$$\frac{\partial(\rho u)}{\partial t} + \nabla \cdot [\rho uu - BB/\mu_e - \overline{\overline{\tau}} + (p + B \cdot B/2\mu_e)\overline{\overline{I}}] = 0 \tag{11.3}$$

$$\frac{\partial(\rho e)}{\partial t} + \nabla \cdot \left[(\rho e)u + Q + u \cdot (p\overline{\overline{I}} - \overline{\overline{\tau}})\right] = |\nabla \times B/\mu_e|^2 /\sigma \tag{11.4}$$

A total of four different waves propagate in a MHD field, the Alfven, slow and fast plasma waves are transverse waves, as well as, a longitudinal acoustic wave to make the wave structure more complex [5,6]. The formulation of the wave dominant equation of motion also becomes very complicated, because eigenvalues of the MHD equations can locally degenerate to coincide with each other, dependent on the relative magnitude and polarity of the magnetic field [1,2,15].

For most aerospace applications, the magnetic Reynolds number ($R_{em} = \mu_m \sigma u L$) is much less than unity, which means the induced magnetic flux density is negligible in comparison with the externally applied field [5,6]. Under this circumstance, Faraday's induction law of the Maxwell equations can be decoupled from the rest. In this formulation, the Lorentz force and Joule heating appear as source terms in the modified Navier-Stokes equations [2,10,12,16].

$$\frac{\partial \rho}{\partial t} + \nabla \cdot (\rho u) = 0 \tag{11.5}$$

$$\frac{\partial \rho u}{\partial t} + \nabla \cdot \left(\rho uu + p\overline{\overline{I}} - \overline{\overline{\tau}}\right) - J \times B = 0 \tag{11.6}$$

$$\frac{\partial \rho e}{\partial t} + \nabla \cdot \left[\rho e u + Q + u \cdot \left(p\overline{\overline{I}} - \overline{\overline{\tau}}\right)\right] - E \cdot J = 0 \tag{11.7}$$

Since the basic partial differential system consists mainly of the compressible Navier-Stokes equations and the electromagnetic variables in source terms that will not modify the eigenvector, all traditional flux splitting formulations and solving procedures are directly applicable. Accordingly, the system of equations can be easily cast into the flux vector form and solved by numerical procedures developed in the CFD community [10,16].

$$\frac{\partial U}{\partial t} + \frac{\partial F_\xi}{\partial \xi} + \frac{\partial F_\eta}{\partial \eta} + \frac{\partial F_\zeta}{\partial \zeta} = 0 \tag{11.8}$$

The boundary conditions for the aerodynamic variables are straightforward, thus will not be repeated here [10,16]. For the electromagnetic variable, in general, the tangential component of the electrical field intensity and the normal component of the magnetic flux density are continuous across the media interface [15].

11.5 Plasma models

Two fundamental mechanisms of charged particle movement are the drift velocity and the ordinary diffusion [7]. This behavior is independent of how the gas discharge is generated. Partially ionized gas produced by thermal collision has been extensively studied in a high temperature environment and the modeling of the plasma has been built on the Lighthill's model for dissociating gas and the Saha equation for ionization [15,17]. The ionization by electron collision is widely used for flow control, the physical model however is not as well known. Part of the reason is that the physical phenomenon is very complex and involves interaction at the atomic structure level of gas and solid [7]. The electron component is the result of the avalanche growth of electrons number density produced by secondary emission from the electrodes. In an AC field, the displacement current increases with frequency, while the importance of the secondary emission diminishes [13].

In self-sustaining plasma, the rate of change for charged number density in a control volume is mainly balanced by generation through ionization and depletion by recombination. The continuity equations for species concentration of the two-component plasma are given by the drift-diffusion theory [12].

$$\frac{\partial n_e}{\partial t} + \nabla \cdot \Gamma_e = \alpha(E,P)|\Gamma_e| - \beta n_e n_i \tag{11.9}$$

$$\frac{\partial n_i}{\partial t} + \nabla \cdot \Gamma_i = \alpha(E,P)|\Gamma_e| - \beta n_e n_i \tag{11.10}$$

The electrical field intensity in the low Magnetic Reynolds number limit must satisfy the charge conservation equation and the global neutral condition to be compatible with the drift-diffusion formulation.

$$\nabla \cdot [(\mu_i + \mu_e^*)nE + (D_e^* - D_e)\nabla n] = 0, \tag{11.11}$$

where $\Gamma_e = -D_e \nabla n_e + n_e \mu_e E$ and $\Gamma_i = -D_i \nabla n_i + n_i \mu_i E$ are the electron and ion flux densities, respectively. In this formulation, $\alpha(E,p)$ and β are the first Townsend ionization coefficient and recombination coefficient, μ_e and μ_i are the electron and ion mobility, and D_e and D_i are the electron and ion diffusion coefficients [7,12]. In most electrodynamic formulations, the electrical field intensity is replaced by an electrical potential function,

$$E = -\nabla \phi. \tag{11.12}$$

The compatibility condition, equation (11.9) and equation (11.10) lead to the well-known Poisson equation for the electrical potential ϕ.

The electrical current density, by definition, is given by:

$$J = e(\Gamma_i - \Gamma_e). \tag{11.13}$$

In the presence of a magnetic field, the Hall effect can be included by explicitly modifying the mobility of ions and electrons, diffusion and ionization coefficients of the drift-diffusion theory [12].

For the drift-diffusion model, the appropriate boundary on the cathode requires enforcement of zero electrical potential and the number density of electrons is proportional to the coefficient of secondary electron emission [7,12]. It is assumed that the anode reflects all ions; the number density of the ion vanishes and electrical potential is prescribed by the difference over the electrodes. On the dielectric surface, the charged number density is negligible and the outward normal gradient of the electrical potential is zero.

In view of the great disparity of the characteristic speeds between the sonic speed of fluid dynamic motion and speed of light of the electrodynamics, the governing equations are solved loosely coupled. In that, the much-short-time-scale electrodynamic equations are solved iteratively by the SOR (Successive Over Relaxation) scheme in a pentadiagonal matrix system [10,12]. All numerical results are obtained with a nominal second-order spatial and temporal resolution.

11.6 Electro-fluid-dynamic interaction

The viscous-inviscid interaction near the sharp leading edge of a slender body reveals that a perturbation to the growth rate of the displacement thickness of a shear layer can be greatly amplified [14]. This non-intrusive electromagnetic perturbation appears to be very attractive, because embedded electrodes with an applied electric field can ignite a glow discharge initiating an electro-fluid-dynamic (EFD) interaction. In fact, the glow discharge produces two distinct perturbations to the structure of a shear layer. First, the electrode heating raises the surface temperature and second, the Joule heating increases the gas temperature above the surface in the glow discharge domain. Both mechanisms produce a thermal perturbation to the shear-layer structure.

A schematic of direct current discharge over a sharp leading edge wedge in a Mach number 5.15 hypersonic stream is displayed in Figure 1. The plasma is generated between two electrodes embedded in the flat plate surface. A total electrical current of 50 mA is maintained by an applied electric field of 1.2 kV in the external circuit. The maximum electron number density of the plasma is $3 \times 10^{12}/cm^3$, and the electrode temperature is estimated to be 600 K [10].

The computed temperature contours of the EFD interaction are depicted in a composite presentation. Figure 2 consists of the computed results of a simple hypersonic boundary layer, the same boundary layer with electrode heating, and finally the glow discharge over a wedge model. It is clearly displayed that the effect of Joule heating is much more pronounced than electrode heating. Since the heating mechanism is vastly different, the characteristic time scale for conduction heating is in seconds and the Joule heating occurs in much shorter time scale.

Figure 3 presents the computed charged particle number density over the embedded electrodes. The computed ion number density distribution above the anode reaches excellent agreement with experimental observation. However, there is a large discrepancy between computational and experimental results for the ion number density profile directly above the cathode. The difference between these two results is confined within the measurement disparity between data obtained by the microwave absorption technique and Langmuir probing [9,11].

The computed stagnation temperature profiles produce reasonable agreement with measurements obtained by a shield stagnation temperature probe [9,11]. In Figure 4, the comparison is given in dimensionless temperature normalized by the free-stream temperature (43 K) over the electrodes. The most outstanding feature of the computational results is that the Joule heating is observed to dominate, in that the gas temperature near the wall is actually hotter than the surface temperatures of both cathode and the anode.

In Figure 5 depicts the pressure distributions on the wedge surface. The effect of Joule heating transforms the perturbation to the boundary-layer structure into an intensified viscous-inviscid interaction. The observation is made easier by a direct contrast with the computed results of a classical hypersonic pressure interaction, in which the induced pressure is concentrated mostly at the leading edge of the wedge. On the other hand, the Joule heating triggers additional compression waves above the cathode and anode. These waves coalesce and eventually merge with the oblique shock wave originating from the sharp leading edge. The resulting shock wave produces a higher-pressure rise over the wedge surface.

The relative effectiveness of the EFD interaction as a flow control mechanism is confirmed by experimental data. In Figure 6, numerical results exhibit reasonable agreement with data. More importantly, the glow discharge exhibits a bona fide electro-aerodynamic interaction that is not possible by electrode heating alone. At a very low-level power supply of 60 watts, the EFD interaction generates an equivalent surface deflection of one degree. The maximum effect of the studied EFD interaction for flow control is equivalent to a flow deflection of 7 degrees. From this result, a scaling of power required for the plasma actuator per electrode length is 18.90 watts per cm per degree of deflection.

11.7 Magneto-fluid-dynamic interaction

The full impact of magneto-aerodynamic interaction requires plasma in the presence of an externally applied magnetic field [3]. A transverse magnetic field exerts a profound effect on plasma and especially alters characteristics of the plasma structure including that of the electrode sheaths [7,12]. The basic phenomenon is governed by reduced electron mobility in a magnetic field. For hypersonic flow control, very little is known about the electrodynamic structure, but can only be analyzed from the fundamental collision process and experimental observations.

The present numerical simulation of magneto-fluid-dynamics (MFD) interaction duplicates experimental efforts [9,11]. In this experiments, the transverse magnetic field generates a Lorentz force, $J \times B$ that either expels ($J \times B > 0$) or restrains ($J \times B < 0$) the charged particles to the electrodes. It is anticipated that for the case where the charged particles are expelled from the plate surface, the inelastic collisions between ions and neutral particles enhances the viscous-inviscid interaction. Conversely, the restrained discharge particle motion suppresses the intensity of the pressure interaction. Preliminary experimental observations have confirmed this finding, but uncertainty of discharge instability is evident over a wide range of magnetic field strengths [9,11].

At the testing condition, the electron Larmor frequency at B=1.0 Tesla is $1.76 \times 10^{11} radian/s$ and the electron-heavy particle collision frequency is estimated to be in the range from $3.9 \times 10^9/s$ to $2.3 \times 10^{10}/s$. Under these conditions, the behavior of the plasma in the hypersonic MHD channel is not collisionally dominant, and the maximum and effective Hall parameters range approximately from 7.6 to 45.1. In order to alleviate the uncertainty, the simulations are limited to a lower intensity of magnetic field, $-0.2T < B < 0.2T$, then the Hall parameter is reduced to the order of unity.

In numerical simulations, this effect of the Lorentz force is easily detected from the computed charged particle density contours over the electrodes as shown in Figure 7. In the absence of an external magnetic field, the highest degree of ionization of the plasma model reaches a value of $8.11 \times 10^{10}/cc$ near the electrodes, and this value is compatible with the measurement. In the presence of an externally applied magnetic field, the discharge pattern is substantially altered. The expelling Lorentz force $(J \times B > 0)$ indeed pushes the charged particles away from the electrodes and enlarges the discharge domain. The degree of ionization is also reduced at the electrodes by 24.1% in comparison with the glow discharge without the external magnetic field. The trend of increasing discharge domain is reversed with the opposite polarity of the magnetic field. The charged particles are strongly constricted to the electrodes with a maximum value of $1.06 \times 10^{11}/cc$ in the cathode and anode layers and the discharge domain is significant reduced.

The effect of the Lorentz force is easily detected from the computed surface pressure over the electrodes in Figure 8. The computed result generally over predicts the pressure measurement, and the data also exhibits an unusually large data scattering band due to the unstable glow discharge pattern in the presence of an applied and uniform magnetic field. The expelling Lorentz force $(J \times B > 0)$ indeed pushes the charged particles away from the electrodes and intensifies the viscous-inviscid interaction. As the consequence, the induced surface pressure is greater than the EFD interaction under an identical electric field strength. The trend of increasing discharge domain is reversed with the opposite polarity of the magnetic field $(J \times B < 0)$. The induced surface pressure is also diminishing accordingly by the suppressing outward deflection of the streamline. However, there is also an increased effort required for maintaining the computational stability. The numerical results using the drift-diffusion model capture the difference between the applied magnetic fields of opposite polarity.

11.8 Concluding remarks

The plasma actuator using electromagnetic perturbation to the shear layer structure and amplified by the viscous-inviscid interaction in hypersonic stream has been fully substantiated by experimental observations. The basic mechanism is the combined effects of the conductive electrode and volumetric Joule heating. The time scale of the Joule heating is estimated in the microsecond range and is much shorter than the conducting electrode heating and ensued convection. The electro-fluid-dynamic interaction by a plasma actuator embedded in a fixed and non-movable plate surface behaves as if the plate is executing a pitching movement. Both experimental and computational results have shown the effectiveness of flow deflection can be scaled as 18 watts per degree per cm of electrode length.

The magneto-fluid-dynamic interaction exhibits even greater amplification with an applied external magnetic field than the electro-fluid-dynamic interaction. The drift-diffusion plasma model has been successfully applied to numerical simulations, and the Hall effect is explicitly included into the formulation. From both experiments and computations, the Hall effect significantly constricts the mobility of charged particles and alters the surface discharge pattern.

Acknowledgment

The invaluable contributions by Prof. S.T. Surzhikov of Russian Academy of Science, Dr. R. Kimmel of Air Force Research Laboratory, and Prof. J. Menart of Wright State University are sincerely acknowledged.

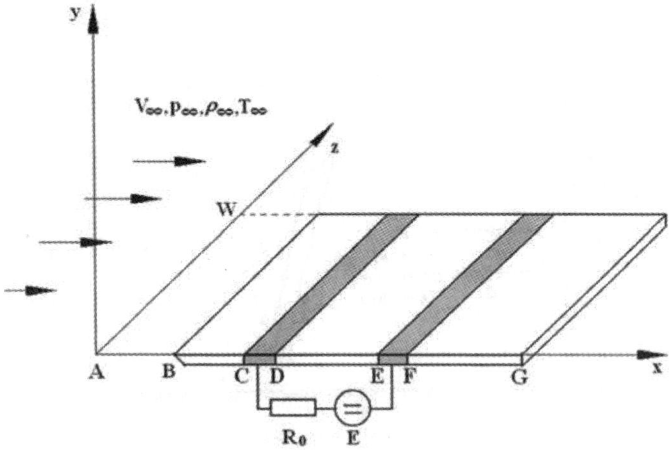

Fig 1. Schematic of embedded electrodes for electro-fluid-dynamic interaction

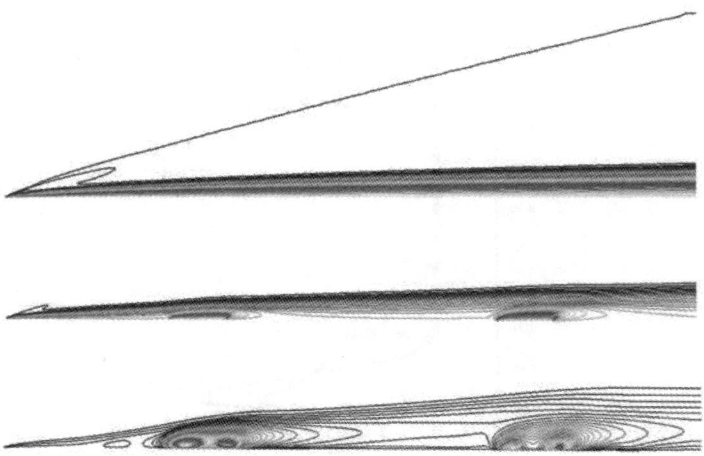

Fig 2. Modified shear layer structure by MFD interaction

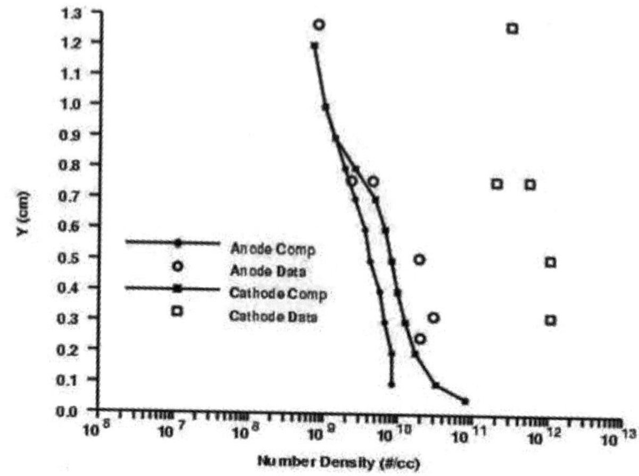

Fig 3. Ion number density distributions over electrodes

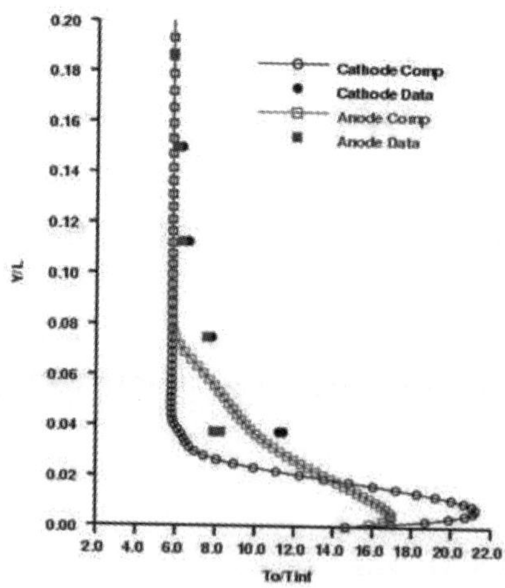

Fig 4. Comparison of stagnation temperature profiles over electrodes

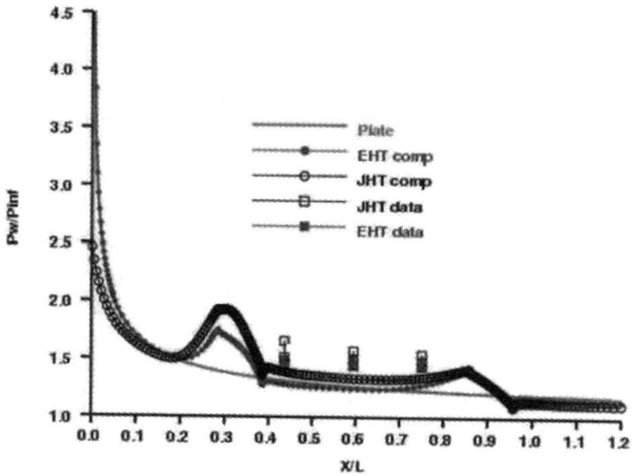

Fig 5. Surface pressure distributions with/without EFD interaction

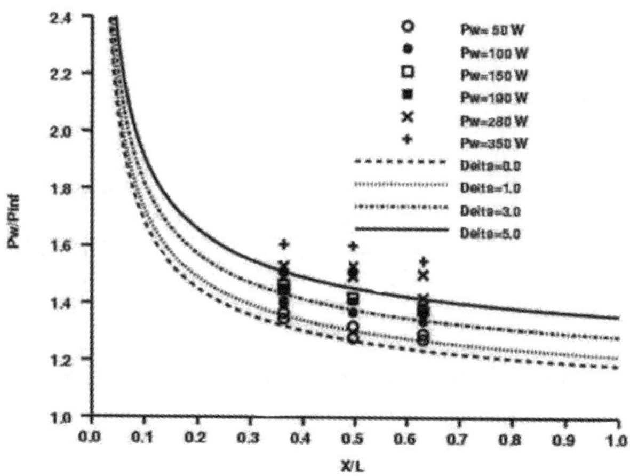

Fig 6. Effectiveness of the EMT interactions for flow control

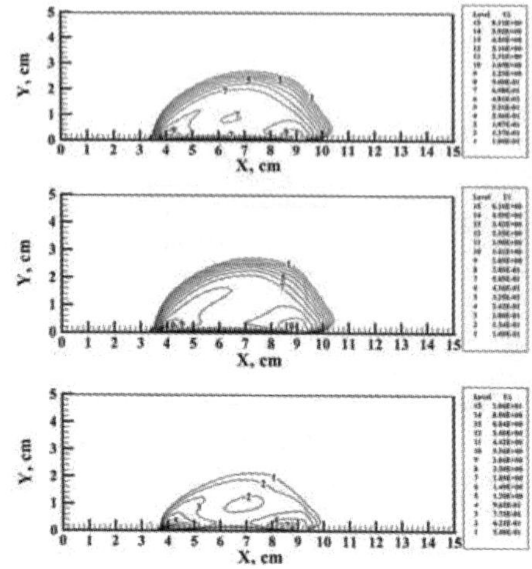

Fig 7. Charged particle density Contours for B=0,B=+0.2T,and B=-0.2T

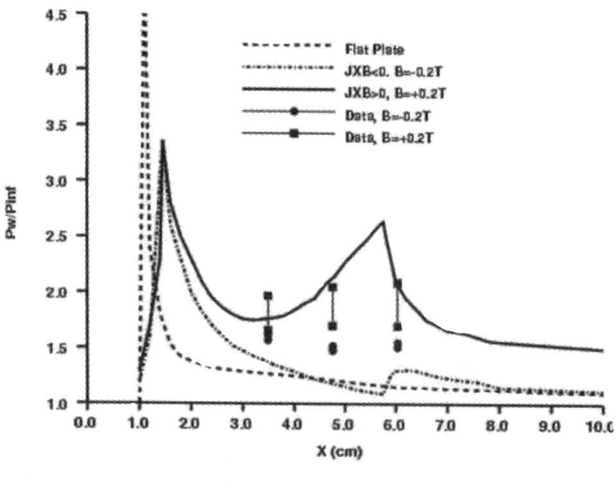

Fig 8. Pressure distributions on wedge surface with MFD interactions

Bibliography

[1] Shang, J. S., Historical Perspective of Magneto-Fluid-Dynamics, Introduction to Magneto-Fluid-Dynamics for Aerospace Applications, Lecture Series 2004-01, von Karman Institute for Fluid Dynamics, Rhode-ST-Genese Belgium, Oct. 2003.

[2] Shang, J.S., MFD Research in US Toward Aerospace Applications, Introduction to Magneto-Fluid-Dynamics for Aerospace Applications, Lecture Series 2004-01, von Karman Institute for Fluid Dynamics, Rhode-ST-Genese Belgium, Oct. 2003.

[3] Resler, E.L., Sears, W.R., The Prospect for Magneto-aerodynamics, J. Aero. Science 1958, Vol. 25, 1958, pp. 235-245 and 258.

[4] Fraishtadt, V.L., Kuranov, A.L., and Sheikin, E.G., Use of MHD Systems in Hypersonic Aircraft, Technical Physics, Vol. 11, 1998, p. 1309.

[5] Sutton, G.W. and Sherman, A., Engineering Magnetohydrodynamics, McGraw-Hill, New York, 1965.

[6] Mitchner, M. and Kruger, C.H., Partially Ionized Gas, John Wiley & Sons, New York, 1973.

[7] Raizer, Yu. P., Gas Discharge Physics, Springer-Verlag, Berlin, 1991.

[8] Macheret, S.O., Shneider, M.N., and Miles, R.B., Magnetohydrodynamics and Electrohydrodynamic Control of Hypersonic Flows of Weakly Ionized Plasmas, AIAA J., Vol. 42, July 2004, pp. 1378-1387.

[9] Menart, J., Shang, J.S., Kimmel, R. and Hayes, J., Effects of Magnetic Fields on Plasma Generated in Mach 5 Wind Tunnel, AIAA 2003-4165, Orlando FL, June 2003.

[10] Shang, J.S and Surzhikov, S.T., Magneto-Aerodynamic Interaction for Hypersonic Flow Control, AIAA 2004-0508, Reno NV, January 5-8, 2004.

[11] Kimmel, R, Hayes, J., Menart, J., and Shang, J.S., Effect of Surface Plasma Discharges on Boundary Layer at Mach 5, AIAA 2004-0509, Reno NV, January 5-8, 2004.

[12] Surzhikov, S.T. and Shang, J.S., Two-Component Plasma Model for Two-dimensional Glow Discharge in Magnetic Field, J. Comp. Physics, Vol. 199, 2004, pp. 437-464.

[13] Elisson, B., and Kogelschatz, U., Nonequilibrium Volume Plasma Chemical Processing, IEEE Trans. Plasma Science, Vol. 19, 1991, pp 1063-1077.

[14] Hayes, W.D., and Probstein, R.F., Hypersonic Flow Theory, Academy Press, 1959.

[15] Shang, J.S., Shared Knowledge in Computational Fluid Dynamics, Electromagnetics, and Magneto-Aerodynamics, Progress in Aerospace Sciences, Vol. 38, 2002, pp. 449-467.

[16] Gaitonde, D., Higher-Order Solution Procedure for Three-dimensional Non-ideal Magnetogasdynamics, AIAA J., Vol. 39, No. 1, 2001, pp. 2111-2120.

[17] Shang, J.S., Numerical Simulation of Hypersonic Flows, Computational Methods in Hypersonic Aerodynamics, Kluwer Academic Publishers, South Hampton UK, 1992, pp. 81-114.

Chapter 12

Interaction between Longitudinal Vortices and Normal and Oblique Shocks

Oliver Thomer, Egon Krause, and Wolfgang Schröder
Aerodynamisches Institut, RWTH Aachen, Germany

Dedicated to Lu Ting on his eightieth birthday

Abstract

Breakdown of a slender vortex caused by normal (NSVI) and oblique shocks (OSVI) is studied using numerical solutions of the EULER and NAVIER-STOKES equations for unsteady, three-dimensional, supersonic flow at free stream MACH numbers of 1.6 and 2.0. At the inflow boundary a slender BURGERS vortex with a given circulation and axial velocity distribution is prescribed. The calculations using approximately 2 million grid points show that breakdown is primarily controlled by pressure forces and is therefore easily initiated when the vortex interacts with a normal shock. Oblique shocks are deformed into an 's'-shaped part near the vortex core where the shock becomes normal. The results also indicate that initiation of breakdown is more sensitive to variations in the axial velocity component than in the circulation of the vortex, and that the flow structure is time-dependent.A simple breakdown criterion based on the axial momentum flux is presented for both NSVI and OSVI.

12.1 Introduction

Recent investigations of flows in turbine engine cascades seem to indicate that shock induced vortex breakdown may be responsible for stall observed at high compressor loads [14]. Although not verified, this conjecture may lead to a new insight and understanding of the stall mechanism. Until now reliable breakdown conditions could not be formulated for incompressible or for compressible flow. Previous investigations mainly dealt with vortices in incompressible flow [1]. For supersonic flow first attempts to derive breakdown criteria for shock-vortex interaction can be found in [10; 4; 9]. Related to this problem is the leading edge vortex breakdown on delta wings, which was studied numerically by KANDIL *et al.* in [6; 7].

A combined experimental and numerical investigation was initiated at the Aerodynamisches Institut to study shock-vortex interaction. The free stream MACH number, strength of the normal and oblique shocks, and the circulation of the free vortex embedded in the flow were varied in the investigation. Although previously obtained results indicate that breakdown is mainly dominated by pressure forces, it was decided to base the investigation on a numerical solution of the EULER and NAVIER-STOKES equations for unsteady, three-dimensional, supersonic flow to determine the influence of the STOKES stresses and the FOURIER heat flux. The study was first purposely restricted to laminar flow, but is presently being extended to turbulent flow by using time-averaged NAVIER-STOKES equations with several higher order closure assumptions for the REYNOLDS stresses and the turbulent heat flux.

The breakdown is initiated by letting either an oblique (OSVI) or a normal (NSVI) shock at free stream MACH numbers ranging from 1.6 to 2.0 intersect a slender vortex. In the experiments the strength of the oblique shock is varied by changing the deflection angle of the shock generating wedge. The circulation of the vortex is varied by modifying the angle of attack of the double wedge causing the vortex. In the present paper we focus on the discussion of the numerical results and use the preliminary experimental findings to confirm the numerical data qualitatively since no quantitative measurements are available yet.

Three interaction modes are found, namely weak, moderate and strong interaction. For a weak shock-vortex interaction the vortex passes through the shock without remarkably changing its shape. The moderate interaction mode yields a change of the shock contour and has a slight influence on the vortex structure. The third mode, when breakdown occurs, is called strong interaction. In the case of NSVI, a region of reversed flow with a free stagnation point is formed at a certain distance downstream from the shock. A bubble-like flow structure emerges, growing in the axial and radial direction until a stable position of the stagnation point is reached. Downstream from the shock, which remains normal near the axis of the vortex and is curved further away from it, the flow is unsteady, slightly oscillating.

For extremely high vortex circulation and/or pronounced wake structure the size of the breakdown bubble exceeds the limits of the computational domain.

In the case of OSVI the computed structure of the burst vortex part differs from that obtained with NSVI. It does not increase as much in the radial direction as it does in the case of normal shock-vortex interaction. For moderate and strong interaction the shock is 's'-shaped in the area of the vortex axis. A small region of subsonic flow is generated downstream of the deformed shock part. For the strong interaction mode a free stagnation point is formed inside the subsonic flow pocket balanced by the recirculation region. The computations show that either a small axial or large azimuthal velocity component can trigger breakdown. The larger the pressure rise or the circulation or the lower the axial velocity component the more likely breakdown will occur. Breakdown is found to be more probable in the case of NSVI than for OSVI. For the latter stability is more dependent on the vortex wake than on the circulation.

Based on the observations made a simple theoretical breakdown criterion for both, NSVI and OSVI is presented on grounds of momentum conservation on the vortex axis and the assumption of a free stagnation point.

12.2 Governing equations

We numerically solve the EULER and the NAVIER-STOKES equations for time-dependent three-dimensional, compressible flow. They read in dimensionless[1] transformed coordinates ξ, η and ζ

$$\frac{\partial \hat{Q}}{\partial t} + \frac{\partial \hat{E}_a}{\partial \xi} + \frac{\partial \hat{F}_a}{\partial \eta} + \frac{\partial \hat{G}_a}{\partial \zeta} = \frac{1}{Re_0}\left(\frac{\partial \hat{E}_d}{\partial \xi} + \frac{\partial \hat{F}_d}{\partial \eta} + \frac{\partial \hat{G}_d}{\partial \zeta}\right) . \tag{12.1}$$

The REYNOLDS number $Re_0 = (\rho_0^* a_0^* r_0^*)/\mu_0^*$ is based on stagnation conditions and $\vec{Q}$ is the solution vector, whose components consist of the density ρ, the momentum fluxes $\rho\vec{v}$ and the specific energy ρe, multiplied by the

[1] In the present investigation variables are nondimensionalized. For example $\Gamma_0 = \Gamma_0^*/(a_0^* r_0^*)$ is the circulation nondimensionalized by the speed of sound at stagnation conditions and the core radius, and $T_\infty = T_\infty^*/T_0^*$ is the free stream temperature nondimensionalized by the stagnation temperature. Additionally the total density ρ_0^*, the reference time $t_{ref}^* = r_0^*/a_0^*$, and the pressure $p_{ref}^* = \rho_0^*\, a_0^{2*}$ are used.

Jacobian determinant

$$\begin{aligned} J &= x_\xi y_\eta z_\zeta + y_\xi z_\eta x_\zeta + z_\xi x_\eta y_\zeta - x_\xi y_\zeta z_\eta - y_\xi z_\zeta x_\eta - z_\xi x_\zeta y_\eta \\ &= (\xi_x \eta_y \zeta_z + \xi_y \eta_z \zeta_x + \xi_z \eta_x \zeta_y - \xi_x \eta_z \zeta_y - \xi_y \eta_x \zeta_z - \xi_z \eta_y \zeta_x)^{-1} \end{aligned} \quad (12.2)$$

that represents the cell volume, $\hat{Q} = J\vec{Q}$. The fluxes in the ξ, η and ζ directions are

$$\hat{E}_a = J \begin{pmatrix} \rho U \\ \rho U u + \xi_x p \\ \rho U v + \xi_y p \\ \rho U w + \xi_z p \\ U(\rho e + p) \end{pmatrix} \quad (12.3)$$

$$\hat{F}_a = J \begin{pmatrix} \rho V \\ \rho V u + \eta_x p \\ \rho V v + \eta_y p \\ \rho V w + \eta_z p \\ V(\rho e + p) \end{pmatrix} \quad (12.4)$$

$$\hat{G}_a = J \begin{pmatrix} \rho W \\ \rho W u + \zeta_x p \\ \rho W v + \zeta_x p \\ \rho W w + \zeta_x p \\ W(\rho e + p) \end{pmatrix} \quad (12.5)$$

respectively

$$\hat{E}_d = J \begin{pmatrix} 0 \\ \xi_x \tau_{xx} + \xi_y \tau_{yx} + \xi_z \tau_{zx} \\ \xi_x \tau_{xy} + \xi_y \tau_{yy} + \xi_z \tau_{zy} \\ \xi_x \tau_{xz} + \xi_y \tau_{yz} + \xi_z \tau_{zz} \\ \xi_x \beta_x + \xi_y \beta_y + \xi_z \beta_z \end{pmatrix} \quad (12.6)$$

$$\hat{F}_d = J \begin{pmatrix} 0 \\ \eta_x \tau_{xx} + \eta_y \tau_{yx} + \eta_z \tau_{zx} \\ \eta_x \tau_{xy} + \eta_y \tau_{yy} + \eta_z \tau_{zy} \\ \eta_x \tau_{xz} + \eta_y \tau_{yz} + \eta_z \tau_{zz} \\ \eta_x \beta_x + \eta_y \beta_y + \eta_z \beta_z \end{pmatrix} \quad (12.7)$$

$$\hat{G}_d = J \begin{pmatrix} 0 \\ \zeta_x \tau_{xx} + \zeta_y \tau_{yx} + \zeta_z \tau_{zx} \\ \zeta_x \tau_{xy} + \zeta_y \tau_{yy} + \zeta_z \tau_{zy} \\ \zeta_x \tau_{xz} + \zeta_y \tau_{yz} + \zeta_z \tau_{zz} \\ \zeta_x \beta_x + \zeta_y \beta_y + \zeta_z \beta_z \end{pmatrix} . \quad (12.8)$$

The quantities $U = \xi_x u + \xi_y v + \xi_z w = \vec{v} \cdot \nabla \xi$, $V = \eta_x u + \eta_y v + \eta_z w = \vec{v} \cdot \nabla \eta$ and $W = \zeta_x u + \zeta_y v + \zeta_z w = \vec{v} \cdot \nabla \zeta$ are the contravariant velocities. On the right-hand side of the energy equation the abbreviation $\beta_j = u_i\tau_{ij} + q_j$ is used, where $\tau_{ij} = \mu[(\frac{\partial u_i}{\partial x_j} + \frac{\partial u_j}{\partial x_i}) - \frac{2}{3}\delta_{ij}\frac{\partial u_k}{\partial x_k}]$ are the STOKES stresses and $q_j = -\frac{\mu}{(\gamma-1)Pr}(\xi_j\frac{\partial T}{\partial \xi} + \eta_j\frac{\partial T}{\partial \eta} + \zeta_j\frac{\partial T}{\partial \zeta})$ is the FOURIER heat transfer including the dynamic viscosity. The quantity Pr is the PRANDTL number.

12.3 Computational setup and boundary conditions

At NSVI the computational domain is a rectangular box containing $99 \times 99 \times 199$ grid points in the x-, y-, and z-direction (Fig. 12.1, $G1$). For the case of OSVI a wedge-like grid was generated ($G2$). Although OSVI can be simulated using $G1$ for the normal shock case in conjunction with the RANKINE-HUGONIOT equations the $G2$ configuration is more similar to the experimental setup. While the wedge generates an oblique shock and a PRANDTL-MEYER expansion stronger vortices (larger wake or higher circulation) are necessary to enforce breakdown. The breakdown depends on the wedge geometry, i.e., the interchange between the shock and expansion fan. Near the vortex axis both grids are clustered such that the vortex core ($|r| < 1$) is resolved with approximately 21 grid points in the radial $r = \sqrt{x^2 + y^2}$ direction.

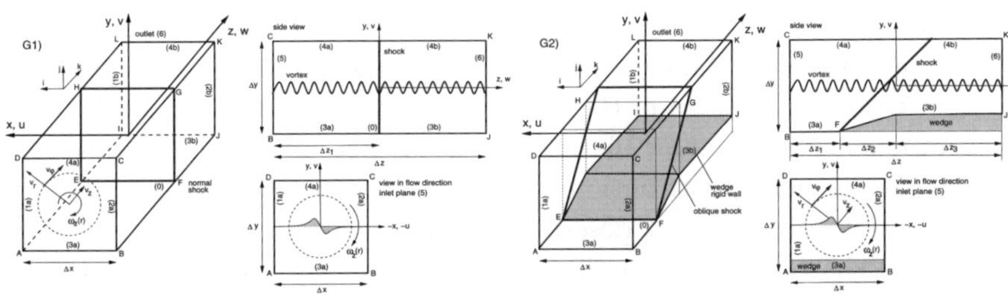

Fig. 12.1 Computational domain and boundary conditions for the NSVI (left) and for the OSVI (right). Flow is in the positive z-direction.

Grid	Δx	Δy	Δz	Δz_1	Δz_2	Δz_3	i_{max}	j_{max}	k_{max}
G1 (NSVI)	40	40	80	40	-	-	99	99	199
G2 (OSVI)	10	10	19	4	5	10	99	99	199

Fig. 12.2 Parameters for grid $G1$ and $G2$. The step sizes are scaled by the vortex core radius r_0^*. The quantities i_{max}, j_{max} and k_{max} are the total number of grid points in the x, y and z direction.

The inflow conditions are assumed to be described by a slender BURGERS vortex with the circumferential velocity distribution $v_\varphi(r) = \Gamma_0\, r\, e^{0.5\,(1-r^2)}/(2\pi)$ and vanishing radial velocity $v_r = 0$. The vortex is centered at x=y=0 together with a radial distribution of the axial velocity component $v_z(r) = v_{z_\infty}\,(1 - \delta\, e^{-\mu_w\, r^2})$ (Fig. 12.1, plane 5). The azimuthal velocity component v_φ reaches its maximum at $|r| = 1$, and decreases exponentially with increasing radius $|r| > 1$. The circulation at $|r| = 1$ is defined as $\Gamma_0 = 2\,\pi\, v_\varphi(|r| = 1)$. Because of the exponential decrease of the azimuthal velocity component in the radial direction disturbances near the axis of the vortex are barely propagated to the lateral boundaries of the computational domain.

For vortices with finite core size the core diameter d is usually defined as twice the radial distance r between the location of the maximum of the azimuthal velocity v_φ and the axis. If the ratio of the diameter d over the length of the vortex l is much smaller than unity, the vortex is called slender implying that the axial velocity $v_z(r)$ is much larger than the radial velocity component $v_r(r)$, i.e., $v_{r\text{max}}/v_\infty = O(d/l) << 1$. For high circulation or large vortex wake it is well known that in the case of breakdown the vortex tends to propagate out of the integration area. Subsequently, the breakdown bubble may reach the inlet causing problems with the boundary condition. Therefore the slenderness condition also implies that the core radius r_0^* is small in comparison to the bursting length l_b^*, defined as the distance between the inflow cross-section and the breakdown point, i.e., the point where the flow stagnates thereby initiating the breakdown [10].

Controlled by the sign of the wake parameter $\delta \in [0, 1[$ the axial velocity component in the inflow plane is modeled as a 'jet-like' or a 'wake-like' profile, and its radial distribution is given by a GAUSSian distribution. The quantity $r_i = \pm\sqrt{1/(2\mu_w)}$ describes the position of the inflection point of the bell shaped velocity profile. The ratio between the azimuthal and the axial velocity components is defined as $\tau = v_\varphi(|r| = 1)/v_z(r = 0) = \left(\Gamma_0\,\sqrt{1 + \frac{\gamma-1}{2} Ma_\infty^2}\right) / \left(2\,\pi\, Ma_\infty\,(1 - \delta)\right)$. The equation shows that τ increases if either the circulation Γ_0 and the wake parameter δ are increased or the MACH number Ma_∞ is decreased. In the following the kinematic parameter τ is used to describe the vortex strength. Strictly speaking the critical ratio τ_{crit}, indicating vortex breakdown, can only be compared for similar vortex types since it does not include any information about the progression of the primitive variables inside the vortex core and it varies for different vortices. This implements the question if the relation between circumferential and axial velocity is adequate to classify vortex stability in general. In our opinion the ratio between the advective flux on the vortex axis and the pressure jump enforced by the shock is more adequate for supersonic flow. Figure 12.6 shows the advective flux in the streamwise direction in the plane perpendicular to the vortex axis. It can be seen that the curves differ remarkably for the given vortex types although the kinematic parameters Γ_0, δ, μ_w and the free stream MACH number Ma_∞ are fixed in this example. The same holds for the axial MACH number.

For the OSVI the vortex was assumed to have constant stagnation enthalpy. The cyclostrophic balance leads under the condition of $dp = (\partial p/\partial r)dr$ to $dp(r)/dr = \rho(r)\, v_\varphi^2(r)/r$ for the pressure distribution perpendicular to the axis of the symmetric vortex. After separation of variables and introduction of the ideal gas law $\gamma\, p = \rho\, T$ the pressure distribution is defined by the integral

$$\int_{p(r)}^{p_\infty} \frac{1}{p(r)}\, dp = \gamma \int_r^\infty \frac{v_\varphi^2(r)}{T(r)\, r}\, dr \tag{12.9}$$

where $T(r) = 1 - (\gamma - 1)\,(v_\varphi^2(r) + v_r^2(r))/2$. The left-hand side can be solved directly. The right-hand side is numerically integrated due to the functions for the circumferential $v_\varphi(r)$ and the axial velocity $v_z(r)$. The pressure distribution is given by

$$I(r)|_{dh=0} = \gamma \int_r^\infty \frac{v_\varphi^2(r)}{T(r)\, r}\, dr \rightarrow p(r)|_{dh=0} = \frac{p_\infty}{\exp\left(I(r)|_{dh=0}\right)}\,. \tag{12.10}$$

For NSVI constant stagnation entropy was prescribed. In this case the radial momentum equation leads after separation of variables for arbitrary tangential velocity profiles $v_\varphi(r)$ to the integral

$$\int_{p(r)}^{p_\infty} \frac{1}{(\gamma\, p)^{\frac{1}{\gamma}}} dp = \int_r^\infty \frac{v_\varphi^2(r)}{r} dr \;\text{ resp. }\; I(r)|_{ds=0} = \int_r^\infty \frac{v_\varphi^2(r)}{r} dr \tag{12.11}$$

In this case the axial velocity $v_z(r)$ has no influence on the thermodynamic variables determined by integration. The pressure-, density- and temperature distributions follow from

$$p(r)|_{ds=0} = \left((\gamma\, p_\infty)^{\frac{\gamma-1}{\gamma}} - (\gamma - 1)\;\, I(r)|_{ds=0}\right)^{\frac{\gamma}{\gamma-1}}\, \frac{1}{\gamma} \tag{12.12}$$

$$\rho(r)|_{ds=0} = \left((\gamma\, p_\infty)^{\frac{\gamma-1}{\gamma}} - (\gamma - 1)\;\, I(r)|_{ds=0}\right)^{\frac{1}{\gamma-1}} \tag{12.13}$$

$$T(r)|_{ds=0} = \left((\gamma\, p_\infty)^{\frac{\gamma-1}{\gamma}} - (\gamma - 1)\;\, I(r)|_{ds=0}\right)\,. \tag{12.14}$$

For the BURGERS vortex the right-hand side of equation (12.11) can be solved analytically. In this case the dimensionless form of the inflow conditions read for constant stagnation entropy

$$T(r) = [T_\infty - \frac{(\gamma-1)\,\Gamma_0^2}{8\pi^2}\,e^{(1-r^2)}]^1 \tag{12.15}$$

$$\rho(r) = [T_\infty - \frac{(\gamma-1)\,\Gamma_0^2}{8\,\pi^2}\,e^{(1-r^2)}]^{\frac{1}{\gamma-1}} \tag{12.16}$$

$$p(r) = [T_\infty - \frac{(\gamma-1)\Gamma_0^2}{8\pi^2}\,e^{(1-r^2)}]^{\frac{\gamma}{\gamma-1}}\,\frac{1}{\gamma} = \frac{p^*(r^*)}{\rho_0^*\,a_0^*} = \frac{1}{\gamma}\,T^{\frac{\gamma}{\gamma-1}}\,. \tag{12.17}$$

The approximation for the lateral boundary conditions is obtained by assuming the boundaries to be undisturbed streamline surfaces. Then, the kinematic flow condition can be implemented. For example, at the upper surface of the computational domain $G1$ (Fig. 12.1, plane $DCKL$) the normal derivatives of the local velocity vector with respect to the slope of the local streamline vanish, if the flow is assumed to be inviscid. The pressure is determined by solving the momentum equation normal to the boundary. Additionally, the LODI-method for kinematic walls [17; 12] holds as a good approximation.

In the case of NSVI the shock is fixed by prescribing the RANKINE-HUGONIOT relations at the bottom (Fig. 12.1, EF) of the computational domain. For OSVI and grid $G2$ the shock is automatically generated by the wedge. The wedge surface is treated as a rigid wall, i.e.

$$\frac{\partial\rho}{\partial\vec{n}} = \frac{\partial p}{\partial\vec{n}} = \frac{\partial T}{\partial\vec{n}} = 0 \quad \text{and} \quad \vec{v} = 0\,(\text{NS}) \quad \text{or} \quad \vec{v}_n = 0\ (\text{Euler})\,. \tag{12.18}$$

For the flow field initialization at $t = 0$ several solutions exist. One possibility is to prescribe the slender vortex for the entire flow field. In this case the shock may locally be generated at the bottom of the computational domain by the boundary condition at $t = 0$ and be propagated into the interior of the flow field during the computation. In an alternative approach the initial conditions are generated using a plane shock that divides the computational domain into two parts. Upstream of the shock a slender vortex is prescribed whereas downstream of it a uniform flow is specified. The computations show that the initial conditions have barely any impact on the vortex structure downstream of the shock. Particularly in the case of breakdown, it can be evidenced that the flow inside the bubble is not affected by the varying initial conditions.

The outflow conditions must be adjusted to the form of the conservation equations used. For the solution of the EULER and NAVIER-STOKES equations at supersonic outflow, a simple gradient condition can be applied since all characteristics exit the computational domain.

Downstream of the normal shock the flow is subsonic. One of the characteristics enters the computational domain at the outlet boundary requiring, e.g., the pressure to be specified. Unfortunately, this necessary physical information is part of the solution (the burst part of the vortex) and therefore not immediately available. By formulating 'soft' (numerical) boundary conditions for this problem special care has to be taken for the pressure formulation because of the wave character of the governing equations. Wrong approximations lead to disturbances from the boundary which can travel upstream into the computational domain causing an artifical vortex breakdown in the worst case.

For subsonic outflow different approximations are available. Assuming that the axial flow near the outflow cross-section does not experience any back flow, one way is to prescribe a TAYLOR series expansion for the EULER or the NAVIER-STOKES equations to determine the properties on the boundary via an analytic extension. Then, the primitive variables on the boundary for the outward going characteristics can be computed since the outgoing waves are given by the solution of the governing equations inside the domain. For the inward directed characteristic this method is not useful because it propagates opposite to the transport out of the computational domain. Especially for a time-dependent solution, a simple extrapolation does not seem to describe the local flow situation properly and appears to be arbitrary. A more promising solution is to eliminate artifical disturbances propagating from the outflow section into the computational domain caused by the insufficient formulation of boundary conditions by a time-dependent, non-reflecting, characteristic approach which is used here for NSVI. THOMPSON [16; 17] introduced non-reflecting boundary conditions for the EULER equations, POINSOT *et al.* [12] extended the method for viscous flows and RUDY & STRIKWERDA [13] developed a pressure relaxation method which approximates the problem of an accurate pressure formulation at the outlet boundary.

The non-reflecting boundary conditions for inviscid flows are based on the Lagrangian form of the Euler equations that read in the ξ-direction

$$\left[\frac{\partial p}{\partial t} - \frac{\rho a}{|\nabla\xi|}\left(\xi_x\frac{\partial u}{\partial t} + \xi_y\frac{\partial v}{\partial t} + \xi_z\frac{\partial w}{\partial t}\right)\right] + \mathcal{L}_1 = 0 \quad (12.19)$$

$$\left[\frac{\xi_x}{|\nabla\xi|}\left(a^2\frac{\partial \rho}{\partial t} - \frac{\partial p}{\partial t}\right) + \frac{\rho a}{|\nabla\xi|}\left(\xi_z\frac{\partial v}{\partial t} - \xi_y\frac{\partial w}{\partial t}\right)\right] + \mathcal{L}_2 = 0 \quad (12.20)$$

$$\left[\frac{\xi_y}{|\nabla\xi|}\left(a^2\frac{\partial \rho}{\partial t} - \frac{\partial p}{\partial t}\right) + \frac{\rho a}{|\nabla\xi|}\left(-\xi_z\frac{\partial u}{\partial t} + \xi_x\frac{\partial w}{\partial t}\right)\right] + \mathcal{L}_3 = 0 \quad (12.21)$$

$$\left[\frac{\xi_z}{|\nabla\xi|}\left(a^2\frac{\partial \rho}{\partial t} - \frac{\partial p}{\partial t}\right) + \frac{\rho a}{|\nabla\xi|}\left(\xi_y\frac{\partial u}{\partial t} - \xi_x\frac{\partial v}{\partial t}\right)\right] + \mathcal{L}_4 = 0 \quad (12.22)$$

$$\left[\frac{\partial p}{\partial t} + \frac{\rho a}{|\nabla\xi|}\left(\xi_x\frac{\partial u}{\partial t} + \xi_y\frac{\partial v}{\partial t} + \xi_z\frac{\partial w}{\partial t}\right)\right] + \mathcal{L}_5 = 0\,. \quad (12.23)$$

The spatial derivatives of the primitive variables ρ, u, v, w, and p are represented by the wave terms $\mathcal{L}_i$

$$\mathcal{L}_1 = \lambda_1\left(\frac{\partial p}{\partial x} - \rho a\frac{\partial u}{\partial x}\right) \quad (12.24)$$

$$\mathcal{L}_2 = \lambda_2\left(a^2\frac{\partial \rho}{\partial x} - \frac{\partial p}{\partial x}\right) \quad (12.25)$$

$$\mathcal{L}_3 = \lambda_3\left(\rho a\frac{\partial w}{\partial x}\right) \quad (12.26)$$

$$\mathcal{L}_4 = \lambda_4\left(-\rho a\frac{\partial v}{\partial x}\right) \quad (12.27)$$

$$\mathcal{L}_5 = \lambda_5\left(\frac{\partial p}{\partial x} + \rho a\frac{\partial u}{\partial x}\right)\,. \quad (12.28)$$

The factors $\lambda_{1,2,3,4,5}$=$u - a$, $u + a$, u, u, u are the eigenvalues of the JACOBIAN matrix in the ξ-direction. A physical interpretation of the eigenvalues can be given by the linearized EULER equations. According to this approximation the velocities λ_1 and λ_5 represent the convection of pressure waves, λ_2 is the convection velocity of entropy waves, whereas λ_3 and λ_4 are advection velocities of the contravariant velocities normal to the control surfaces of the finite volume scheme.

From the above equations (12.24) the *LODI*–Relations (***L**ocal* **One**–**D**imensional **I**nviscid) can be determined

$$\left.\frac{\partial p}{\partial t}\right|_{\xi}^{LODI} = -\frac{1}{2}(\mathcal{L}_1 + \mathcal{L}_5) \tag{12.29}$$

$$\begin{aligned}\left.\frac{\partial \rho}{\partial t}\right|_{\xi}^{LODI} &= -\left(\mathcal{L}_2 \frac{\xi_x}{|\nabla\xi|} + \mathcal{L}_3 \frac{\xi_y}{|\nabla\xi|}\right. \\ &\left. + \mathcal{L}_4 \frac{\xi_z}{|\nabla\xi|} + \frac{1}{2}(\mathcal{L}_1 + \mathcal{L}_5)\right)\frac{1}{a^2}\end{aligned} \tag{12.30}$$

$$\left.\frac{\partial u}{\partial t}\right|_{\xi}^{LODI} = -\left(\frac{\xi_x}{2\rho a|\nabla\xi|}(\mathcal{L}_5 - \mathcal{L}_1) + \frac{1}{\rho a|\nabla\xi|}(\xi_y\mathcal{L}_4 - \xi_z\mathcal{L}_3)\right) \tag{12.31}$$

$$\left.\frac{\partial v}{\partial t}\right|_{\xi}^{LODI} = -\left(\frac{\xi_y}{2\rho a|\nabla\xi|}(\mathcal{L}_5 - \mathcal{L}_1) + \frac{1}{\rho a|\nabla\xi|}(\xi_z\mathcal{L}_2 - \xi_x\mathcal{L}_4)\right) \tag{12.32}$$

$$\left.\frac{\partial w}{\partial t}\right|_{\xi}^{LODI} = -\left(\frac{\xi_z}{2\rho a|\nabla\xi|}(\mathcal{L}_5 - \mathcal{L}_1) + \frac{1}{\rho a|\nabla\xi|}(\xi_x\mathcal{L}_3 - \xi_y\mathcal{L}_2)\right). \tag{12.33}$$

If the flow can be considered locally one-dimensional and the pressure forces are small the time-dependent change of the primitive variables on the boundary can be determined using the wave amplitudes $\mathcal{L}_i$. The expansion to three-dimensional, viscous heat conducting flows can be done as follows. The EULER fluxes normal to the outflow boundary are calculated using the characteristic approach described above, and the tangential EULER fluxes, the viscous terms and the heat transfer are calculated using the original governing equations. The assumption of locally one-dimensional flow is still valid. The conservative variables at the outflow boundary can then be written as follows

$$\begin{aligned}\frac{\partial \rho}{\partial t} &= \left.\frac{\partial \rho}{\partial t}\right|_{\xi} - \eta_x \frac{\partial \rho u}{\partial \eta} - \eta_y \frac{\partial \rho v}{\partial \eta} \\ &- \eta_z \frac{\partial \rho w}{\partial \eta} - \zeta_x \frac{\partial \rho u}{\partial \zeta} - \zeta_y \frac{\partial \rho v}{\partial \zeta} - \zeta_z \frac{\partial \rho w}{\partial \zeta}\end{aligned} \tag{12.34}$$

$$
\begin{aligned}
\frac{\partial \rho u}{\partial t} = & \; u \left. \frac{\partial \rho}{\partial t} \right|_{\xi} + \rho \left. \frac{\partial u}{\partial t} \right|_{\xi} \\
& - \eta_x \frac{\partial \rho u u + p}{\partial \eta} - \eta_y \frac{\partial \rho v u}{\partial \eta} - \eta_z \frac{\partial \rho w u}{\partial \eta} \\
& - \zeta_x \frac{\partial \rho u u + p}{\partial \zeta} - \zeta_y \frac{\partial \rho v u}{\partial \zeta} - \zeta_z \frac{\partial \rho w u}{\partial \zeta} \\
& + \xi_x \frac{\partial \tau_{xx}}{\partial \xi} + \xi_y \frac{\partial \tau_{yx}}{\partial \xi} + \xi_z \frac{\partial \tau_{zx}}{\partial \xi} + \eta_x \frac{\partial \tau_{xx}}{\partial \eta} + \eta_y \frac{\partial \tau_{yx}}{\partial \eta} + \eta_z \frac{\partial \tau_{zx}}{\partial \eta} \\
& + \zeta_x \frac{\partial \tau_{xx}}{\partial \zeta} + \zeta_y \frac{\partial \tau_{yx}}{\partial \zeta} + \zeta_z \frac{\partial \tau_{zx}}{\partial \zeta}
\end{aligned} \tag{12.35}
$$

$$
\begin{aligned}
\frac{\partial \rho v}{\partial t} = & \; v \left. \frac{\partial \rho}{\partial t} \right|_{\xi} + \rho \left. \frac{\partial v}{\partial t} \right|_{\xi} \\
& - \eta_x \frac{\partial \rho u v}{\partial \eta} - \eta_y \frac{\partial \rho v v + p}{\partial \eta} - \eta_z \frac{\partial \rho w v}{\partial \eta} \\
& - \zeta_x \frac{\partial \rho u v}{\partial \zeta} - \zeta_y \frac{\partial \rho v v + p}{\partial \zeta} - \zeta_z \frac{\partial \rho w v}{\partial \zeta} \\
& + \xi_x \frac{\partial \tau_{xy}}{\partial \xi} + \xi_y \frac{\partial \tau_{yy}}{\partial \xi} + \xi_z \frac{\partial \tau_{zy}}{\partial \xi} + \eta_x \frac{\partial \tau_{xy}}{\partial \eta} + \eta_y \frac{\partial \tau_{yy}}{\partial \eta} + \eta_z \frac{\partial \tau_{zy}}{\partial \eta} \\
& + \zeta_x \frac{\partial \tau_{xy}}{\partial \zeta} + \zeta_y \frac{\partial \tau_{yy}}{\partial \zeta} + \zeta_z \frac{\partial \tau_{zy}}{\partial \zeta}
\end{aligned} \tag{12.36}
$$

$$
\begin{aligned}
\frac{\partial \rho w}{\partial t} = & \; w \left. \frac{\partial \rho}{\partial t} \right|_{\xi} + \rho \left. \frac{\partial w}{\partial t} \right|_{\xi} \\
& - \eta_x \frac{\partial \rho u w}{\partial \eta} - \eta_y \frac{\partial \rho v w}{\partial \eta} - \eta_z \frac{\partial \rho w w + p}{\partial \eta} \\
& - \zeta_x \frac{\partial \rho u w}{\partial \zeta} - \zeta_y \frac{\partial \rho v w}{\partial \zeta} - \zeta_z \frac{\partial \rho w w + p}{\partial \zeta} \\
& + \xi_x \frac{\partial \tau_{xz}}{\partial \xi} + \xi_y \frac{\partial \tau_{yz}}{\partial \xi} + \xi_z \frac{\partial \tau_{zz}}{\partial \xi} + \eta_x \frac{\partial \tau_{xz}}{\partial \eta} + \eta_y \frac{\partial \tau_{yz}}{\partial \eta} + \eta_z \frac{\partial \tau_{zz}}{\partial \eta} \\
& + \zeta_x \frac{\partial \tau_{xz}}{\partial \zeta} + \zeta_y \frac{\partial \tau_{yz}}{\partial \zeta} + \zeta_z \frac{\partial \tau_{zz}}{\partial \zeta}
\end{aligned} \tag{12.37}
$$

$$
\begin{aligned}
\frac{\partial \rho e}{\partial t} = & \frac{1}{\gamma-1}\,\frac{\partial p}{\partial t}\bigg|_\xi + \frac{u^2+v^2+w^2}{2}\,\frac{\partial \rho}{\partial t}\bigg|_\xi \\
& + \rho u\,\frac{\partial p}{\partial t}\bigg|_\xi + \rho v\,\frac{\partial p}{\partial t}\bigg|_\xi + \rho w\,\frac{\partial p}{\partial t}\bigg|_\xi \\
& - \eta_x \frac{\partial u(\rho e+p)}{\partial \eta} - \eta_y \frac{\partial v(\rho e+p)}{\partial \eta} - \eta_z \frac{\partial w(\rho e+p)}{\partial \eta} \\
& - \zeta_x \frac{\partial u(\rho e+p)}{\partial \zeta} - \zeta_y \frac{\partial v(\rho e+p)}{\partial \zeta} - \zeta_z \frac{\partial w(\rho e+p)}{\partial \zeta} \\
& + \xi_x \frac{\partial \beta_x}{\partial \xi} + \xi_y \frac{\partial \beta_y}{\partial \xi} + \xi_z \frac{\partial \beta_z}{\partial \xi} + \eta_x \frac{\partial \beta_x}{\partial \eta} + \eta_y \frac{\partial \beta_y}{\partial \eta} + \eta_z \frac{\partial \beta_z}{\partial \eta} \\
& + \zeta_x \frac{\partial \beta_x}{\partial \zeta} + \zeta_y \frac{\partial \beta_y}{\partial \zeta} + \zeta_z \frac{\partial \beta_z}{\partial \zeta}\,. \qquad (12.38)
\end{aligned}
$$

Further details on the formulation of the boundary conditions may be found in reference [12]. Due to the pressure relaxation method of RUDY & STRIKWERDA [13] the magnitude of reflexion caused by the inward directed characteristic $\lambda_{1\xi} = U - a|\nabla\xi|$ can be adjusted by the basic approach

$$
\begin{aligned}
\mathcal{L}_{1\xi} &= \lambda_{1\xi}\left[\frac{\partial p}{\partial \xi} - \frac{\rho a}{|\nabla \xi|}\left(\xi_x \frac{\partial u}{\partial \xi} + \xi_y \frac{\partial v}{\partial \xi} + \xi_z \frac{\partial w}{\partial \xi}\right)\right] \\
&= K\,(p_I - p_F) \text{ and } K = \sigma\,(1 - Ma_I)\,\frac{a_I}{L} \qquad (12.39)
\end{aligned}
$$

where L is the relaxation length, a_I and Ma_I are the speed of sound and the MACH number, resp., at the first inner grid point and the index F characterizes values from the free stream. The quantity $\sigma = [0,1]$ describes the reflexion magnitude.

12.4 Numerical method and vortex identification

The differentials in time and space are discretized to second-order accuracy. The temporal integration is performed using an optimized 5-step RUNGE-KUTTA scheme

$$
\begin{aligned}
\hat{Q}^{(n+1)} &= \hat{Q}^n \\
&+ \Delta t(\alpha_{m1} RHS(\hat{Q}^n) \\
&+ \alpha_{m2} RHS(\hat{Q}^1) + \cdots + \alpha_{mm} RHS(\hat{Q}^{m-1})) \qquad (12.40)
\end{aligned}
$$

where $\alpha_{11} = 0.059$, $\alpha_{22} = 0.145$, $\alpha_{33} = 0.273$, $\alpha_{44} = 0.5$ and $\alpha_{55} = 1.0$ are the non-zero coefficients and RHS is the right-hand side of the governing equations including the spatial derivatives.

A finite volume node-centered scheme is used to approximate the inviscid and viscous fluxes. Viscous fluxes are computed using a modified cell-vertex scheme. The convective fluxes are discretized using an explicit finite-volume scheme based on the $AUSM+$ approach by LIOU [8]. The fluxes $\mathcal{F}^K_\pm$ are determined using variables interpolated via the $MUSCL$ approach from the left (+) and right (–) side of the interface, and the split MACH numbers $(\widetilde{Ma_\pm})$ are weighted by the interpolated MACH numbers $(\widehat{Ma_\pm})$ according to $\widetilde{Ma_\pm} = \frac{1}{2}\widehat{Ma_\pm}$. The split pressure $\widetilde{p}_\pm$ is given by $\widetilde{p}_\pm = \frac{1}{2}p_\pm$.

There exist numerous methods to detect a vortex pattern in an unsteady flow field. In the following, we briefly describe the most widely used definitions. JEONG & HUSSAIN [5] introduced an eigenvalue criterion based on the decomposition of the velocity gradient tensor in a symmetric and an asymmetric part

$$\nabla\vec{v} = \begin{pmatrix} u_x & u_y & u_z \\ v_x & v_y & v_z \\ w_x & w_y & w_z \end{pmatrix} = S + \Omega\,. \tag{12.41}$$

The tensor S corresponds to the stresses and Ω contains the cartesian parts of the fluid rotation vector $\vec{\omega} = \nabla \times \vec{v}$

$$S = \frac{1}{2}\begin{pmatrix} 2u_x & u_y + v_x & u_z + w_x \\ v_x + u_y & 2v_y & v_z + w_y \\ w_x + u_z & w_y + v_z & 2w_z \end{pmatrix} \text{ and} \tag{12.42}$$

$$\Omega = \frac{1}{2}\begin{pmatrix} 0 & u_y - v_x & u_z - w_x \\ v_x - u_y & 0 & v_z - w_y \\ w_x - u_z & w_y - v_z & 0 \end{pmatrix}. \tag{12.43}$$

JEONG & HUSSAIN found that the second negative eigenvalue of the symmetric matrix $M = S^2 + \Omega^2$ corresponds to a local pressure minimum and closed streamline pattern. Both properties are also characteristic for a vortex structure. After calculating the elements of M it is easy to find the real eigenvalues λ_1, λ_2, λ_3, which hold $\lambda_1 \geq \lambda_2 \geq \lambda_3$, from the characteristic polynomial $det(M) = 0$, e.g., via CARDAN's equations and the casus irreduziblis of VIETA's root equation.

Another vortex identification criterion was developed by CHONG *et al.*. They propose that the vortex core is a region of complex eigenvalues of

$det(\nabla\vec{v}) = 0$ corresponding to closed or spiral streamline patterns. The characteristic polynomial reads $\sigma^3 - \mathcal{P}\sigma^2 + \mathcal{Q}\sigma - \mathcal{R} = 0$ with the coefficients $\mathcal{P} = \mathrm{tr}(\nabla\vec{v})$, $\mathcal{Q} = \frac{1}{2}\left(\mathcal{P}^2 - \mathrm{tr}\left[(\nabla\vec{v})^2\right]\right)$ and $\mathcal{R} = \det(\nabla\vec{v})$. For $\Delta = \left(\frac{1}{3}\mathcal{Q}\right)^3 + \left(\frac{1}{2}\mathcal{R}\right)^2 > 0$ all eigenvalues are complex.

HUNT *et al.* defined a vortex as a region with positive second invariant $\mathcal{Q}$ of $det(\nabla\vec{v}) = 0$ where $\mathcal{Q} = (||\Omega^2|| - ||S^2||)/2$ using $||\Omega^2|| = \sqrt{tr(\Omega\,\Omega^t)}$ and $||S^2|| = \sqrt{tr(S\,S^t)}$. The number $\mathcal{Q}$ therefore represents the local balance between shear strain rate and vorticity magnitude.

TRUESDELL introduced a kinematic vorticity number to describe the 'quality of rotation'. The ratio $N_k = ||\Omega||/||S||$ gives the quality of rotation regardless of the vorticity magnitude.

At last it is prevalent to plot the vorticity magnitude $|\vec{\omega}| = |\nabla \times \vec{v}|$ to detect vortical structures. This method is used here as a standard definition in comparison with the above outlined methods. The disadvantage of the $|\vec{\omega}|$-method is that the vorticity magnitude doesn't correspond to vortex tubes in general.

12.5 Results

The results for NSVI are presented first. The specified vortex identification criteria are compared. Subsequently, the numerical results for OSVI are discussed, and some general remarks will be made concerning the interaction of slender vortices with normal and oblique shocks. Finally, a simple breakdown criterion for slender, axisymmetric vortices in supersonic flow is introduced.

In general vortex breakdown, characterized by the occurrence of a free stagnation point on the vortex axis, can be enhanced by either a small axial or large azimuthal velocity component. In the case of OSVI the flow downstream of the shock has to become locally subsonic, i.e., a locally strong shock solution is a necessary condition for vortex breakdown. Since the flow downstream of normal shocks is subsonic, vortex breakdown is more probable in NSVI than in OSVI. Additionally it is well known that the larger the pressure rise the more likely breakdown will occur.

Since the flow field downstream from a normal shock differs from that downstream from an oblique shock, the structure of the burst part of the vortex is also different. Either for a sufficiently high circulation and uniform axial velocity or for a small circulation and an adequate vortex wake the shock migrates upstream in the course of time, yielding a cone-like shock shape in the case of NSVI. For OSVI the vortex wake triggers the shape of the breakdown bubble, which is much smaller than for NSVI. In this case it is very unlikely to enhance vortex breakdown for vortices with uniform axial flow since in general the flow field downstream of the interaction remains supersonic.

Figure 12.7 shows instantaneous pictures of NSVI ($Ma = 1.6$, $\Gamma_0 = 2.5$, $\delta = 0.1$, $\mu_w = 2.0$, $Re = 15000$) using the vortex identification criteria outlined above. At $t = 139$ the slender isentropic BURGERS vortex broke up through the normal shock. A bubble type breakdown and the beginning of spiral breakdown emerges in addition with vortical ring like structures (VRLS) traveling parallel to a small longitudinal vortex tube in the main flow direction. The λ_2 criterion identifies seperate longitudinal and ring like vortex structures. The surface of the longitudinal vortex even shows a spiraling contortion which is typical for the vortex tube when the transition to spiral type breakdown is initiated. From the iso-vorticity contours the same vortex structures are identifiable while the delimitation of the different structures is not as clear as for the λ_2 criterion. The resolution of the vortex pattern is less crisp.

The Δ criterion resolves the longitudinal vortex and the dividing process. Unfortunately, it identifies only strongly pronounced vortices like the embedded vortex tube and a small number of VRLS generated inside the breakdown bubble downstream of the normal shock.

The N_k criterion identifies additional to the breakdown bubble and the vortex tube only the strong VRLS. The VRLS is less pronounced than that for the λ_2 or the $|\vec{\omega}|$ criterion. Their diameters are much smaller than those visible for the λ_2 and ω criteria. The same is true for the $\mathcal{Q}$ criterion. However, there is an improvement in the resolution of vortical structures compared to the N_k and Δ criteria.

To discuss the temporal behavior of NSVI Fig. 12.8 shows a sequence of instantaneous pictures of the interaction for various time steps $t = t^*/(l_0^* a_0^*) = 60..200$. In the first column the streamline pattern and in the second column the vortex lines in the x-y center plane of the computational domain are plotted. The projection of the three-dimensional stream- and vortexlines into the two-dimensional center plane evidences the intricacy of the breakdown process. Since the flow for the shown time interval is nearly axisymmetric, the third column shows the thermodynamic flow variables, the velocity component, and the MACH number in the streamwise direction on the vortex axis. On the very right the spatial vortex structures are visualisized using the λ_2 criterion are shown.

At the beginning of the breakdown process the streamline pattern shows a small region of reversed flow with a free stagnation point on the axis which is confirmed by the plot of the axial velocity component. The first stagnation point is formed at a certain distance downstream from the shock followed by a region of reversed flow, which is terminated by a second stagnation point inside the bubble. Directly downstream of the shock near the vortex axis the flow is strongly decelerated. The maximum deceleration occurs directly on the vortex axis (Fig. 12.6) yielding the first stagnation point. It is decreasing for increasing radius. Further downstream on the vortex axis a pressure minimum exists which is accompanied by the development of RLVS shown by the λ_2 visualization.

At increasing time the second stagnation point is shifted in the z-direction. The reverse flow causes the shock to move upstream, and a bubble-like flow structure emerges, growing in the axial and radial direction until the upstream stagnation point reaches a stable position. Further downstream from the shock, which remains normal near the axis of the vortex but is curved further away from it, the flow is unsteady, slightly oscillating. The deformed shock resembles that of a bow shock observed at supersonic flow around blunt bodies.

The reversed flow inside the vortex bubble can be explained using several small identifiable ring-like vortical structures (RLVS) (Figure 12.8, 4th column) traveling downstream with progressing time. The velocity and MACH number distributions evidence a strong axial reverse flow along the axis of the vortex, and the RLVS inside the bubble are evidenced by the streamline pattern. At the beginning of the calculations the streamline pattern yields several adjacent clockwise rotating RLVSs, generated inside the bubble, which merge further downstream. After some time the flow shows a periodic shedding of single RLVS in conjunction with a stagnation point inside the upstream part of the breakdown bubble. With progressing time the flow deviates more and more from its initial axial symmetry.

Figure 12.9 shows an artificial case of NSVI at Ma_∞=1.6, and time level t=200 for which the axial velocity profile is assumed to be uniform and the circulation is highly increased to $\Gamma = 5.0$. Nearly uniform axial velocity profiles can be generated experimentally in two steps. First, the vortex is originated in subsonic flow by e.g. a delta wing. After that the longitudinal vortex is accelerated to supersonic conditions by a divergent-convergent nozzle like shown by DELERY [3]. The discrepancy between that method and the vortex investigated in this study is that the vortex circulation in the experiments was much smaller. On the other hand, ERLEBACHER *et al.*

[4] selected such a vortex which is why we analyze this case.

The left column shows the results of the EULER calculations and the right column those of the NAVIER-STOKES computations at Re=15000. In the upper pictures the three-dimensional vorticity distribution is exhibited. In these figures the MACH number distribution of the x-z symmetry plane is also projected onto the boundary surface of the computational domain. It can be seen that the inviscid flow computation (left column) yields larger vortical structures than the viscous flow computation. The MACH number contours exhibit a multiple shock structure. The deformed shock resembles that of a bow shock observed at supersonic flow around blunt bodies. Further away from the axis the shock angle decreases until a triple point is encountered. The calculation for inviscid flow yields two adjacent clockwise rotating VRLS inside the bubble, while for the viscous flow computations multiple vortex structures can be observed. An analysis of the temporal development of the flow shows a periodic shedding of the VRLS. A detailed investigation of the inviscid flow pattern evidences that the two vortices are separated by a shock which is caused by a supersonic jet-like flow in the upstream direction along the axis. This shock is perpendicular to the z-axis. The streamlines in the recirculation region show a convergent-divergent nozzle-like flow structure that enables the upstream flow to become supersonic inside the bubble. Planes inside the breakdown bubble, which are

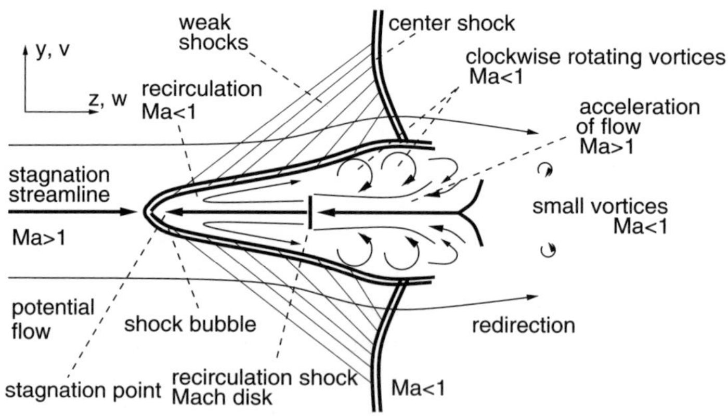

Fig. 12.3 Flow schematic for very strong NSVI.

perpendicular to the main flow direction, show a fan like flow regime with multiple shock structures, which also was noticed by BRILLANT *et al.* in their experiments [2]. A schematic of the flow in the bubble is given in

Fig. 12.3 to clarify the details of the flow in the x-z-plane.

Figure 12.10 shows the color schlieren and differential interferometry pictures for the OSVI at a free stream MACH number of $Ma_\infty = 2.0$ and figure 12.11 the flow field in the x-z-middle plane of the computational domain for a strong OSVI ($\Gamma = 3.5$, $\delta = 0.345$, $\mu = 0.4$ and $\beta = 11.5°$ wedge angle). The instantaneous pictures at $t = 200$ indicate that the flow field differs remarkably from the NSVI case. The MACH number distribution shows that in the vicinity of the vortex core a normal shock is observed. The originally straight oblique shock is distorted into an 's'-shaped shock front near the vortex core. Downstream of the shock a region of subsonic flow occurs. With progressing time a small cone like shape emerges which is of the order of the vortex diameter. Compared to the normal shock case this cap size is smaller. Since breakdown of the vortex is necessarily connected with the formation of a stagnation point on or near the axis, a locally subsonic flow downstream of the shock has to occur and as such the increased shock angle is a must for a bursting vortex. As a second condition the pressure rise across the deformed part of the shock has to be sufficient to initiate breakdown. The streamlines inside the burst part of the vortex indicate recirculation and a stagnation point.

The parameter study for OSVI shows that the shock deformation in the vicinity of the vortex core depends primarily on the magnitude of the axial velocity deficit in conjunction with the free stream shock strength $Ma_\infty \sin\sigma$ and the wedge angle. Secondly, the vortex circulation seems to determine the pressure on the vortex axis. For a slender isentropic vortex the radial momentum equation and the energy equation are decoupled. In this case the influence of circulation is only linear on the axial momentum while the wake is of quadratic order.

The velocity vectors evidence a recirculation region inside the burst part of the vortex. It is moved upward away from the axis of the undisturbed vortex. The distribution of the λ_2 contours, which are not shown here, show separate vortex tubes downstream from the burst part of the vortex, indicating the change to the spiral type breakdown.

In addition to the numerical and experimental investigations a simple breakdown criterion is derived. The stability region for NSVI is compared with the results of the literature in Fig. 12.5. Based on the axial momentum equation for inviscid flow and the RANKINE-HUGONIOT relations and the energy equation, the onset of breakdown can be predicted by requiring a stagnation point on the vortex axis. Dividing the axial momentum equation by the scaled free stream pressure p_{1_∞} upstream of the shock leads to the relation $\Phi_I + \Phi_{II} \geq \Phi_{III}$ with its stability variables $\Phi_I = p_{1_A}/p_{1_\infty}$, $\Phi_{II} = \rho_{1_A}\, v_{z1_A}^2/p_{1_\infty}$ and $\Phi_{III} = p_{2_\infty}/p_{1_\infty}$. The momentum flux on the vortex axis has to be sufficient to pass the pressure jump enforced by the shock without the occurrence of a stagnation point. The term $\Upsilon = (\Phi_I + \Phi_{II})/\Phi_{III}$ expresses the stability of an arbitrary longitudinal streamwise vortex. The numerator contains upstream pressure and momentum forces, the denominator comprises the downstream pressure forces determined by the shock. $\Upsilon >> 1$ is required for stability, $\Upsilon \leq 1$ denotes the possibility of vortex breakdown depending on the local shock properties, i.e., whether or not a stagnation point develops. Additionally, for OSVI the local shock angle has to be greater than that necessary for the weak shock solution. In other words, the flow downstream of the shock has to be locally subsonic. In the model of the axisymmetric slender iso-energy or isentropic RANKINE [9], BURGERS [4] and LAMB-OSEEN [3] vortices the pressure, density and MACH number distribution in a plane perpendicular to the undisturbed vortex tube has a minimum on the axis and so does the pressure across the shock. Because of the wake-like axial velocity profile and the low pressure and density on the vortex axis the momentum flux is reduced compared to the outer region far away from the vortex center. To estimate the stability properties for a given vortex, Υ has to be determined by calculating the quantities Φ_I, Φ_{II} and Φ_{III}.

Φ_I describes the relation between the pressure on the vortex axis and in the outer region $p_{1\infty} = T_{1\infty}^{\frac{\gamma}{\gamma-1}}/\gamma$. In conjunction with the free stream temperature $T_{1\infty} = (1 + \frac{\gamma-1}{2} Ma_\infty^2)^{-1}$ the relation reads $\Phi_I = \frac{p_{1_A}}{p_{1_\infty}} = \gamma\, p_{1_A}\,(1 + \frac{\gamma-1}{2} Ma_\infty^2)^{\frac{\gamma}{\gamma-1}}$, whereas the pressure on the vortex axis has to be determined by integrating the radial momentum equation for a given vortex model. Φ_{II} contains the relation between the axial momentum flux and the free stream pressure. For all supersonic flow $(Ma_\infty(1-\delta) > 1.0)$ the following relation holds: $\Phi_{II} = \frac{\rho_{1_A}\, v_{z1_A}^2}{p_{1_\infty}} = \rho_{1_A}\,\gamma\,(1-\delta)^2\, Ma_\infty^2\,(1 + \frac{\gamma-1}{2} Ma_\infty^2)^{\frac{1}{\gamma-1}}$. The free stream velocity $v_{z_\infty} = Ma_\infty\,\sqrt{T_\infty}$ can be determined by the free stream MACH number and temperature. In combination with the energy equation and the perfect gas law one obtains for the pressure p_{1A} on the axis

$$\Phi_{II} = p_{1_A} \left[\frac{\gamma^2\,(1-\delta)^2\, Ma_\infty^2\,(1 + \frac{\gamma-1}{2} Ma_\infty^2)^{\frac{1}{\gamma-1}}}{1 - \frac{\gamma-1}{2}(1-\delta)^2\, Ma_\infty^2\,(1 + \frac{\gamma-1}{2} Ma_\infty^2)^{-1}} \right] . \qquad (12.44)$$

$\Phi_{III} = p_{2_\infty}/p_{1_\infty} = 1 + \frac{2\,\gamma}{\gamma+1}(Ma_\infty^2\,\sin(\sigma)^2 - 1)$ describes the pressure jump over the shock and follows from the RANKINE-HUGONIOT relations.

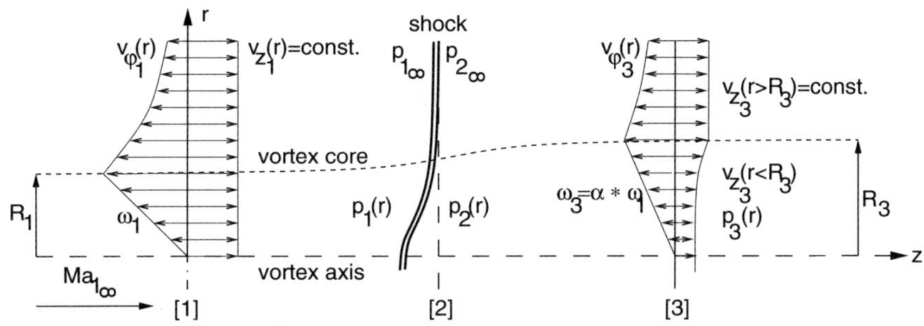

Fig. 12.4 Theoretical model to predict vortex breakdown.

As already shown, for iso-energy vortices the axis pressure p_{1A} can be determined by integrating the radial momentum equation for slender vortices in conjunction with the ideal gas law, $\int_{p_A}^{p_\infty} \frac{1}{p}\, dp = \gamma \int_0^\infty \frac{v_\varphi^2(r)}{T(r)\, r}\, dr$. Separation of variables leads to the integral to determine the axis pressure $p_{1A}|_{dh=0} = \frac{p_\infty}{exp(I|_{dh=0})}$, where $I|_{dh=0} = \gamma \int_0^\infty \frac{v_\varphi^2(r)}{T(r)\, r}\, dr$. The energy equation provides the axis temperature $T_{1A}|_{dh=0} = 1 - \frac{\gamma-1}{2} \frac{(1-\delta)^2\, Ma_\infty^2}{1+\frac{\gamma-1}{2}\, Ma_\infty^2}$ for a given wake $v_{z_{1A}} = (1-\delta)\, v_{z_\infty}$. Note that $T_A|_{dh=0}$ is independent of the vortex model, i.e., a given vortex circulation. For isentropic vortices the radial momentum equation and the LAPLACE-POISSON law leads to an integral for the axis pressure $\int_{p_A}^{p_\infty} (\gamma p)^{-\frac{1}{\gamma}}\, dp = \int_0^\infty (v_\varphi^2(r)/r)\, dr$. The minimum pressure follows directly from $p_A|_{ds=0} = ((\gamma\, p_\infty)^{\frac{\gamma-1}{\gamma}} - (\gamma - 1)\;\; I|_{ds=0})^{\frac{\gamma}{\gamma-1}}/\gamma$ with $I|_{ds=0} = \int_r^\infty (v_\varphi^2(r)/r)\, dr$. For a slender iso-energy or isentropic RANKINE vortex with uniform axial velocity the pressure-, density- and temperature at the vortex axis can be determined analytically. For a the slender BURGERS vortex the isentropic pressure-, density- and temperature distribution can also be found analytically, while in the iso-energy case the thermodynamic variables must be determined numerically. The same is true for the LAMB-OSEEN vortex.

The comparison of the numerical results obtained for NSVI with the relation $\tau = \tau(Ma_\infty)$ and with numerical and experimental data from [3; 4; 9] in Fig. 12.5 shows good agreement in the range of $1.5 \leq Ma_\infty \leq 2.0$.

It predicts breakdown, when the ratio of the maximum azimuthal and the axial velocity components reaches a critical value.

In the following the introduced breakdown criterion is briefly discussed for OSVI. For oblique shocks and wedge angles $\beta < \beta_{max}$ the EULER equa-

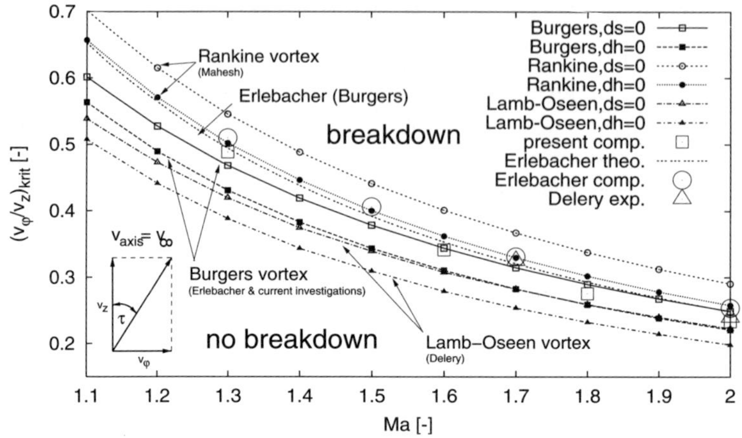

Fig. 12.5 Breakdown map for NSVI. $\tau = v_\varphi(r=1)/v_z(r=0)$ characterizes the critical ratio for which vortex breakdown occurs at normal shocks. Either a small axial or large azimuthal velocity component can trigger breakdown. The larger the pressure rise or the circulation the more likely breakdown will occur.

tions provide two possible real shock angles σ, namely the weak and strong solution. In the latter case the flow downstream of the shock is subsonic. If the wedge angle exceeds the limit $\beta > \beta_{max}$ the shock will be detached producing subsonic flow, which is required for vortex breakdown. The MACH number has its minimum on the vortex axis and decreases for increasing wake, e.g., $Ma_A = v_{z_A}/a_A = (1-\delta)\, v_{z_\infty}/\sqrt{T_\infty + \frac{\gamma-1}{2}\, v_{z_\infty}^2\, (1-(1-\delta)^2)}$. In our model the MACH number on the vortex axis needed for vortex breakdown has to be smaller than that for no breakdown. To establish breakdown, the following conditions must be satisfied. First, the local MACH number on the vortex axis has to be lower than the MACH number for the weak solution, $Ma_A < Ma_{A_{min}} \rightarrow \beta > \beta_{max}(Ma)$. Second, the local pressure jump has to be sufficient to ensure $\Phi_1 + \Phi_2 < \Phi_3 \rightarrow \Upsilon < 1$. The problem is to predict the deflection angle on the vortex axis, which is not necessarily the same as the wedge angle. From the computations it is known that for increasing distance between wedge surface and vortex axis the likelihood of breakdown is attenuated. As a first approximation it is assumed that the flow angle corresponds to the known wedge angle. This correlates to enforced redirection, e.g., by a fan blade in a turbine.

In this manner it is possible to catalogue the already explained OSVI modes: $\beta_A < \beta_{crit}(Ma_A) \wedge \Upsilon < 1$ or $\beta_A < \beta_{crit}(Ma_A) \wedge \Upsilon > 1$ leads to a weak shock vortex interaction mode. Both shock and vortex are only slightly influenced. The flow field remains supersonic everywhere. $\beta_A > \beta_{crit}(Ma_A) \wedge \Upsilon > 1$ stands for a moderate interaction mode. The shock resembles a strong solution in the vortex core vicinity. A small subsonic pocket develops downstream of the shock. Elsewhere the flow remains supersonic. The vortex is influenced without being destroyed. No stagnation point is observed. In the computational investigation of NEDUNGADI & LEWIS the streamwise vortex tube was divided by the shock. $\beta_A > \beta_{crit}(Ma_A) \wedge \Upsilon < 1$ denotes a strong interaction. The shock has a similar form to that observed for moderate interaction. Additionally a larger subsonic region is formed and a stagnation point can occur. NEDUNGADI & LEWIS [11] investigated OSVI at MACH numbers of $Ma_\infty = 3.0$ and 5.0 for different shock- and vortex strengths. The results match with the presented criterion.

12.6 Conclusions

Breakdown of slender vortices in supersonic flow caused by NSVI and OSVI was investigated using numerical solutions of the EULER and NAVIER-STOKES equations. For OSVI the numerical results were compared with preliminary experimental results. The calculations show, that NSVI and OSVI does not necessarily lead to breakdown. If the circulation of the vortex is small and if the axial velocity profile is uniform, the vortex does not change its overall shape. Increasing the vortex strength, the free stream MACH number, and a wake-like axial flow profile can lead to a free stagnation point on or near the axis of the vortex. This can be considered as the initiation of vortex breakdown. The results show that a reduction of the axial velocity or an increase of the azimuthal velocity component enhance breakdown. In other words, the stronger the shock and hence the pressure jump, the less the circulation has to be to cause breakdown, and for weaker shocks a larger circulation is necessary for breakdown. For wake-like profiles breakdown occurs earlier, since the necessary deceleration of the axial flow to reach a stagnation point is smaller than for a uniform or even a jet-like profile. While breakdown of vortices with uniform axial flow is possible for normal shocks, a wake-like velocity profile is required for oblique shocks. Although an axisymmetric vortex is initialized it can be concluded that vortex breakdown is a truly three-dimensional process. For NSVI several ring-like, slightly oscillating vortex structures are formed in the beginning of vortex breakdown immediately downstream from a stagnation point on the axis. For certain inviscid flow conditions strong upstream flow motion, even supersonic flow may occur near the axis. The numerical results obtained indicate that the flow structure of the burst part of the vortex agrees well with experimental observations. A breakdown criterion was derived that shows good agreement with the numerical and experimental findings and other data from the literature.

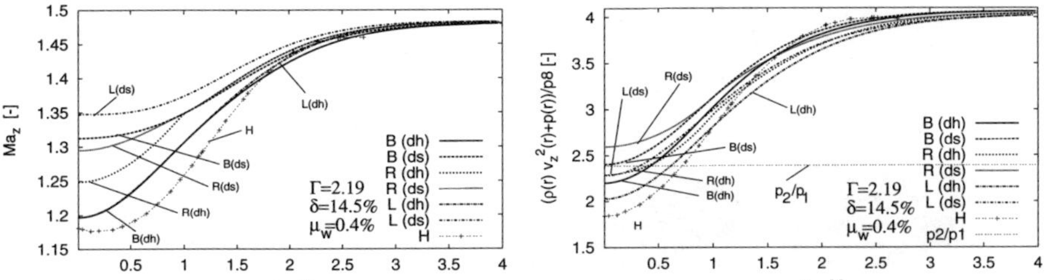

Fig. 12.6 MACH number in the streamwise direction Ma_z and advective flux $(\rho(r)\ v_z^2(r) + p(r))/p_\infty$ perpendicular to the vortex axis for the **R**ankine, **B**urgers, **L**amb-Oseen and a fan tip vortex (priv. comm. W. **H**ofmann, LFM, RWTH Aachen). ($Ma_\infty = 1.48$, $\Gamma_0 = 2.19$, $\delta = 0.145$, $\mu_w = 0.4$). Indices *dh*: constant enthalpy, *ds*: constant entropy.

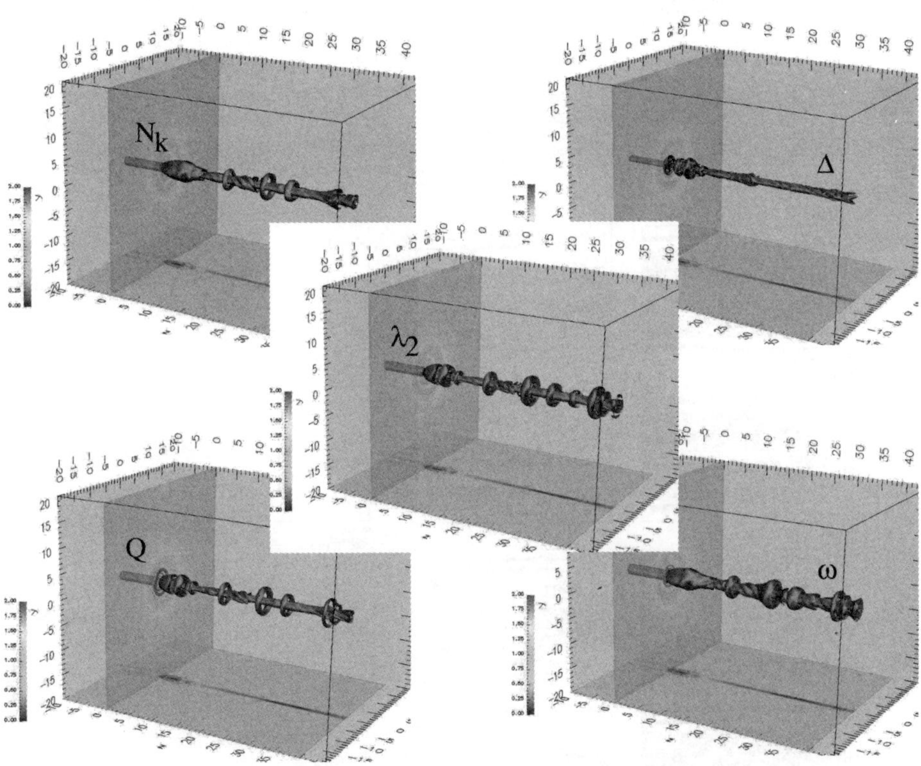

Fig. 12.7 Comparison of different vortex criteria. The pictures show NSVI for the BURGERS vortex at $Ma_\infty = 1.6$, $\Gamma_0 = 2.5$, $\delta = 0.1$, $\mu = 2.0$, $t = 139$. The vortex identification criteria are denoted by N_k: TRUESDELL-, Δ: CHONG & PERRY & CANTWELL-, Q: HUNT *et al.*, ω vorticity, and λ_2: JEONG & HUSSAIN [5].

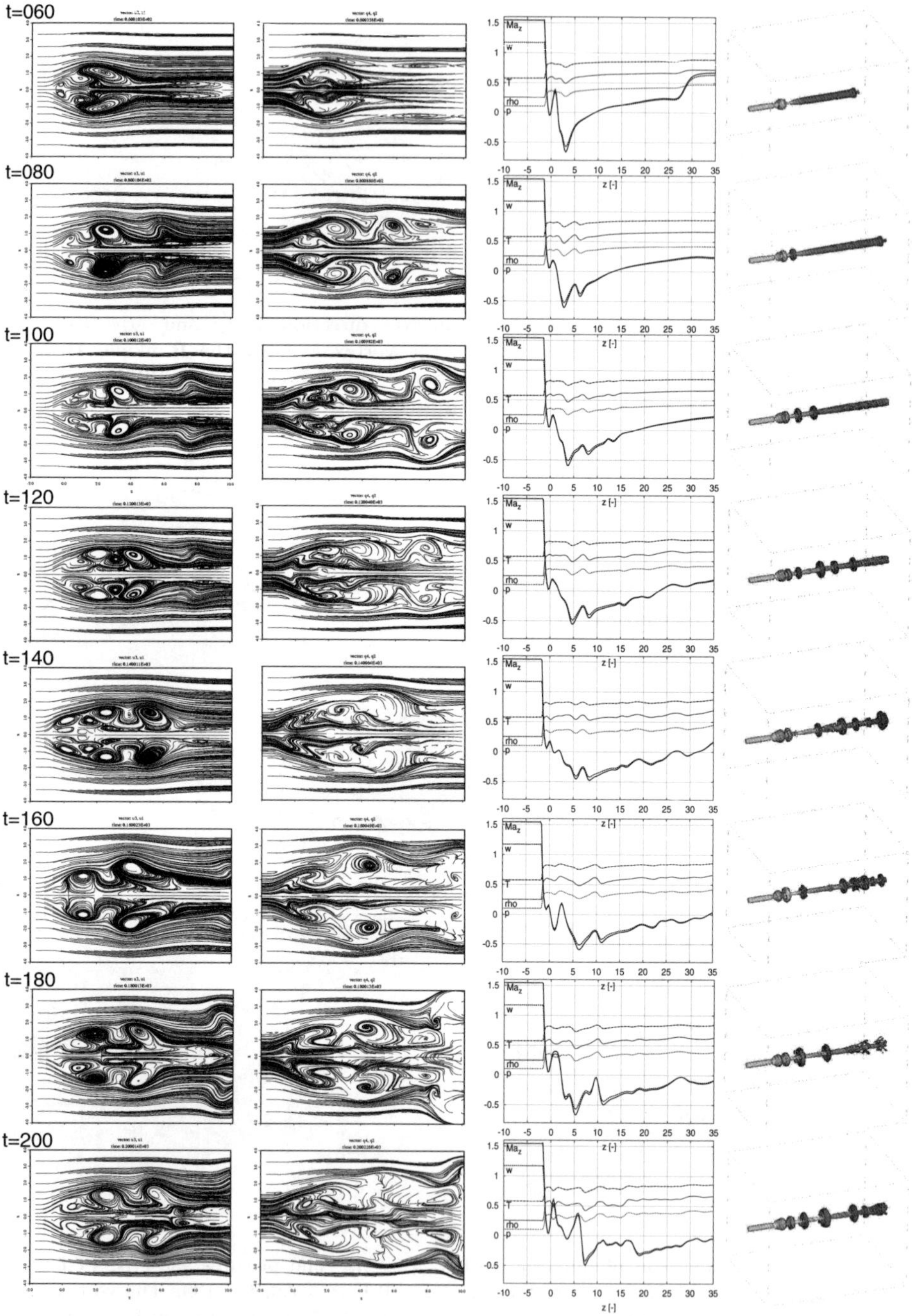

Fig. 12.8 Temporal evolution of streamlines (1st column), and vortex lines (2nd column) in the center plane $y = 0$; primitive variables and MACH number (3rd column) and λ_2-iso surface (4th column, color=local density) for certain time steps in the case of NSVI ($Ma_\infty = 1.6$, $\Gamma_0 = 2.5$, $\delta = 0.1$, Re=15000).

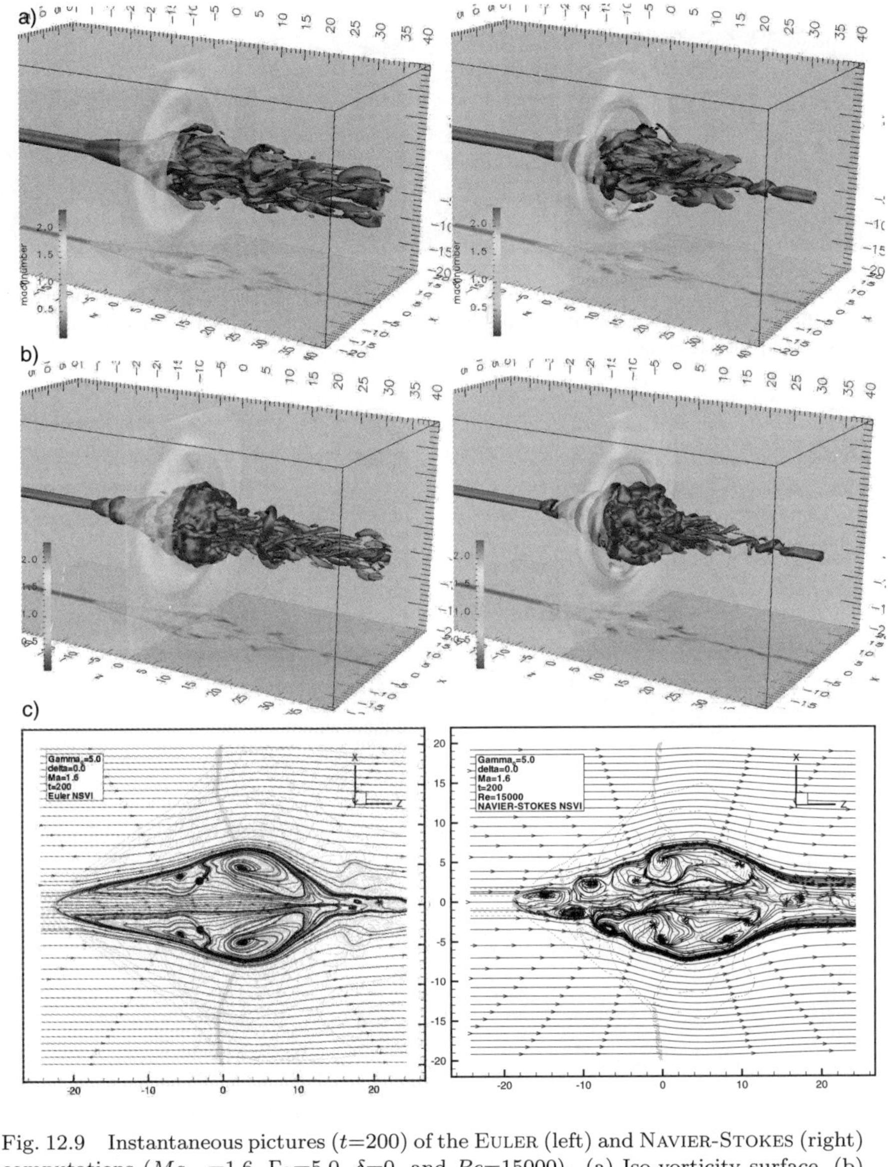

Fig. 12.9 Instantaneous pictures (t=200) of the EULER (left) and NAVIER-STOKES (right) computations (Ma_∞=1.6, Γ_0=5.0, δ=0, and Re=15000). (a) Iso-vorticity surface, (b) iso-λ_2 surface combined with the MACH number distribution in the center plane of the computational domain mapped onto the surface plane, and (c) streamline pattern in the y-z-plane at x=0 of the bubble.

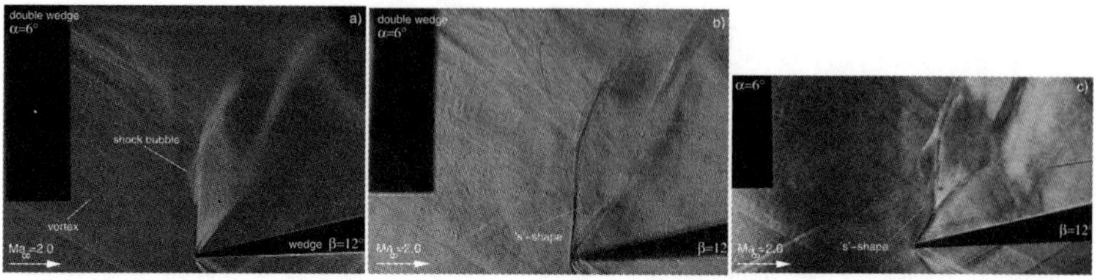

Fig. 12.10 OSVI for $Ma_\infty = 2.0$, Re=15000 at a $\beta = 12°$ wedge and an angle of attack $\alpha = 6°$ of the vortex generator (double wedge). (a) Color schlieren picture, (b)+(c) differential interferometry.

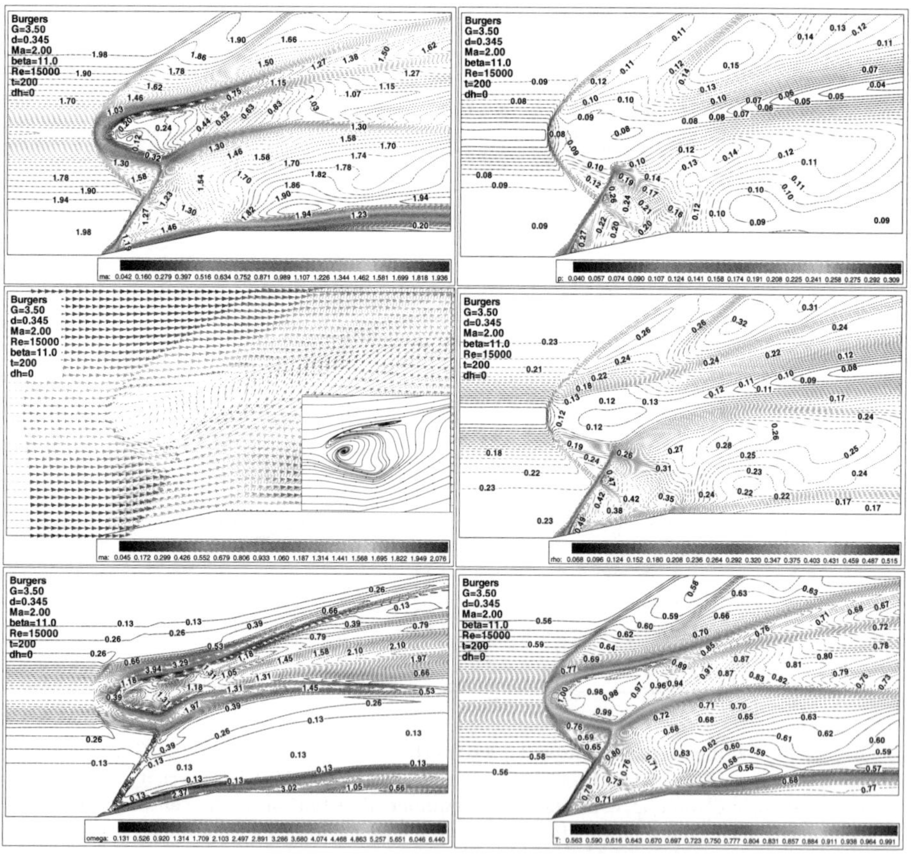

Fig. 12.11 OSVI (Ma = 2.0, Re = 15000, Γ = 3.50, δ = 0.345, μ = 0.4) at a thin wedge (β = 11.5°). The instantaneous pictures (t=200) show the $Ma-$, $p-$, $T-$, $\rho-$, iso-vorticity lines $|\vec{\omega}|$, velocity vectors, and streamlines in the y-z middle plane (x=0) of the computational domain.

Bibliography

[1] W. ALTHAUS, M. WEIMER, Review of the Aachen Work on Vortex Breakdown, IUTAM Symposium on Dynamics of Slender Vortices, Kluwer Academic Publishers, pp. 331-344, 1997.

[2] P. BILLANT *et al.*, Instabilities and Vortex Breakdown in Swirling Jets and Wakes. IUTAM Symposium on Dynamics of Slender Vortices, Kluwer Academic Publishers, pp. 267-286, 1997.

[3] J. DELERY, Aspects of Vortex Breakdown, Prog. Aerospace Sci., vol. 30, pp. 1-59, 1994.

[4] G. ERLEBACHER, M. HUSSAINI, C. SHU, Interaction of a Shock with a Longitudinal Vortex, ICASE Report, No. 96-31, 1996.

[5] J. JEONG, F. HUSSAIN, On the identification of a vortex, JFM, vol. 285, pp. 69-94, 1995.

[6] O. KANDIL, H. KANDIL, C. LIU, Computation of Steady and Unsteady Compressible Quasi-Axisymmetric Vortex Flow and Breakdown, AIAA Paper 91-0752, 1991.

[7] O. KANDIL, H. KANDIL, C. LIU, Shock/Vortex Interaction and Vortex-Breakdown Modes, IUTAM Symposium of Fluid Dynamics of High Angle of Attack, Tokyo, Springer Verlag, 1992.

[8] M. LIOU, A Sequel to AUSM: $AUSM^+$, JCP, vol. 129, pp. 364-382, 1996.

[9] K. MAHESH, A Model for the Onset of Breakdown in an Axisymmetric Compressible Vortex, Physics of Fluids, vol. 8, no. 12, pp. 3338-3345, 1996.

[10] S. MENNE, Rotationssymmetrische Wirbel in achsparaller Strömung, PhD Thesis, Aerodynamisches Institut, RWTH Aachen, 1986.

[11] A. NEDUNGADI, M. LEWIS, Computational Study of the Flow-Fields Associated with Oblique Shock/Vortex Interactions, AIAA Journal, 34-12, 1996.

[12] T. POINSOT, S. LELE, Boundary Conditions for Direct Simulations of Compressible Viscous Flows, JCP, vol. 101, pp. 104-129, 1992.

[13] D. RUDY, J. STRIKWERDA, A Nonreflecting Outflow Boundary Condition for Subsonic Navier-Stokes Calculations, JCP, vol. 36, pp. 55-70, 1980.

[14] S. SCHLECHTRIEM, M. LÖTZERICH, Breakdown of tip leakage vortices in compressors at flow conditions close to stall, IGTI-ASME Conference, Orlando Florida, June 1997.

[15] O. THOMER *et al.*, Computational Study of Normal and Oblique Shock-Vortex Interactions, Conference proceedings ECCOMAS 2000 Barcelona, 2000.

[16] K. THOMPSON, Time Dependent Boundary Conditions for Hyperbolic Systems, JCP, vol. 68 no. 1, pp. 1-24, 1987.

[17] K. THOMPSON, Time Dependent Boundary Conditions for Hyperbolic Systems II, JCP, vol. 89, pp. 439-461, 1990.

Chapter 13

Analysis of Rotor Vortex Wake Structure Using 3-C PIV Measurements

Chee Tung and Joon Lim
Army/NASA Rotorcraft Division
Aeroflightdynamics Directorate, AMRDEC
US Army Research, Development, and Engineering Command
Ames Research Center, Moffett Field, Ca 94035

Dedicated to Lu Ting on his eightieth birthday

Abstract

Three-component (3-C) PIV measurements of a model rotor wake in descent flight are analyzed. The measuring PIV plane is always cutting the vortex with an angle. Another difficulty is the convective velocity of the vortex wake is not known accurately either. The measured vortex wake geometry has been used to determine the convection velocity. For three closely adjacent points along the curved vortex, one may fit a circular arc. The orientation of the PIV measuring plane to the plane that is normal to the curved vortex at the middle point can be easily determined by the geometry of these three points. Both vortex core radius and peak swirl velocity are smaller in the resolved normal plane than the one obtained in the measuring plane. Vortex formation, structure and decay are also discussed.

13.1 Introduction

The current civil helicopter has been used extensively because of its excellent hover and low-speed forward-flight capability, but the helicopter fleet continues to be too noisy for community acceptance, preventing it from reaching its full potential of utilization. One of the major noise sources comes from the rotor blade cutting through its own wake. This phenomenon is known as blade-vortex interaction (BVI). It occurs mostly during descent flight for landing. Over the last two decades many experimental studies from full-scale flight tests [Sakowsky (1976)], [Boxwell (1980)] to model-rotor tests [Boxwell (1983)], [Dadone (1989)], [Lorber (1990)] have been performed to investigate this BVI phenomenon. At the some time, several analytical methods have been developed with limited success to predict the BVI noise. It turns out that in order to correctly predict the aerodynamics and acoustic pressure for a helicopter operating in power decent flight, it is of utmost importance to model the geometry of the wake vortices shed from the blade tips. The wake geometry model should contain information on the vortex location, structure, strength, and core size.

In 1994, an international cooperative test, called the higher-harmonic-control aeroacoustic rotor test (HART) was conducted by the Army Aeroflightdynamics Directorate, Deutsche Forschunganstalt fur Luft-und Raumfahrt (DLR), Duits-Nederlandse Wind Tunnel (DNW), NASA, and the Office National d'Etudes et de Recherches Aerospatiales (ONERA) at DNW to measure not only the blade surface pressure distributions, the blade deformation, and the acoustic signatures, but also the wake geometry using the laser light sheet technique and the vortex strength using laser Doppler velocimetry method. However, the wake information such as vortex location, structure, strength, and core size is still incomplete to validate the prediction tools. Therefore, a second entry (HART II -consisting of the same five organizations) to DNW was carried out in 2001. In this test, advanced flow measurement techniques like particle image velocimetry (PIV) were used to obtain 3-component (3-C) velocity field as well as wake geometry.

The advantages and difficulties associated with PIV measuring technique applied to rotor vortex properties have been reported in [Heineck (2000)], [McAlister (2001)], [Raffel (1998)], [Yu (2002)]. One of main difficulties in deriving the vortex structure from 3-C velocity measurement is that the plane normal to the vortex axis is unknown a priori. The measuring plane is always cutting the vortex with an angle. Another difficulty is the convective velocity of the vortex wake is not accurately known either. The wake moves downstream with free stream velocity plus the induced velocity by other vortices. [Burley (2003)] presented two different but elaborate methods to determine the orientation of the vortex axis with respect to the PIV measuring plane. Specifically, one method uses the axial flow within the vortex core and the other uses the tangential (or spin) velocity within the vortex core to define the orientation. In this paper, a rather simple geometry method based on the measured vortex location is presented to determine the convective velocity and the orientation of the vortex axis.

13.2 Test set-up and 3-C PIV measurements

Like the HART I test, the Mach-scaled rotor is a 40% model of a hingeless BO-105 main rotor. The radius of the rotor is 2 meters with a pre-cone of 2.5 degrees at hub. The rectangular blade has a -8 degrees twist with NACA-23012 airfoil and a chord length of 0.121 m. The rotor is operated at a nominal 1041 rpm and a thrust coefficient of 0.0044. The hover tip speed is 218m/s and an advance ratio of 0.15. The baseline case has a rotor shaft angle of 5.3 degrees aft. Detailed information about the model and test conditions is given in [Yu (2002)] .

All the PIV equipment including 3 double-pulsed Nd:YAG lasers and 5 digital cameras was mounted onto a remote controlled traverse system as shown in Fig. 13.1. This traverse system was located outside of the shear layer of the wind tunnel in the open jet configuration and moved in the direction of flow so the distance between the camera and light sheet plane remains constant. A total of four cameras have been used to collect 3-C PIV data. One pair of cameras has 100mm lenses (DNW measurement window) for a larger observation area and the other pair has 300mm lenses (DLR measurement window) to investigate the tip vortex with high spatial resolution. The remaining lasers and cameras have been used to determine the location of wake vortices before the 3-C PIV cameras taking data, and the locations and orientations of the measuring cut-planes over the rotor disk are shown in Fig. 13.2. To avoid the interference of the blades with the measuring plane, a reference blade located at an azimuth of 20 degrees and 70 degrees alternatively is used for different wake measurements in Figs. 13.2(a) and 13.2(b). The entire traverse system has to move to the opposite side of the wind tunnel to measure the wake on the retreating side. Further details of test set-up and measuring technique are given in [Raffel (2002)] , [Richard (1995)].

13.3 PIV data process

In each measuring plane (black line segment in Fig. 13.2), one hundred instantaneous PIV images are obtained. This line segment is oriented at angle of 30.7 degrees with respect to wind tunnel x-axis. The measuring procedure is to measure the wake vortex from the same blade to avoid blade-to-blade differences. Figure 13.3 shows one individual image taken by the camera for Dpt 995 (Position 21). The upper image (0.45m by 0.38m) is the global view taken from DNW camera and shows the apparent center of swirl motion in the flow field. The close-up image (0.15m by 0.13m) from the DLR camera is not clear whether or not the identified swirl center is in the image since it does not show a black spot. In addition, the vortex structures move around from image to image due to vortex wandering.

An instantaneous 3-component velocity (U_e, V_e and W_e) contour plot at the position 18 in the DLR measuring window is shown in Fig. 13.4 and the derived vorticity contour in the direction normal to the measuring plane is also accompanied. The PIV measuring window coordinate system is represented with the positive x-axis to the right direction from the observer, the positive y-axis up and the positive z-axis toward the observer. This vortex was generated initially by the blade tip vortex at an azimuth of 136 degrees, and reached this position after a wake age of 118 degrees. The 3-C velocity vectors shown in Fig. 13.4 were unfiltered. In this figure, the inplane velocity vectors, $U_e\ \vec{i} + V_e\ \vec{j}$ are overlaid in the cross flow (W_e) contour plot, and the free stream velocity contributes mostly to these inplane vectors. The vorticity, γ was derived using the measured PIV velocity vectors by $\gamma = 2\omega = \vec{k} \cdot (curl\ \vec{V})$ where ω is the angular velocity and $\vec{V} = U_e\ \vec{i} + V_e\ \vec{j} + W_e\ \vec{k})$, and shows the maximum vorticity clearly and noises of the measurement.

In order to eliminate vortex turbulence and noise of the measurements, an averaging of the all hundred PIV images is necessary, and a simple way of averaging the 3-C velocity is using the simple mean by purely overlaying all the PIV images. This method would be a good choice for the measurements where the vortex centers are mostly collocated, which is typically observed in the 2nd quadrant of the rotor disk where there are few vortex interactions with the younger-age vortices. But conditional averaging obtains the averaged PIV image by aligning the individual image with its vortex centers as shown in Fig. 13.5, and is found to be often effective for the analysis of vortex properties [Jeong (1995)].

There are many methods for conditional averaging of the velocity field. The centroid of vorticity method is employed in the present study. Figure 13.6 shows the centroids of all the hundred PIV images for the positions 17 and 21 of the baseline DLR data. Due to vortex wandering, the centroids are not at the same points the younger vortex has smaller scatter of the

Table 13.1 Comparison of 3-C convective velocities calculated using vortex center and vortex distance methods for the DLR data, position 18.

	Vortex Center (% ΩR)			Vortex Distance (% ΩR)		
Pos 17	-9.46	1.58	12.17	-10.85	2.22	13.47
Pos 18	-9.94	3.11	12.47	-10.45	2.02	13.81
Pos 19	-10.03	0.09	13.61	-8.97	2.99	14.65

vortex center. The averaged domain of measuring is slightly smaller after conditional averaging due to vortex wandering. Convective velocity of the vortex is obtained from the measured velocity data at the vorticity center, and the vorticity field is derived after conditionally averaging of all the instantaneous PIV image. Since the distances of the vortex core centers and the elapsed time to reach to the next positions are known, the convective velocity can be easily calculated using the finite difference scheme. This alternative approach using the vortex distance is tabulated in Table 13.1, and interestingly this simple approach allows 3 to 9% relative errors in terms of the convective velocity magnitude, except for the vertical velocity component at the position 19, which shows 17% error. Figure 13.7 compares the cross flow and the vorticity contours after subtracting the 3-C convective velocities calculated by the vortex distance method with the ones with vortex center method. The cross flow field shows the shift of the zero velocity point (.x. mark) by the difference of the convective velocities, but the contours of the cross flow and the vorticity are identical between two methods.

A composite of averaged vorticity field from the positions 17 to 19 (2nd quadrant) and 20 to 23 (1st quadrant) is given in Fig. 13.8. The evolution of the tip vortex is seen. The wake age for the position 17 is 28 degrees, and 90 degrees apart from the positions 17 to 19. Similarly, the wake age for the position 20 is 248 degrees, and 90 degrees in increment from the positions 20 to 23. As the vortex wake ages, the vortex sheet get fainted and disappears.

13.4 Rotation of PIV measuring plane

The method used in [Burley (2003)] to correct the orientation of the vortex is quite complicated. A simple geometrical method is proposed here. The location of the vortex center at each PIV measuring plane is known. For any three adjacent points such as the positions 9, 17 and 24, one can fit a circular arc through these points. As shown in Fig. 13.9, the tangent vector (k) at the position 17 is parallel to the vector connecting the positions 9 and 24. The cross product of the tangent vector to the vector between the positions 9 and 17 gives the normal vector (j) at the position 17. Then the cross product of the tangent vector to the normal vector gives bi-normal vector (i). The plane contains the normal and bi-normal vectors in the plane normal the circular arc and thus the corrected plane normal to the vortex. Two rotational angles can be easily computed.

Figure 13.10 shows the vorticity and cross flow contours for the position 17 before and after rotation. Because the wake age for the position 17 is only 28 degrees, the vortex remains in the rotor disk plane. Rotations of the measuring plane normal to the vortex trajectory are 0.6 and -20.8 degrees along the horizontal (x) and vertical (y) axes respectively. The 3-C PIV velocities are referenced with the vortex core center (.x. mark) by subtracting the velocity at the vortex core center. The vorticity contours between the unrotated and rotated images reveal insignificant difference in terms of the maximum vorticity and the contour shape, but unbalanced cross flow distribution is significant in the unrotated image.

Figure 13.11 compares the unrotated and rotated circulation and swirl velocity distribution along the vortex core coordinate for the position 17 of the baseline DLR data. After rotation of the PIV measuring plane, the peak swirl velocity and the vortex core radius (defined as the distance at the maximum swirl velocity) become smaller. For a younger vortex such as in this position, the peak swirl velocity after rotation is reduced by 21%, and the vortex core radius is reduced by 39% from 0.62 to 0.38 of the rotor radius (2m). The growth of the corresponding circulations appears to slow down as the vortex radius increases. In fact, one chord length is equivalent to the 6% of the rotor radius and the circulation growth could continue to grow even after the vortex radius increases to a multiple of the chord lengths.

For older vortices such as the one measured at the position 22, the vorticity and cross flow contours are given in Fig. 13.12. The vortex appears fully grown after aging of 428 degrees, and the rotations of the measuring plane normal to the vortex trajectory are 6.3 and -33.4 degrees along the horizontal (x) and vertical (y) axes respectively. Due to relatively large rotation, the projected area for the vorticity contour has been reduced. Calculated two rotation angles are shown in Fig. 13.13 in terms of wake age (from position 17 to 23). The rotations become larger as the wake matures, and the rotation along the vertical axis (qy) turns out to be significant, ranging in the 20 to 40 degrees of the clockwise rotation.

The peak swirl velocities and vortex core radii are compared in Fig. 13.14 with the spin-velocity method results by [Burley (2003)]. The peak of the vortex swirl velocity becomes smaller as wake ages (first 118 degrees) while the vortex core radius increases, then the reverse trend is observed for wake age greater than 118 degrees. It seems that the vortex has higher swirl velocity after formation, and then the core radius decays fast. At later times the vortex sheet feeds into the vortex to make it grow. This phenomenon is in contrast to the formation of the vortex for a hovering rotor [McAlister (2001)]. Both predictions are in relatively good agreement, and the peak swirl velocity shows significant difference at the wake age of 518 degrees (Baseline, position 23), and it is known that at this position it is extremely difficult to accurately analyze the vortex property due to interactions from multiple vortices.

Figure 13.15 shows that the swirl velocity and circulation distribution as a function of the vortex radius after the rotation. Since the analysis of the PIV image is based on the data observed by the 300mm lenses, the field view is limited to about 3 to 4 percent of the blade radius (about 1/2 to 2/3 chord length). The swirl velocity and the circulation show reasonably good trends up to the wake age of 118 degrees. After that, the circulation seems to not reach a constant value, and the swirl velocity no longer shows 1/r decay and does not die out.

13.5 Concluding remarks

There are many benefits in applying PIV techniques to study the rotor vortex formation, structure and aging. But there are equal amounts of difficulty to analyze the PIV data. Notably the method on how to average all instantaneous PIV frames, the elimination of convection velocity, and the orientation of the averaged PIV plane normal to the vortex trajectory are typical problem areas. In this study, the 3-component of convection velocity based on vortex distance method is computed. It does not change as compared to the velocity averaging method over the PIV measuring plane. The method of vorticity centroid is selected to determine the conditional average. The proposed re-orientation of the PIV measuring plane based on vortex geometry is simple, but the effectiveness requires further study due the domain of PIV measuring plane is limited to the camera pixels.

This study also suggests for future PIV measurement of rotor wakes that the convective velocity and re-orientation of the measuring plane should be done in closer distance than the one used in the HART II test.

Acknowledgments

The authors are deeply indebted to Lu Ting for sharing, over the course of many communiations, his insights on vortex wake structures. These proved indispensable in the writing of this paper.

We also acknowledge the support of our partners from DLR, DNW, NASA, and ONERA with special thanks to Dr. Berend van der Wall of DLR and Casey L. Burley of NASA Langley Research Center.

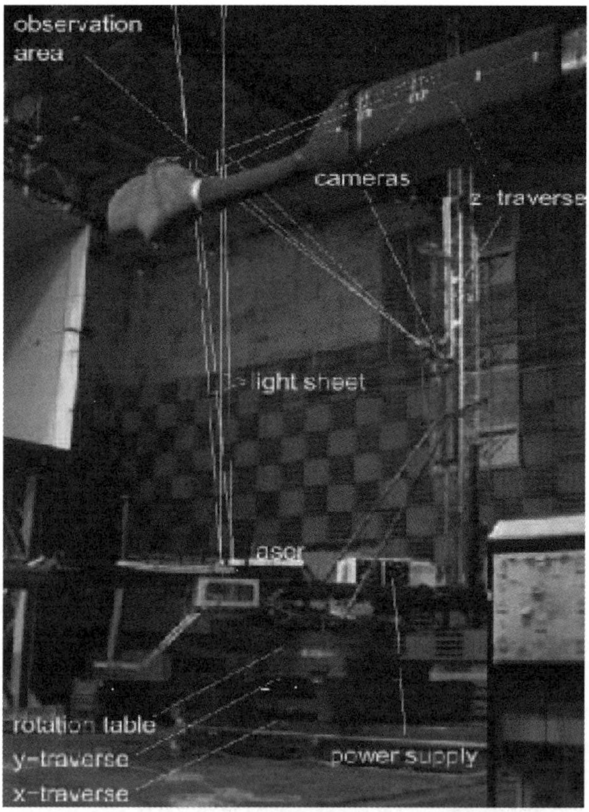

Fig. 13.1 HART-II rotor and 3-component PIV test setup at DNW facility

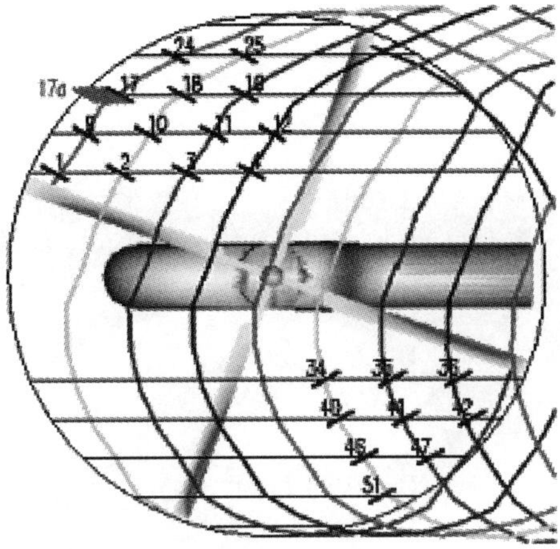

(a) $\epsilon_0 = 70^o$

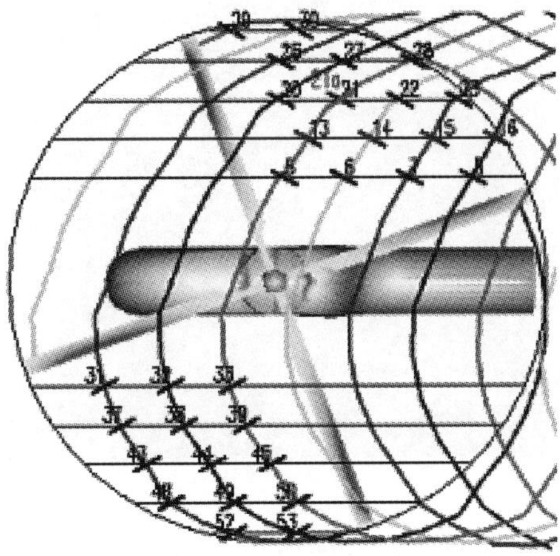

(b) $\epsilon_0 = 20^o$

Fig. 13.2 Schematic diagram for PIV measuring plane at two blade positions

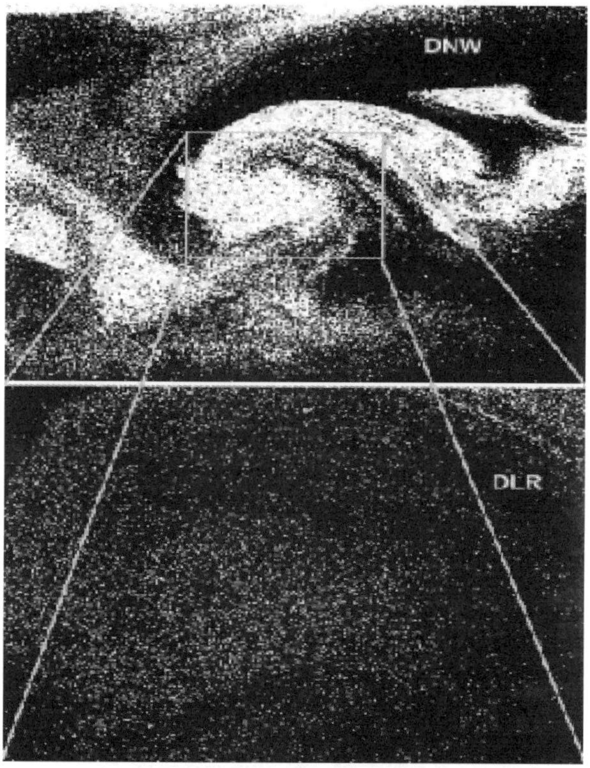

Fig. 13.3 Raw data images from two different cameras

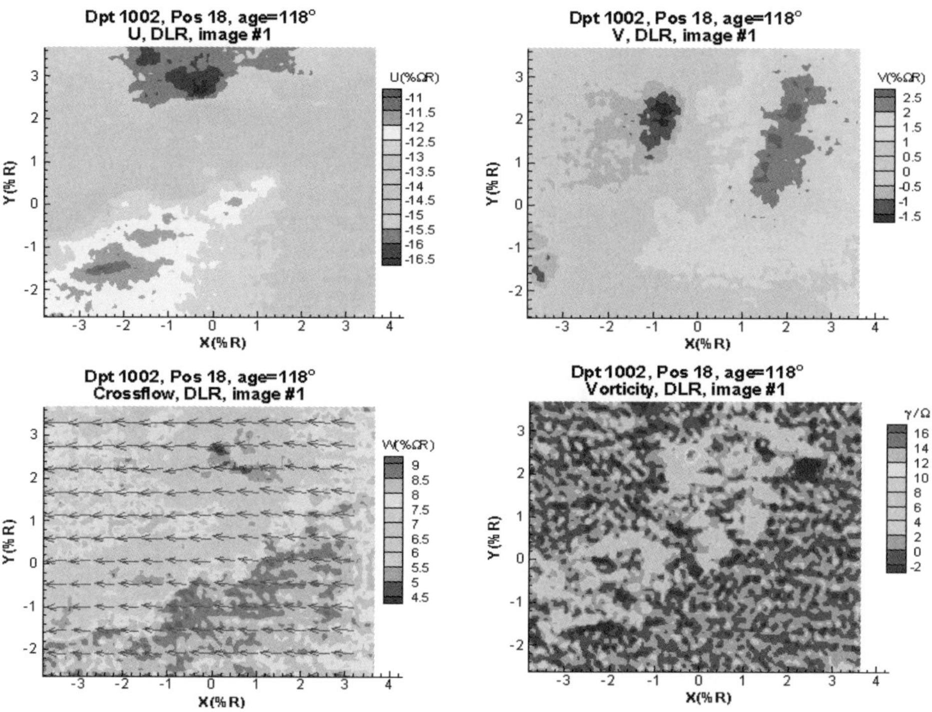

Fig. 13.4 Single PIV image of 3-component velocity and out-of-plane vorticity contour plots

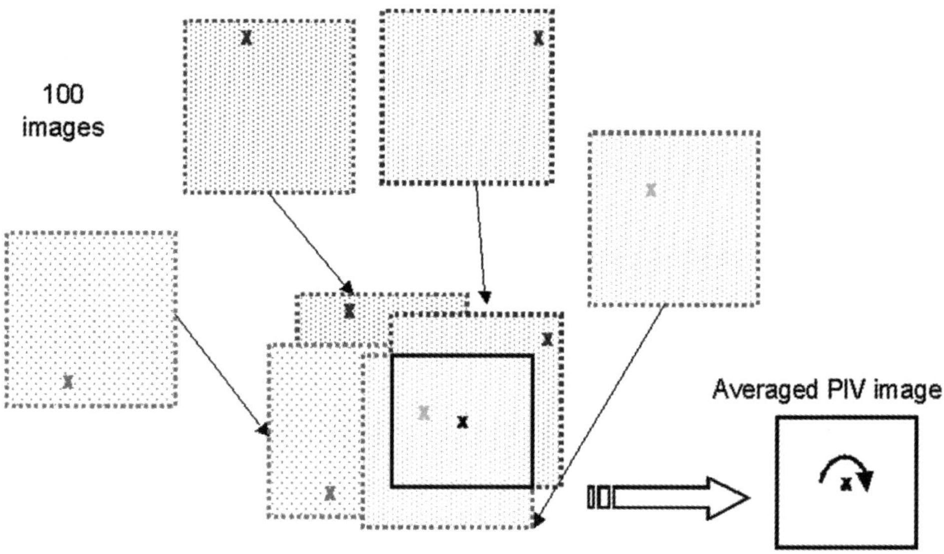

Fig. 13.5 Image averaging procedure

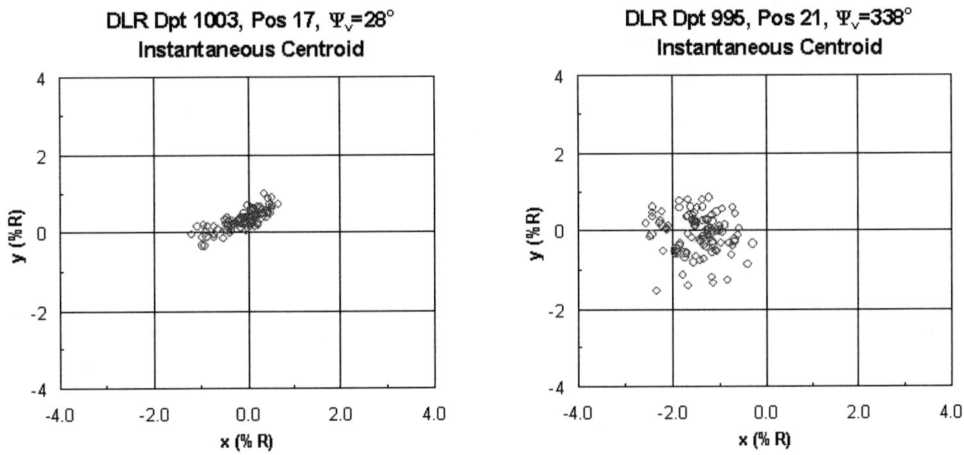

Fig. 13.6 Variation of vortex center for positions 17 and 21 of the DLR data

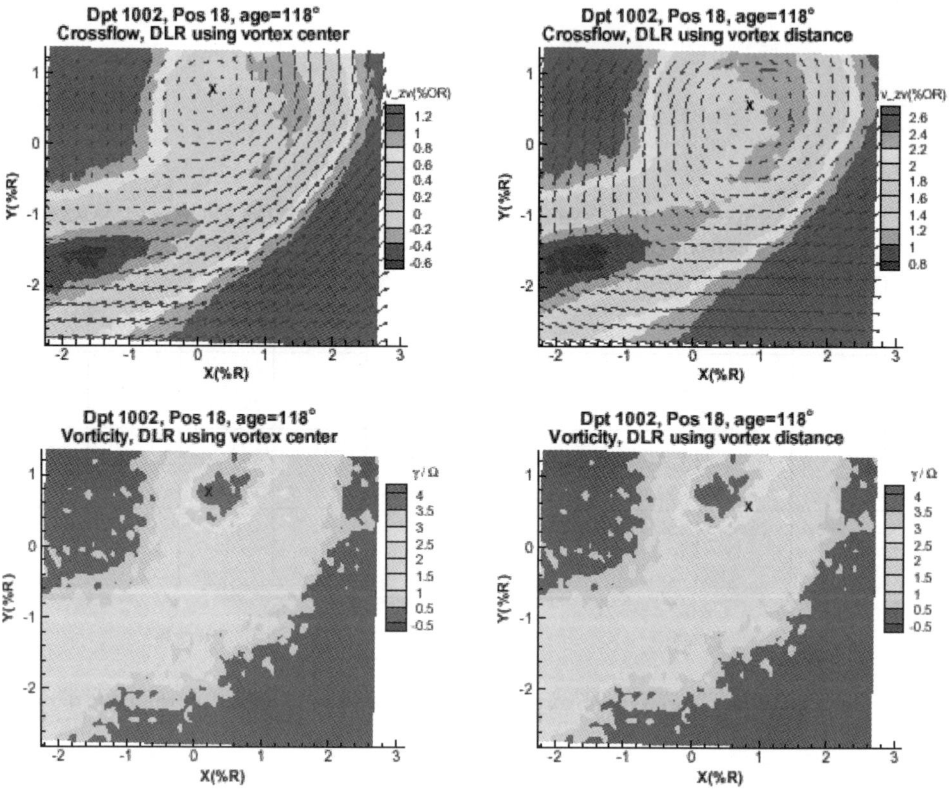

Fig. 13.7 Comparison of 3-component velocity and out-of-plane vorticy contour plot for Point 18 using vortex center method and vortex distance method

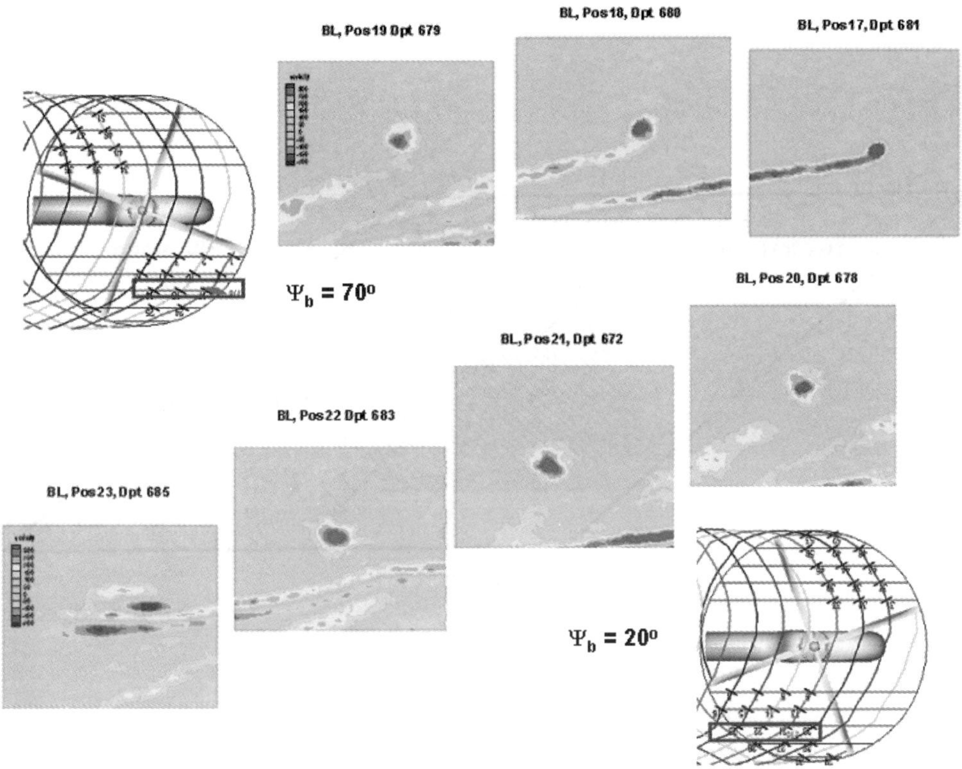

Fig. 13.8 Processed vorticity contour images

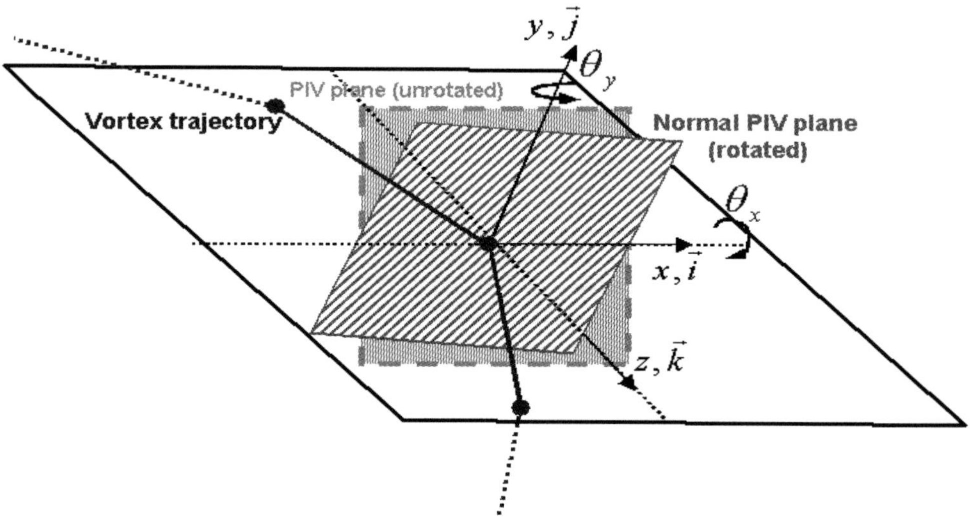

Fig. 13.9 Re-orientation of the measuring plane

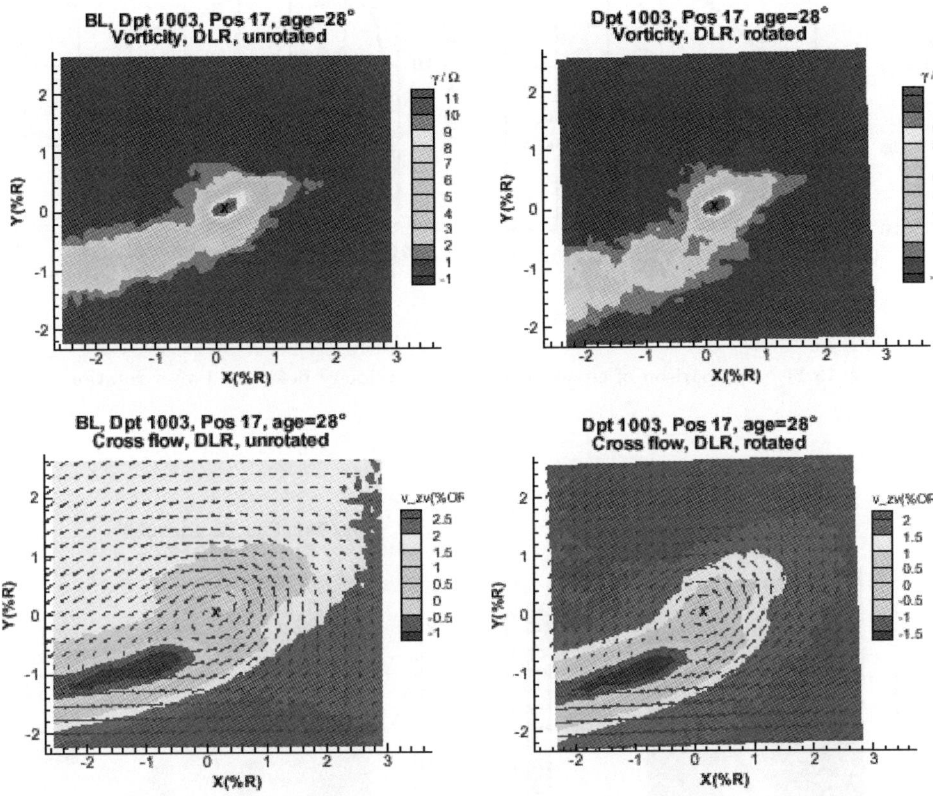

Fig. 13.10 3-component velocity and out-of-plane vorticity contour plot after re-orientation

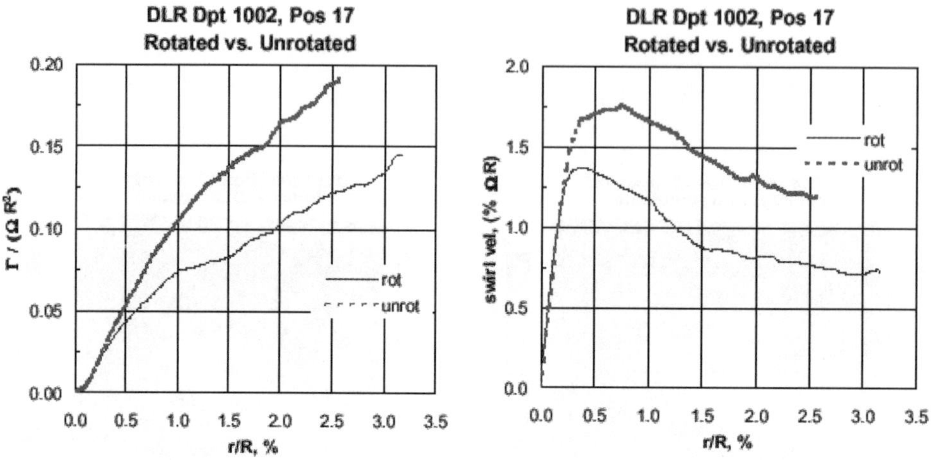

Fig. 13.11 Comparison of circulation and swirl velocity before and after rotation

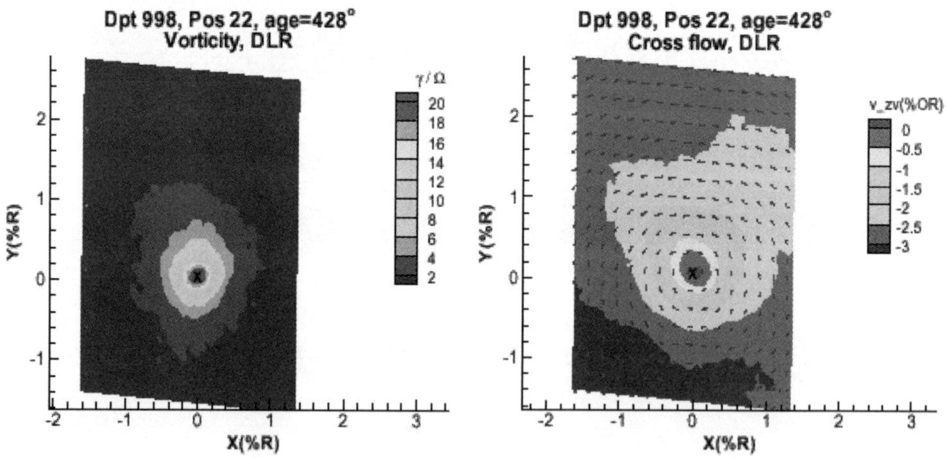

Fig. 13.12 3-component velocity and out-of-plane vorticity contours, DLR Dpt 998, Pos 22

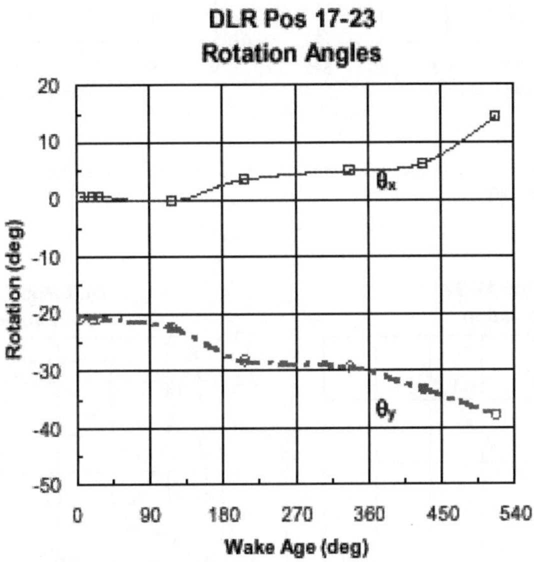

Fig. 13.13 Rotation angles of measuring plane as a function of wake age

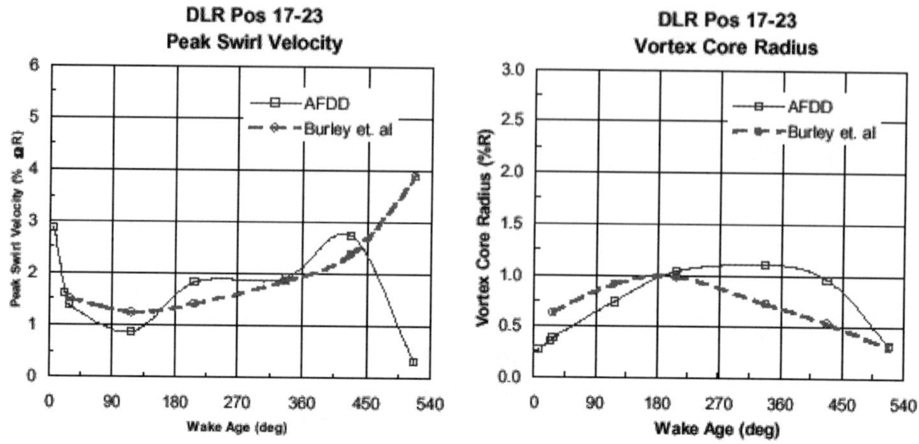

Fig. 13.14 Comparison of peak swirl velocity and vortex core radius with spin velocity method used in Ref. 10

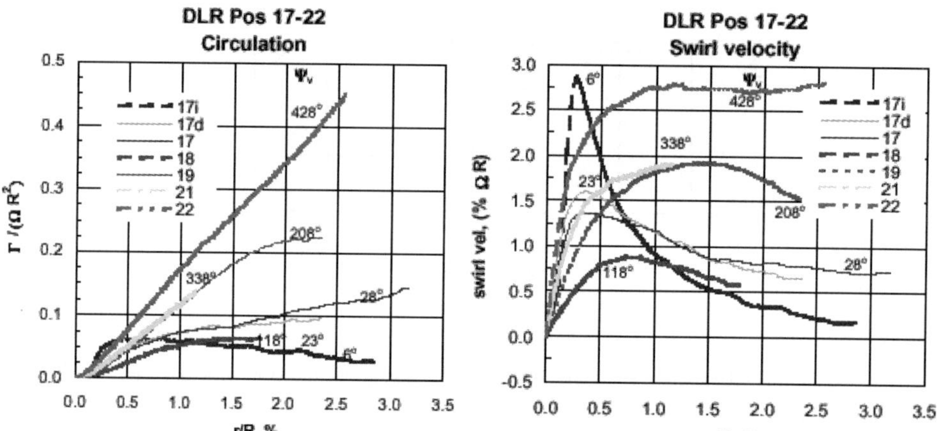

Fig. 13.15 Circulation and swirl velocity distribution as a function of wake age

Bibliography

Sakowsky, P. C., and Charles, B. D. (1976). Noise Measurement test Results for AH-1G Operational Loads Survey, Vol. I and II, Bell Helicopter Company, Rept. 299-099-831, Ft Worth, TX.

Boxwell, D. A., and Schmitz, F. H. (1980). Full-Scale Measurements of Blade-Vortex Interaction Noise, *American Helicopter Society 36th Annual Forum*, Washington, D. C.

Boxwell, D. A., Schmitz, F. H., Splettstoesser, W. R., and Schultz, K. J. (1983). Model Helicopter Rotor High Speed Impulsive Noise: Acoustics and Blade Pressures, *9th European Rotorcraft Forum*, Paper 17, Stressa, Italy.

Dadone, L. U., Caradonna, F. X., Ramachandra, K., Silva, M. J., and Poling, D. (1989). The Prediction of Loads on the Boeing Helicopter Model 360 Rotor, *American Helicopter Society 45th Annual Forum*, Boston, MA.

Lorber, P. F. (1990) Aerodynamics Results of a Pressure-Instrumented Model Rotor test at DNW, *American Helicopter Society 46th Annual Forum*, Washington, D. C.

Heineck, J. T., Yamauchi,G. K., Wadcock, A. J., Lourenco, L., and Abrego, A. (2000) Application of Three-Component PIV to a Hovering Rotor Wake, *American Helicopter Society 56th Annual Forum*, Virginia Beach, VA.

McAlister, K. W., Tung, C., and Heineck, J. T. (2001) Forced diffusion of Trailing Vorticity from a Hovering Rotor, *American Helicopter Society 57th Annual Forum*, Washington, D. C.

Raffel, M., Seelhorst, U., and Willert, C. (1998) Vortical Flow Structure at a Helicopter Model Rotor measured by LDV and PIV, The Aeronautical Journal, Paper No. 2227.

Yu, Y. H., Tung, C., van de Wall, B. G., Pausder, J., Burley, C. L., Brooks, T. F., Beaumier, P., Delrieux, Y., Mercker, E., and Pengel, K., .The HART II Test: Rotor Wake and Aeroacoustics with Higher-Harmonic Control Inputs, *American Helicopter Society 58th Annual Forum*, Montreal, Canada.

Burley, C. L., Brooks, T. F., van de Wall, B. G., Richard, H., Raffel, M., Beaumier, P., Delrieux, Y., Lim., J. W., Yu, Y. H., Tung, C., Pengel, K., and Mercker, E. (2003) Rotor Wake Vortex Definition Using 3C-PIV Measurements - Corrected for Vortex Orientation, AIAA 2003-3175,*9th AIAA/CEAS Aeroacoustics Conference*, AIAA Paper No. 2003-3175, Hilton Head, SC.

Raffel, M., Richard, H., Schneider, G., Klinge, F., Ehrenfried, K., Pengel,K., and Geenstra, G. (2002) Recording and Evaluation Methods of PIV Investigations on a Helicopter Rotor Model, *11th International Symposium on Applications of Laser Techniques to Fluid Mechanics*, Lisbon, Portugal.

Richard, H., and Raffel, M. (1995) Rotor Wake Measurements: Full-Scale and Model Tests, *American Helicopter Society 58th Annual Forum*, Montreal, Canada.

Jeong, J., Hussain, F. (1995) On the Identification of a Vortex, Journal of Fluid Mechanics, vol. 285, pp. 69-94.

Chapter 14

Typical Vortex Phenomena in Flow Fields Past Space Vehicles

Claus Weiland
EADS Space Transportation
81663 Munich, Germany

Dedicated to Lu Ting on his eightieth birthday

Abstract

Every flow field around flight vehicles is governed by vortex phenomena. The main questions are how these vortices influence the aerodynamic performance and the thermal loads of a space vehicle and how the generation of these vortices can be controlled. Further, the question arises under which conditions an optimization of the design with respect to flight reliability and controllability of a space transportation system could be achieved. To give an answer to these questions requires tools which could be experimentally – windtunnel facilities and advanced measurement systems - or theoretically – numerical simulation methods- . The investigations reported in this paper are mainly carried out by employing a numerical simulation tool solving the full Navier-Stokes equations. Since for space applications the flight vehicle has to be operable in a wide range of Mach number, pressure and density, the simulation methods applied must be able to deal with either laminar and turbulent flows as well as various thermodynamic states of the air–perfect, equilibrium and non-equilibrium real gas. Some details of the governing equations and the numerical approximation methods used are discussed. Finally, the physical influence of the simulated vortical flow field examples on the efficiency of the flight vehicle is investigated.

14.1 Introduction

A great challenge for the future consists in the aerodynamic design and development of reusable launch vehicles (RLV). These vehicles have to be flyable and controllable along the whole re-entry trajectory covering the Mach number regime from 0 to approx. 30. Consequently, numerous activities over the world aim to develop concepts for reusable or partly reusable space transportion systems. Conceptual design studies have covered single stage to orbit (SSTO) and two stage to orbit (TSTO) systems. Examples for a TSTO are Germany's SÄNGER or French's STAR-H concept (Fig. 14.1) and for SSTO the U.S. X-33 (Fig. 14.2) and the Suborbital Hopper SOH (Fig. 14.3), with the 1:7 demonstrator PHOENIX developed, designed and tested in some free-flight conditions by Germany's industry. The flight of every of these lifting vehicles is only possible in the presence of vortical flow. Vorticity is produced mainly in boundary layers, but also by curved shocks in supersonic or hypersonic flows and by shear layers available due to shock-shock or jet interactions. Wherever boundary layers separate from surfaces of flight vehicles the separated flow tends to roll up into vortices.

The reasons for flow separation could be quite different, for example due to a too strong positive pressure gradient, or due to shock boundary layer interaction or due to severe curvature of shapes, *etc.*. The goal is to physically understand and to control the behaviour of the separated vortical flows, in particular with regard to the optimization of the aerothermodynamic performance of space vehicles. Excellent surveys about these problems are given by Peake & Tobak (1983) and Stollery (1990). There exists a wide range in which vortical flows effect the characteristic values of flight vehicles. The influence of vortical flow on lift, drag, pitching moment and heat flux will be described qualitatively and quantitatively and experimental data are taken for comparison wherever possible. The special aspect is also considered that a space vehicle during re-entry comes in the situation to have a hot surface also in lower flight altitudes with the consequence that heat is transferred from the wall into the ambient air, influencing the profiles of the boundary layer and the vortical structure of the flow. By that, the characteristic aerodynamic values are changed.

14.2 Numerical simulation tool

All the numerical simulations reported in this paper are conducted by using the EADS Space's Navier-Stokes solver DAVIS-VOL. This method was developed over a period of approx. 15 years and addresses the specific requirements of vehicle flight to and from space [Weber *et al.* (2000), Schröder & Mergler (1993)]. Keypoints of the physical modelling are

- Favre-averaged Navier-Stokes equations for gases in thermodynamic non-equilibrium
- two equation turbulence models ($k - \omega$, SST and variants)
- view factor approach for concave surface elements
- treatment of wall radiation

The numerical approximation technique can be described by

- implicit finite volume technique
- symmetric TVD methods for inviscid fluxes
- central differences for diffusion terms
- line Gauss-Seidel method for Navier-Stokes equations
- point Gauss-Seidel method for turbulence and species equations
- multiblock capability

For completeness the set of governing equations are listed here [Gross (2000)].

a) Navier-Stokes equations for turbulent flow including non-equilibrium thermodynamics

$$\frac{\partial Q}{\partial t} + \frac{\partial E}{\partial x} + \frac{\partial F}{\partial y} + \frac{\partial G}{\partial z} = \frac{\partial E_\nu}{\partial x} + \frac{\partial F_\nu}{\partial y} + \frac{\partial G_\nu}{\partial z}, \tag{14.1}$$

where

$$Q = \begin{bmatrix} \rho \\ \rho u \\ \rho v \\ \rho w \\ \rho e \end{bmatrix}, \quad E = \begin{bmatrix} \rho u \\ \rho u^2 + p \\ \rho u v \\ \rho u w \\ (\rho e + p)u \end{bmatrix},$$

and

$$E_\nu = \begin{bmatrix} 0 \\ \tau_{xx} \\ \tau_{xy} \\ \tau_{xz} \\ u\tau_{xx} + v\tau_{xy} + w\tau_{xz} + \kappa_T \frac{\partial T}{\partial x} + \sum_i \rho D_i h_i \frac{\partial c_i}{\partial x} - q_{Rx} + \mu' \frac{\partial k}{\partial x} \end{bmatrix},$$

and similarly for F, G and F_ν, G_ν. Furthermore, we have the definitions

$$k = \frac{1}{2\rho}\left(\overline{\rho u'^2} + \overline{\rho v'^2} + \overline{\rho w'^2}\right), \ \tau_{ij} = \mu\left(\frac{\partial \nu_i}{\partial x_j} + \frac{\partial \nu_j}{\partial x_i} - \frac{2}{3}\delta_{ij}\frac{\partial \nu_i}{\partial x_i}\right) - \frac{2}{3}\rho k \delta_{ij}$$

$$e = \varepsilon + \frac{1}{2}\left(u^2 + v^2 + w^2\right) + k$$

$$\mu' = \mu_l + \sigma^* \mu_T, \qquad \mu = \mu_l + \mu_T$$

$$\kappa_T = \kappa_l \left(1 + \frac{\Pr \mu_T}{\Pr_T \mu}\right), \qquad D_i = D_{l,j} + \frac{Sc\mu_T}{Sc_T \rho}$$

where μ_l, κ_l, Pr and Sc denote the laminar, and μ_T, κ_T, $\Pr_T$ and Sc_T the turbulent viscosity, heat conduction, Prandtl and Schmidt numbers. Further, σ^*represents the coefficient of turbulence and D_l of laminar diffusion.

b) Equations of turbulence models (two equation models on $k - \omega$ basis)

$$\frac{\partial Q_T}{\partial t} + \frac{\partial E_T}{\partial x} + \frac{\partial F_T}{\partial y} + \frac{\partial G_T}{\partial z} = \frac{\partial E_{\nu,T}}{\partial x} + \frac{\partial F_{\nu,T}}{\partial y} + \frac{\partial G_{\nu,T}}{\partial z} + S_T \qquad (14.2)$$

with

$$Q_T = \begin{bmatrix} \rho k \\ \rho\omega \end{bmatrix}; \ E_T = \begin{bmatrix} \rho u k \\ \rho u \omega \end{bmatrix}; \ S_T = \begin{bmatrix} P_k - \beta^* \rho k \omega \\ \alpha \frac{\rho}{\mu_T} P_k - \beta \rho \omega^2 \end{bmatrix};$$

and

$$E_{\nu,T} = (\mu_l + \sigma^* \mu_T) \begin{bmatrix} \partial k / \partial x \\ \partial \omega / \partial x \end{bmatrix}$$

(and similarly for F_T, G_T and $F_{\nu,T}, G_{\nu,T}$) with the relations

$$\mu_T = \rho \frac{k}{\omega}; \quad P_k = \tau_{ij} \frac{\partial \nu_i}{\partial x_j}$$

where ω represents the turbulent dissipation and α, β, β^* and σ^* are constants of the turbulence model selected. In compressible flow it could happen that the normal stresses in the Reynolds stress tensor become lower than zero, which must be prevented. To overcome this problem, either a compressible dissipation or a realizability correction is employed.

c) Equations of species continuity

$$\frac{\partial Q_S}{\partial t} + \frac{\partial E_S}{\partial x} + \frac{\partial F_S}{\partial y} + \frac{\partial G_S}{\partial z} = \frac{\partial E_{\nu,S}}{\partial x} + \frac{\partial F_{\nu,S}}{\partial y} + \frac{\partial G_{\nu,S}}{\partial z} + S_S \tag{14.3}$$

where

$$E_S = \begin{bmatrix} \rho_1 u \\ \rho_2 u \\ \dots \end{bmatrix};\ E_{\nu,S} = \begin{bmatrix} \rho D_1 \frac{\partial c_1}{\partial x} \\ \rho D_2 \frac{\partial c_2}{\partial x} \\ \dots \end{bmatrix};\ Q_S = \begin{bmatrix} \rho_1 \\ \rho_2 \\ \dots \end{bmatrix};\ S_S = \begin{bmatrix} \dot{\omega}_1 \\ \dot{\omega}_2 \\ \dots \end{bmatrix}$$

The following relations hold

$$\dot{\omega}_i = M_i \frac{d\,[X_i]}{dt}; \quad [X_i] = \frac{\rho_i}{M_i}; \quad X_i = \frac{M}{M_i} c_i; \quad \sum\nolimits_i X_i = 1$$

where the reaction rate $\frac{d[X_i]}{dt}$ depends on the gas composition and the chemical model selected. The diffusion coefficients D_i are computed by the binary diffusion coefficients D_{ij}

$$D_i = \frac{1 - X_i}{\sum_j (X_i / D_{ij})};\ D_{ij} = \frac{1}{p} \exp\left(d_{1jk} + d_{2jk} T_L + d_{3jk} T_L^2 + d_{4jk} T_L^3\right)$$

and the laminar viscosity is given by

$$\mu = \frac{1}{2}\left(\sum\nolimits_i X_i Z_i + \left(\sum\nolimits_i (X_i / Z_i)\right)^{-1}\right);$$
$$Z_i = \exp\left(d_{1i} + d_{2i} T_L + d_{3i} T_L^2 + d_{4i} T_L^3\right); \tag{14.4}$$

with $T_L = \ln T$, and the relations for the heat conduction look similar to Eq. (14.4). The difference consists in the magnitude of the coefficients d_{li}.

$$\kappa = \frac{1}{2}\left(\sum\nolimits_i X_i Z_i + \left(\sum\nolimits_i (X_i / Z_i)\right)^{-1}\right);$$

$$Z_i = \exp\left(d_{1i} + d_{2i} T_L + d_{3i} T_L^2 + d_{4i} T_L^3\right).$$

14.3 Vortical flow effects on space vehicles

14.3.1 *General*

Flight vehicles which have to move with supersonic or hypersonic speeds have mainly delta-type wings. The sweep angle and the form of the leading edge of these wings depend on the Mach number range and the area of operation of the vehicle. For combat aircraft, sharp leading edges and moderate sweep angles are used, while for space vehicles, blunt leading edges and higher sweep angles are brought into use.The vortical structure essentially on the leeward side and the wing tip is responsible for the reliable flyability and operability of the vehicle. The next four figures demonstrate the problem. The very famous water channel figure [Dyke v. (1982)] shows leeward side vortices and its breakdown (Fig. 14.4). The complicated flow field over the wing of a combat aircraft with angle of attack $\alpha = 24^o$ is displayed in Fig. 14.5 [Hitzel (1993)], and Fig. 14.6 exhibits on the blunt delta wing the leeside vortices and the trace of its on the wall by the plotted wall streamlines [Riedelbauch (1992)]. Measurements of the turbulence intensity indicating the strength and the position of the vortex on a fighter aircraft are shown in Fig. 14.7 [Breitsamter (1997)].

14.3.2 *Influence on lift*

Since the lift is mainly produced by the wing, the dynamics of the vortices created by these wings are one of the major concerns. Let us consider a delta wing with blunt leading edges and a distinct wing tip. For low angles of attack α the flow remains attached over the entire wing surface and rolls up in the wake forming together with the separated tip flow a vortex with a center close to the wing tip. Increasing the incidence lets the flow separate ahead of the trailing edge forming a vortex core more located inboard. Finally, by further increase of α the flow will separate at the leading edge composing the leading edge vortices. In case of sub- and transonic flow this leading edge vortex may break down with the consequence of a drastically reduced lift and a strong change of the pitching moment. The next figures demonstrate these phenomena on the basis of numerical flow field simulations (DAVIL-VOL code) around the German SSTO demonstrator PHOENIX. Fig. 14.8 contains the streamlines emanating from the wing leading edge and the tip for $M_\infty = 0.2$, $\alpha = 15^o$, where the flow remains attached along the leading edge. For $\alpha = 20^o$ it can clearly be seen, that the flow separates at the leading edge (Fig. 14.9). Vortex breakdown happens for $M_\infty = 0.2$, at $\alpha \approx 15\%$ and $M_\infty = 0.5$, $\alpha \approx 22^\circ$ as it can be extracted from the diagram Fig. 14.10, where experimental data of three different windtunnels are plotted. Of course, the accuracy of the numerical results depends on the capability of the method to proper resolve flow turbulence. On the one hand, there is no doubt that the application of advanced two equation models ($k - \omega$ + SST or + realizable) might have deficiencies, but the results obtained are at least for industrial purposes (generation of aerothermodynamic data basis) satisfactory, while on the other hand LES or - in the extreme case - DNS solutions for complete configurations seems to be a goal for the far future.

14.3.3 *Influence on drag*

Winged reusable launch vehicles differ from conventional civil and military aircrafts amongst others by large base areas. The drag of such vehicles, in particular for lower Mach numbers depends strongly on the flow situation at these bases. The flow in the wake of RLVs is characterized by separated boundary layers, vortices, shear layers and in super- and hypersonic regimes by embedded shock systems. All these phenomena interact strongly with each other and constitute the source for the pressure and the heat flux distribution at the base. A good picture of the flow physics in the wake may only be obtained by knowing the dynamics of the vortices either by experimental investigation or numerical simulation. Often it is not evident under what conditions the vortex dominated wake flow is either steady or unsteady and further what the reason is for a non-symmetrical unsteady vortex shedding.

Despite the fact that there were in the last two decades strong improvements in the numerical simulation methods and in the turbulence modeling the prediction of wake flows is still a severe task. Nevertheless some solutions obtained recently show a considerable progress even in the case of wake flows dominated by turbulence.

Figure 14.11 shows the flow field in the wake of the PHOENIX vehicle obtained by numerical simulation (laminar assumption; [Bernaudin (2001)]). 17.5% of the total length L behind the base formation of the vortices starts generated by the separated boundary layer along the edges of the base (Fig. 14.11a, left), which converges downstream (25% of L) to a single vortex with a centerline positioned approximately in parallel to the main stream (Fig. 14.11a, right). Figure 14.11b gives an insight into the extension of the recirculation region, which is defined by iso-surfaces $u = 0$, giving the bounds between downstream and reverse for two trajectory points ($M_\infty = 1.72$, $\alpha = 5^o$; $M_\infty = 2.31$, $\alpha = 10^o$). Wind tunnel results are available for these flow cases (Table 1a). The drag and the lift compare excellently, while for the pitching moment some deviations occur, which seems to only be possible if the pressure forces have different distributions in the experiment and the simulation. Two conclusions can be drawn from this result. First, the flow in the windtunnel was mainly laminar and second, the predicted laminar wake flow produces a pressure distribution at the base, which is not far from reality. In Table 1b, laminar and turbulent flow assumptions are compared for lift and drag. As expected, the lift quantity is only slightly affected, while the deviation in the drag value is considerable.

Table 1a. Comparison of experimental and numerical results

M_∞ =2.31, α=10°, Re = 12$\cdot$10^6	Drag C_D	Lift C_L	P. mom. C_m
SST Windtunnel (NLR)	0.101	0.202	-0.00430
Navier-Stokes solution	0.101	0.203	-0.00409

Table 1b. Comparison of laminar and turbulent simulations

M_∞=3.95, α =10°, Re = 25.1$\cdot$10^6	Drag C_D	Lift C_L
Laminar	0.06950	0.01371
Turbulent	0.07255	0.01375

14.3.4 *Influence on heat flux*

During re-entry space vehicles undergo high thermal loads in particular at the nose and along the leading edges of wings and fins. This happens mainly due to thermodynamic recompression of the air. Another reason for high heat flux rates is given by vortex-boundary layer interactions, which means vortex-driven flow separations and reattachments. Boundary layers are becoming thick where separation occurs and thin in case of reattachment. With the attentuation of the boundary layer the gradient of the flow variables normal to the wall grows considerably. For gases with relatively high enthalpies this is also valid for the temperature with the consequence of locally high heat fluxes and temperatures (hot spots). Flow separation with vortex formation takes place on leeward sides of the vehicle, on deflected aerodyamic controls (e.g. body flaps), in base areas, in cavities and elsewhere. The design of the thermal protection system (TPS) has to allow for these effects.

The flow on the rear part of a capsule moving with a high angle of attack behaves unexpectedly complex. A general flow separation takes place just behind the shoulder due to the large curvature there. Further, the strong crossflow in the rear part produces a primary and a secondary vortex with the corresponding separation and reattachment (Fig. 14.12, left). As explained above a heat flux pattern exists where the separation lines coincide with regimes of low heat fluxes and the reattachment lines with those of high heat fluxes as well (Fig. 14.12, right). Every space vehicle which accomplishes a controlled re-entry in earth's atmosphere needs a Reaction Control System (RCS) which consists of a set of thrusters installed on specified positions of the vehicle. The jet of a firing thruster interacts with the external airflow, where, depending on some geometrical and physical parameters a vortical flow pattern is generated. It seems that the basic vortical structure consists on a ring vortex and a horse shoe vortex which can be observed in the flow field for an exemplary flat plate – jet arrangement (Fig. 14.13, [Zeiss (2002)]). Thrusters installed on a realistic shape like the X-38 vehicle generate vortical structures which are degenerated in the above sense (Fig. 14.14, [Weiland (2003)]). In the regime where the flow reattaches a strong increase of the temperature - here 1000K- compared to the non-firing case can be identified (Fig. 14.11, right). The magnitude of the temperature increase for cases with and without firing thrusters depends on the position of the jets on the vehicle surface, the state of the boundary layer, the flight velocity, the angle of attack and the flight altitude.

14.3.5 *Hot wall effects*

During re-entry space vehicles experience very severe heat transfers that might lead - depending on the flight trajectory - to peak temperatures in the limit of radiation cooled surfaces of approx. 1800K at stagnation points, wing leading edges, deflected aerodynamic controls and others. For example, the maximum temperature distribution for the U.S. Shuttle is reached at an altitude between 55 and 60 km, where the flight speed is still 4500 to 5000 m/s ($M_\infty \approx 15$). For lower altitudes the convective heat transfer from the air to the wall reduces drastically, so that the already heated structure of the vehicle transfers heat to the air on the basis of relatively high wall temperatures. This situation is often not considered in the course of the establishment of the aerothermal data base.

In a numerical experiment the flow around the PHOENIX configuration is calculated for a moderate Mach number (M_∞=3.2) once with a radiation adiabatic wall condition (cold) and in a second case with a prescribed temperature distribution (hot) taken from a flight Mach number M_∞=14.4 [Häberle (2004)]. The stagnation point values for M_∞=14.4 are $T_{\text{rad}} = 1800$K and q_w=519 kW/m2 while for M?=3.2 the radiation adiabatic values are Trad=593K and qw=6 kW/m2. Hot walls reduce the skin friction (despite an increase of the viscosity !) and with that the drag what is due to the decrease of the gradient $\partial u/\partial n_{\text{wall}}$. The vortical structure of the flow is strongly different in the body flap area owing to the remarkable change in the separation and reattachment regimes. Figure 14.15 shows for a body flap deflection of $\eta_{\text{bf}} = 40^\circ$ the skin friction lines and the pressure distribution for the radiation adiabatic wall (left) and the hot wall (right). A presentation of the complete flow field, where cold (upper half) and hot (lower half) wall conditions are compared is displayed in Fig. 14.16.

The general attenuation of the friction over the whole vehicle shape diminishes the drag, but the local change in the vortical structure (separations and reattachments) around the body flap regime alter the pitching moment and with that the trim conditions (Table 2).

The change of the drag is approximately 5% and that of the pitching moment 8.2%. This must be borne in mind for every flight mechanical system.

Table 2. Change of drag and pitching moment depending on wall temperature conditions

M_∞=3.2, α=15°, Re = $2.31 \cdot 10^6$	Drag C_D	Pitching moment C_m
cold wall	0.1963	-0.02515
hot wall	0.1870	-0.02309

14.4 Conclusions

The production of vortices can be critical for the aerodynamic performance and the thermal loads of space vehicles. Vortical dynamics can be investigated in windtunnels by advanced measurement systems and by evaluation of numerical flow field simulations. With these tools the knowledge about vortical flows has grown rapidly in the past with the consequence of reducing the uncertainties during the design process of space vehicles.

Fig. 14.1 Two stage to orbit concept SÄNGER

Fig. 14.2 Single stage to orbit concept X-33

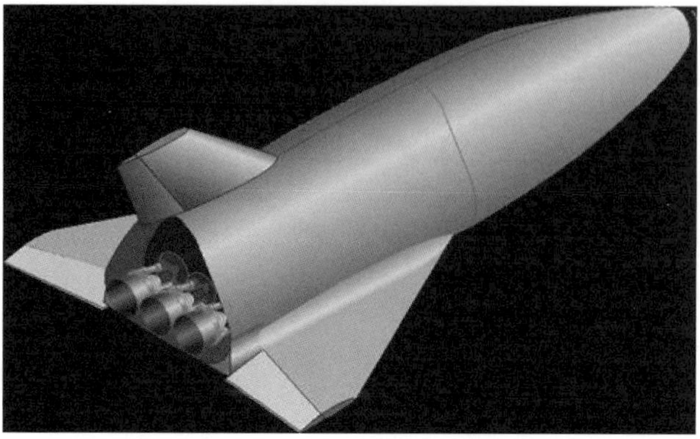

Fig. 14.3 Europe's suborbital Hopper concept

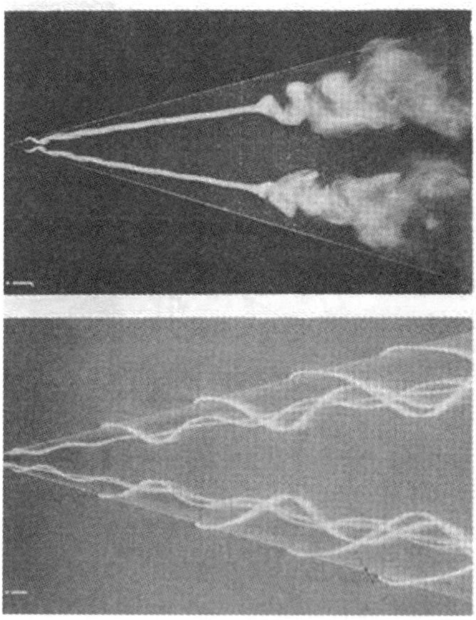

Fig. 14.4 Visualized leading edge vortices on delta wing ($\alpha = 20^{\circ}$, top) and vortex breakdown ($\alpha = 30^{\circ}$, bottom) (experiments Werle in [Dyke v. (1982)])

Fig. 14.5 Water tunnel visualization of vortex structure, $\alpha = 24^o$, slatangle $= -20^o$ ([Hitzel (1993)])

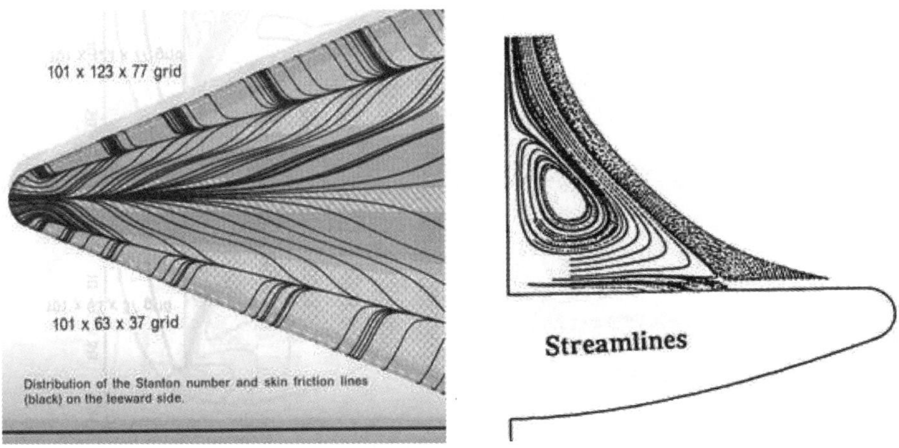

Fig. 14.6 Vortex formation on blunt delta wing, $M_\infty = 8.7$, $\alpha = 30^o$ ([Riedelbauch (1992)])

Fig. 14.7 Measurement of turbulence intensity on fighter aircraft, $M_\infty = 0.12$, $\alpha = 15^o$, $Re = 0.97 \times 10^6$ ([Breitsamter (1997)])

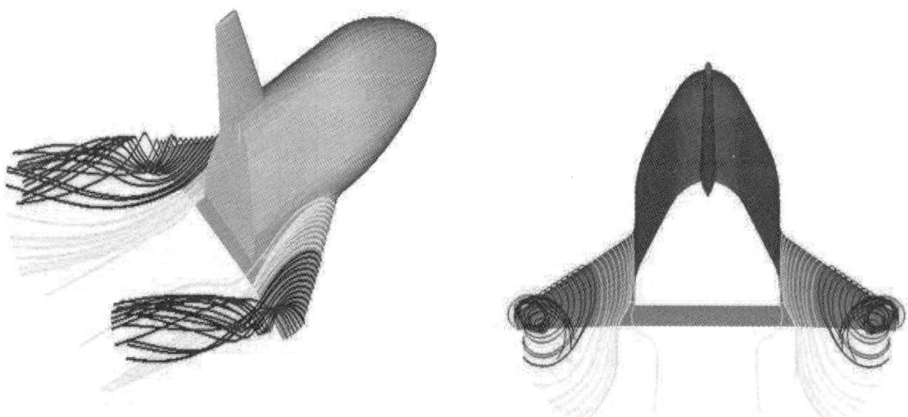

Fig. 14.8 PHOENIX configuration, $M_\infty = 0.2$, $\alpha = 15^o$, $Re = 3.8 \times 10^7$; turbulent flow

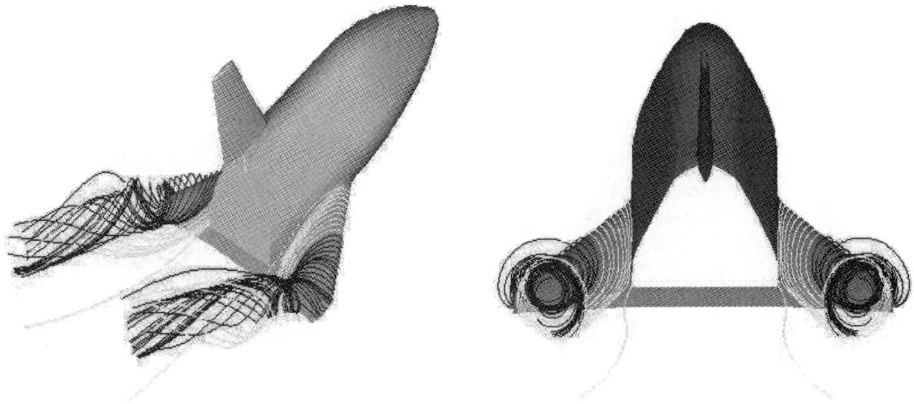

Fig. 14.9 PHOENIX configuration, $M_\infty = 0.2$, $\alpha = 20°$, $Re = 3.8 \times 10^7$; turbulent flow

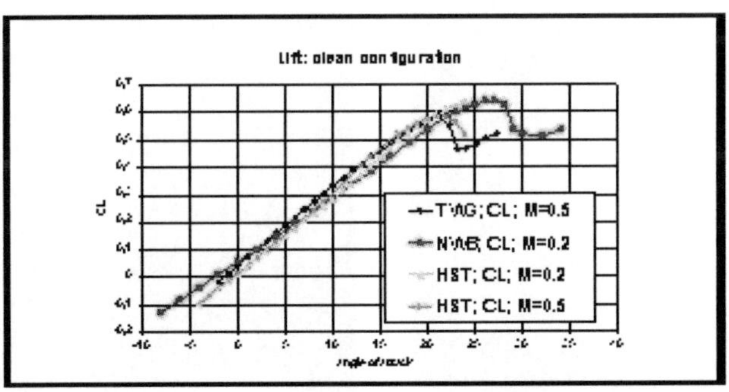

Fig. 14.10 Wind tunnel measurement on PHOENIX configuration, $Re = 3.8 \times 10^7$; note the clear vortex breakdown effect

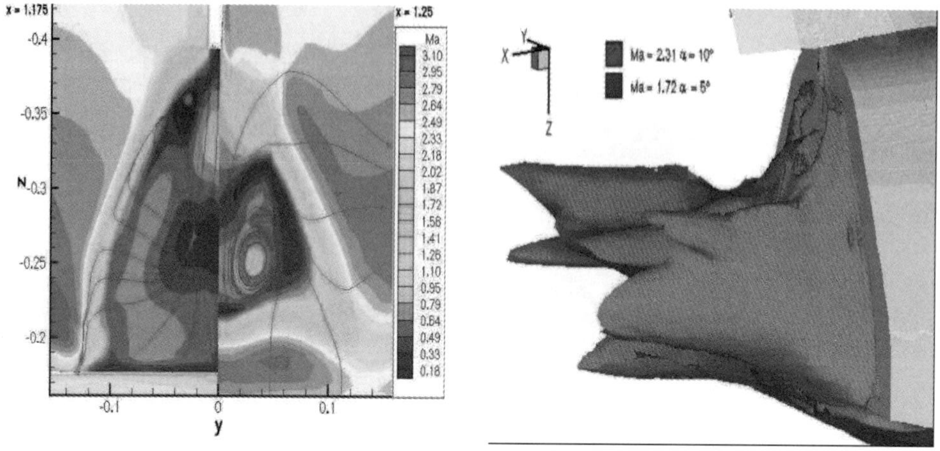

Fig. 14.11 a,b PHOENIX; $M_\infty = 2.31$, $\alpha = 10^o$, $Re = 1.2 \times 10^7$; laminar NS solution

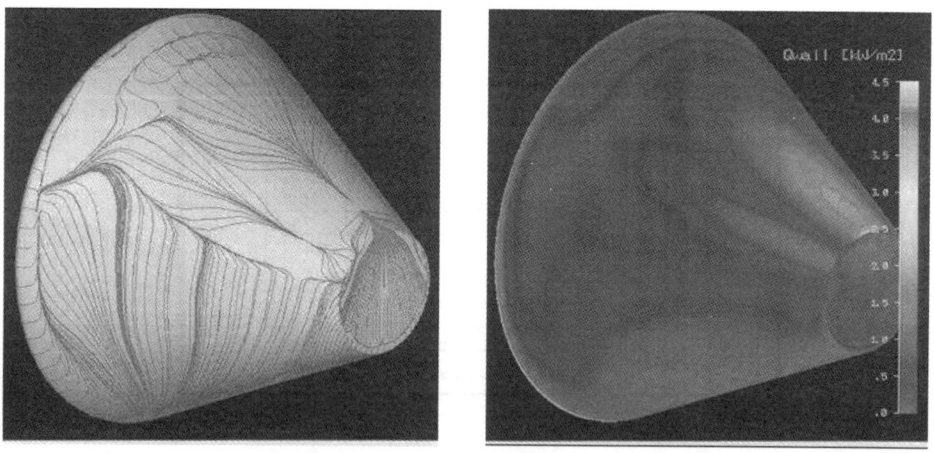

Fig. 14.12 Viking type shape, $M_\infty = 3.0$, $\alpha = 25^o$. Vortex induced heat loads. Wall streamline pattern (left); heat flux pattern (right)

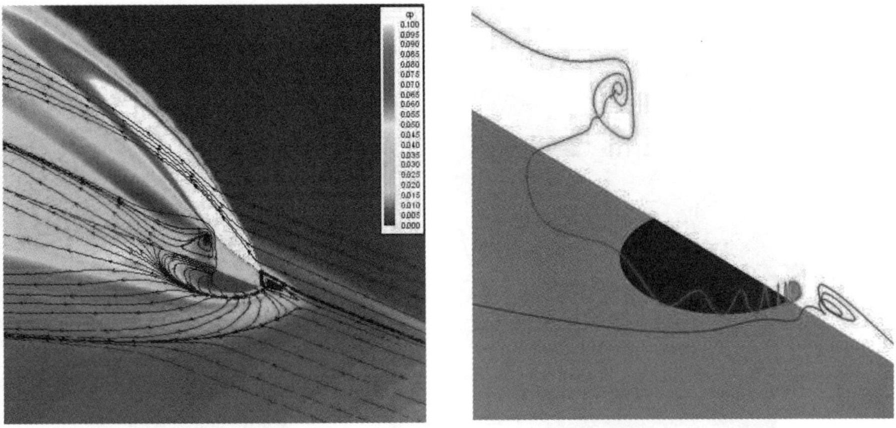

Fig. 14.13 Flat plate-jet arrangement demonstrating jet/external airflow interaction, $M_\infty = 17.5$, $\alpha = 0^o$. Jet: $M_{jet} = 5.99$, $p_{jet} = 192pa$, $T_{jet} = 728K$

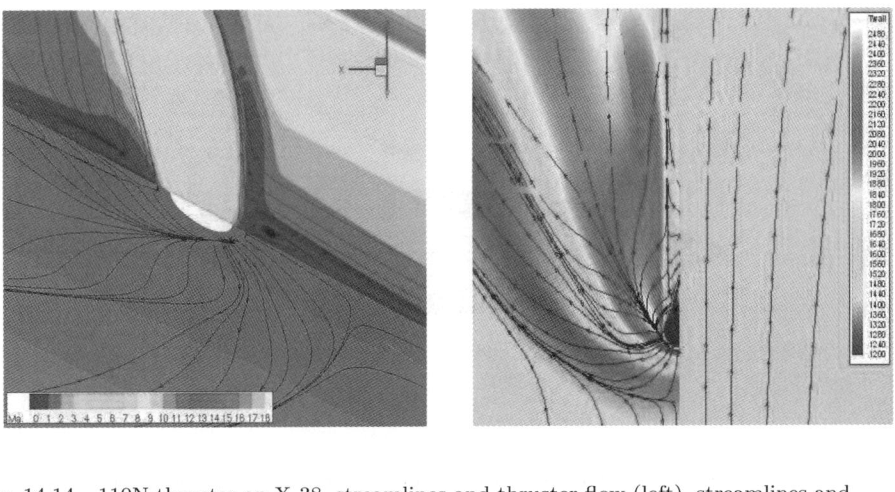

Fig. 14.14 110N thruster on X-38, streamlines and thruster flow (left), streamlines and wall temperature (right); $M_\infty = 17.5$, $\alpha = 0^o$. Jet: $M_{jet} = 4.25$, $p_{jet} = 16210pa$, $T_{jet} = 57K$

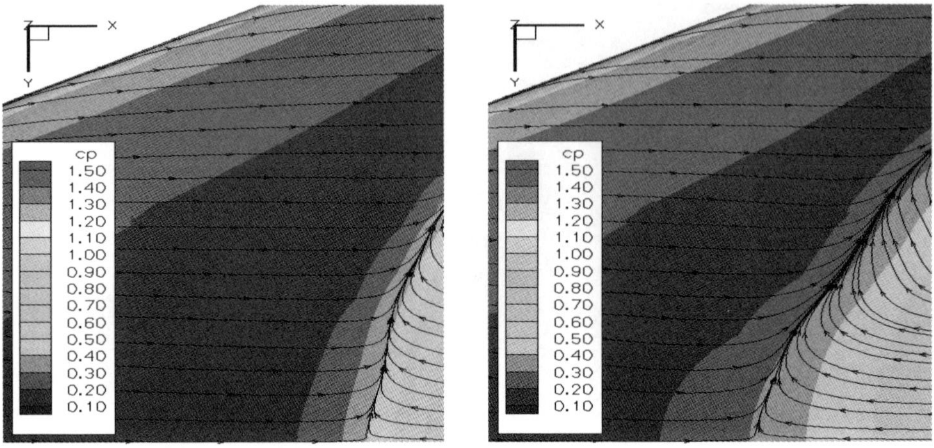

Fig. 14.15 PHOENIX: skin friction lines and pressure distribution in the body flap regime for cold wall (left) and hot wall (right) conditions (left), $M_\infty = 3.2$, $\alpha = 15^o$, Re $= 2.31 \times 10^6$, $\eta_{bf} = 40^o$, turbulent flow

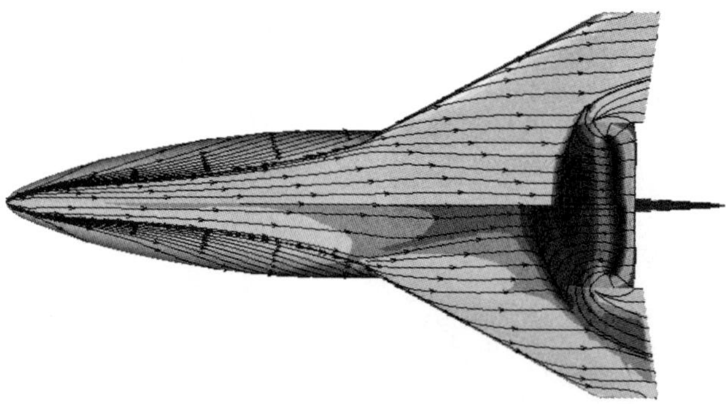

Fig. 14.16 PHOENIX skin friction lines and c_f distribution on the windward side for cold (upper half) and hot wall (lower half) conditions, $M_\infty = 3.2$, $\alpha = 15^o$, Re $=$ 2.31×10^6, $\eta_{bf} = 40^o$, turbulent flow

Bibliography

Bernaudin, J.B. (2001). Aerodynamics of the Phoenix Demonstrator, Diploma Thesis, Ecole Nationale Superieure de L' Aeronautique et de L'Espace - ENSAE-, Toulouse.

Breitsamter, Ch. (1997). Turbulente Strömungsstrukturen an Flugzeugkonfigurationen mit Vorderkantenwirbel, Department of Fluimechanics, Univ.of Munich, 335p. (in German).

Dyke v., M. (1982). An Album of Fluid Motion, The Parabolic Press, Stanford, CA.

Gross, A. (2000). Numerische Untersuchung abgelösterDüsenströmungen, PhD Thesis, Department of Combustion and Propulsion, RWTH Aachen.

Häberle, J.(2004). Einfluss heisser Oberflächen auf aerothermodynamische Flugeigen-schaften von Hopper/Phoenix, Diploma Thesis, IAG, Univ. of Stuttgart, 106p. (in German)

Hitzel, S.M. (1993). Zur numerischen Simulation dreidimensionaler kompressibler Flügelvorderkantenwirbelsysteme mit dem Eulerverfahren, Doctoral Thesis, University of Stuttgart.

Peake, D.J., Tobak, M. (1983). On Issues Concerning Flow Separation and Vortical Flow in Three Dimensions, AGARD-CP-342, Paper No.1.

Riedelbauch, S. (1992), Hypersonic Flow over a Delta Wing at High Angle of Attack, in: Hypersonic Flows for Reentry Problems, Vol III, Springer, pp. 865-880.

Schröder, W. and Mergler, F. (1993). Investigation of the Flow Field over Parallel-Arranged Launch Vehicles, AIAA Paper 93-3060.

Stollery, J.L. (1990). Glancing Shock-Boundary Layer Interactions, AGARD-R-764, Paper No. 6.

Weber, C., Behr, R. and Weiland, C. (2000). Investigation of Hypersonic Turbulent Flow over the X-38 Crew Return Vehicle, AIAA Paper 2000-2601.

Weiland, C. (2003). Vortex phenomena in flows for reusable launch vehicles, Computational Fluid and Solid Mechanics 2003, Elsevier, pp.1197-1200.

Zeiss, W. (2002). Jet/Airflow Interaction, DLR-TETRA Programm, TET-DASA-21-TN-2409.

Zeiss, W. (2002). Baseflow / Rocketjet Interaction, DLR-TETRA Programm, TET-DASA-21-TN-2410.